International Series in Operations Research & Management Science

Founding Editor

Frederick S. Hillier, Stanford University, Stanford, CA, USA

Volume 332

Series Editor

Camille C. Price, Department of Computer Science, Stephen F. Austin State University, Nacogdoches, TX, USA

Editorial Board Members

Emanuele Borgonovo, Department of Decision Sciences, Bocconi University, Milan, Italy

Barry L. Nelson, Department of Industrial Engineering & Management Sciences, Northwestern University, Evanston, IL, USA

Bruce W. Patty, Veritec Solutions, Mill Valley, CA, USA

Michael Pinedo, Stern School of Business, New York University, New York, NY, USA

Robert J. Vanderbei, Princeton University, Princeton, NJ, USA

Associate Editor

Joe Zhu, Foisie Business School, Worcester Polytechnic Institute, Worcester, MA, USA

The book series **International Series in Operations Research and Management Science** encompasses the various areas of operations research and management science. Both theoretical and applied books are included. It describes current advances anywhere in the world that are at the cutting edge of the field. The series is aimed especially at researchers, advanced graduate students, and sophisticated practitioners.

The series features three types of books:

• Advanced expository books that extend and unify our understanding of particular areas.

• Research monographs that make substantial contributions to knowledge.

• Handbooks that define the new state of the art in particular areas. Each handbook will be edited by a leading authority in the area who will organize a team of experts on various aspects of the topic to write individual chapters. A handbook may emphasize expository surveys or completely new advances (either research or applications) or a combination of both.

The series emphasizes the following four areas:

Mathematical Programming : Including linear programming, integer programming, nonlinear programming, interior point methods, game theory, network optimization models, combinatorics, equilibrium programming, complementarity theory, multiobjective optimization, dynamic programming, stochastic programming, complexity theory, etc.

Applied Probability: Including queuing theory, simulation, renewal theory, Brownian motion and diffusion processes, decision analysis, Markov decision processes, reliability theory, forecasting, other stochastic processes motivated by applications, etc.

Production and Operations Management: Including inventory theory, production scheduling, capacity planning, facility location, supply chain management, distribution systems, materials requirements planning, just-in-time systems, flexible manufacturing systems, design of production lines, logistical planning, strategic issues, etc.

Applications of Operations Research and Management Science: Including telecommunications, health care, capital budgeting and finance, economics, marketing, public policy, military operations research, humanitarian relief and disaster mitigation, service operations, transportation systems, etc.

This book series is indexed in Scopus.

Yacob Khojasteh • Henry Xu • Saeed Zolfaghari
Editors

Supply Chain Risk Mitigation

Strategies, Methods and Applications

Editors
Yacob Khojasteh
Sophia University
Tokyo, Japan

Henry Xu
University of Queensland
Brisbane, QLD, Australia

Saeed Zolfaghari
Toronto Metropolitan University
Toronto, ON, Canada

ISSN 0884-8289   ISSN 2214-7934 (electronic)
International Series in Operations Research & Management Science
ISBN 978-3-031-09185-8   ISBN 978-3-031-09183-4 (eBook)
https://doi.org/10.1007/978-3-031-09183-4

© The Editor(s) (if applicable) and The Author(s), under exclusive license to Springer Nature Switzerland AG 2022

This work is subject to copyright. All rights are solely and exclusively licensed by the Publisher, whether the whole or part of the material is concerned, specifically the rights of translation, reprinting, reuse of illustrations, recitation, broadcasting, reproduction on microfilms or in any other physical way, and transmission or information storage and retrieval, electronic adaptation, computer software, or by similar or dissimilar methodology now known or hereafter developed.

The use of general descriptive names, registered names, trademarks, service marks, etc. in this publication does not imply, even in the absence of a specific statement, that such names are exempt from the relevant protective laws and regulations and therefore free for general use.

The publisher, the authors, and the editors are safe to assume that the advice and information in this book are believed to be true and accurate at the date of publication. Neither the publisher nor the authors or the editors give a warranty, expressed or implied, with respect to the material contained herein or for any errors or omissions that may have been made. The publisher remains neutral with regard to jurisdictional claims in published maps and institutional affiliations.

This Springer imprint is published by the registered company Springer Nature Switzerland AG
The registered company address is: Gewerbestrasse 11, 6330 Cham, Switzerland

Preface

Globalization has made supply chains increasingly complex, interconnected, and interdependent that rely heavily on superior performance and efficient operational systems. Such supply chains are often influenced by factors beyond the power and purview of global companies. As supply chains have become more global and complex, supply chain disruptions have become more frequent and severe. Thus, it induces supply chain vulnerabilities and increases the supply chain network's disruption risk exposure. In recent years, supply chain disruptions caused by unexpected events have occurred more frequently, and these disruptions have had both short- and long-term negative impacts on supply chain operations as well as on corporate profitability.

The COVID-19 outbreak, which began in late 2019 and spread worldwide, exposed the vulnerability of supply chains as never before. Other than its social impacts, it caused a significant disruption in global supply chains impacting production and logistics operations. Less than six months into the pandemic, 94% of Fortune 1000 companies experienced disruptions due to COVID-19, while 75% were negatively affected. In fact, the pandemic brought to the forefront the importance of supply chain risk mitigation strategies. Regardless of size and type, it is crucial for companies and enterprises to develop and implement suitable strategies to prevent supply chain disruptions from occurring and recover quickly from them when they occur. This can be done by establishing an effective supply chain risk management (SCRM) that identifies, assesses, and proposes strategies to manage and monitor supply chain risks. SCRM identifies the potential sources of risks and implements appropriate actions in order to mitigate supply chain disruptions. Therefore, firms need to build resilient supply chains to overcome any kind of disruptions.

This book presents a set of strategies, methods, and analyses that are essential for mitigating supply chain risks. It provides a practical and contemporary view on how companies can manage supply chain risks and disruptions in order to create a resilient supply chain. The book provides the state-of-the-art developments in mitigating supply chain risks from various perspectives. It can be used as an essential source for students and scholars who are interested in pursuing research or teaching courses in supply chain management. It also provides an interesting and informative read for managers and practitioners who need to deepen their knowledge on effective supply chain risk management.

Structured in a modular fashion, each chapter in this book presents an analysis of a specific topic on supply chain risk mitigation, allowing readers to identify the chapters that relate to their interests. More specifically, the book is presented in four parts: (1) Review on Supply Chain Risk Management, (2) Supply Chain Risk Strategies and Developments, (3) Supply Chain Sustainability and Resilience, and (4) Supply Chain Analysis and Risk Management Applications.

The first part of the book presents literature reviews on supply chain risk management. The second part focuses on risk and disruption management in supply chains by providing newly developed models and strategies, while the third part focuses on sustainability and resiliency in supply chains and a firm's ability to return to its original state after a disruption occurs. The last part presents supply chain analysis and some applications of the risk management. In the first chapter of the book, Jestin Johny and Amulya Gurtu, by focusing on the existing literature on supply chain risk management, present a review on definitions of supply chain risk, uncertainty, and vulnerability. They identify the supply chain risk characteristics and discuss their managerial and theoretical implications. They contribute by presenting a discussion on the existing literature on how risks are associated with supply chain integration and perceptions toward different types of supply chain risks. In the chapter entitled "Supply Chain Risk Management: An Enterprise View and a Survey of Methods," Mikaella Polyviou, George Ramos, and Eugene Schneller review the literature and the recent research on behavioral influences in the context of SCRM. They review the SCRM process, describe its four stages of risk identification, risk assessment, risk management, and risk monitoring, and propose practices and strategies that help enterprises identify, assess, manage, and monitor supply chain risks. They also discuss factors, such as industry and supply chain complexity, which can constrain an enterprise in implementing specific supply chain risk mitigation approaches. Moreover, they provide some examples of how enterprises across industries have implemented SCRM and identify key technologies employed within its process.

The second part of the book continues on managing supply chain risks by focusing on strategies and newly developed models. Lan Luo and Charles Munson begin this part with a chapter that develops a supply chain stress test. They propose a stress test approach to determine the ability of a supply chain to deal with crises under extreme, but plausible, scenarios. Using predictive global sensitivity analysis, they develop a single predictive structural equation that managers can use to estimate the percentage loss given a disruption scenario. It can be used by managers to assess the risk impacts of their respective supply chains. This enables managers to evaluate the risk of their current supply chain with a few simple calculations, and provide them with some managerial insights for mitigating supply chain risk. In the chapter entitled "Retail Supply Chain Risk and Disruption: A Behavioral Agency Approach," Raul Partyka, Fernando Picasso, and Ely Paiva address the retail supply chain risk management and study the retail consumer behavior during the COVID-

19 pandemic and the unexpected consumer demands experienced by retailers from the perspective of behavioral agency theory. They describe how disruptions in the retail supply chain during COVID-19 can be explained by the mechanisms of behavioral agency theory.

The chapter entitled "Mitigation of Supply Chain Vulnerability Through Collaborative Planning, Forecasting and Replenishment (CPFR)" focuses on vulnerability assessment in supply chains and its mitigation by applying a unique method. In this chapter, Leonardo de Carvalho Gomes addresses vulnerability mitigation in supply chains and shows the application of collaborative methods such as collaborative planning, forecasting, and replenishment (CPFR) in mitigating vulnerability effects. The author proposes a framework that clarifies the relationship between resilience and robustness, and supply chain vulnerability. The chapter entitled "The Boom and Bust of Medical Supplies During the COVID-19 Pandemic" investigates the causes of hoarding behavior toward essential products during the COVID-19 pandemic and explores how this behavior affects organizational decisions on the production rate. In this chapter, Tung Nguyen presents an insightful analysis of strategies to resolve the boom and the bust of medical supplies during the pandemic. The author discusses how hoarding behavior influences the stock management model and proposes collaboration among all actors in the supply chain to address the hoarding problem.

The next five chapters presented in the third part of the book focus on sustainability and resiliency in supply chains that evaluate firms' ability to return to their original states, within an acceptable period of time, after a disruption occurs. In the first chapter of this part (Sustainability Practices for Enhancing Supply Chain Resilience) Alejandro Ortiz-Perez, Elena Mellado-Garcia, and Natalia Ortiz-de-Mandojana analyze the different dimensions of resilience (pre-adversity, in-crisis organizing, and post-crisis resilience) and the typology of disruptions in the supply chain (internal, social, environmental, and global disruptions), and present a discussion on how sustainability practices can improve the resilience of the supply chain against different types of disruptions. They propose resilient practices through which sustainability commitment can foster relationships with suppliers and customers, by creating mechanisms to deal with a disruption in the supply chain. In the chapter entitled "A Theoretical Framework for Supply Chain Resilience Planning," Jennifer Helgeson and Alfredo Roa-Henriquez develop a theoretical framework for supply chain resilience planning based on the theory of planned behavior. They discuss risk management at the firm level and review the need for research that addresses planning for resilience in addition to risk management in the context of the need to plan ahead and recovery post-disaster.

In the chapter entitled "Towards a Resilient Supply Chain," Sandeep Ramachandran and Ganesh Balasubramanian explain the processes and best practices that can help firms manage their supply chain risks and move toward supply chain resiliency. They discuss three critical aspects of a firm—people, process, and technology—in designing and implementing an effective risk plan, and provide industrial applications and best practices for building resilient supply chains. In the next chapter (Integrated Optimization of Resilient Supply Chain Network Design and Operations Under Disruption Risks), Zhimin Guan, Jin Tao, and Minghe Sun develop an integrated optimization approach for the design and operations decisions of resilient supply chain networks with disruption risk considerations. They formulate a mixed binary integer programming model based on some stochastic scenarios by considering uncertainties in supplies, demands, and prices. They use an illustrative example to demonstrate the validity and effectiveness of the proposed model, and report the sensitivity analysis results to examine the effects of important parameters on the performance of the resulting resilient supply chain networks. In the chapter entitled "Balancing Sustainability Risks and Low-Cost in Global Sourcing," the last chapter of this part, Gbemileke Ogunranti and Avijit Banerjee develop an integrated framework as a decision support system for simultaneously addressing the supplier selection and order allocation problems, considering the trade-off between supply chain sustainability-related risks and procurement cost. They formulate a bi-objective mixed-integer programming model for the specific supplier selection and order allocation decisions to maximize the supply chain's sustainability performance while minimizing the procurement cost. They also present an illustrative example of outsourcing contract manufacturers in the apparel industry to demonstrate the applicability of their proposed framework in practice.

The last part in the book presents more supply chain analyses and some applications of the risk management. The first chapter of this part discusses approaches to incorporate risk aversion in supply chain network design. In this chapter (A Bi-objective, Risk-Aversion Optimization Model and Its Application in a Biofuel Supply Chain), Krystel Castillo-Villar and Yajaira Cardona-Valdés propose a mathematical formulation of a bi-objective two-stage stochastic mixed-integer linear programming model that measures the trade-off between the expected cost and the conditional value at risk. The proposed model can be used to determine the tactical and strategic decisions of a biofuel supply chain network considering the biomass quality variability issues and the risk of investment. They also conduct a real-life data-driven case study at a state level to obtain pragmatic and managerial insights that enable the investigation of solutions for different levels of risk aversion. In the chapter entitled "Conceptualizing and Modeling Supply Chains in the Hazard Context," Douglas Thomas and Jennifer Helgeson discuss the impact of natural and human-made hazards in the US economy as a whole and the US manufacturing industry in particular. They present a macro-level analysis and examine how establishments—at the total economy level and within the manufacturing industry in particular—are affected by a disruption in supplies, to determine the magnitude of the downstream effect, which is often referred to as the ripple effect.

In the chapter entitled "Developing Predictive Risk Analytic Processes in a Rescue Department," Mika Immonen, Jouni Koivuniemi, Heidi Huuskonen, and Jukka Hallikas present a data-driven risk management case application in the service supply chains and analyze predictive risk analytic processes and practices in rescue departments. These practices include preventing, detecting, responding, and recovering from issues by using capable information systems in the supply chain. They present a case study of the use of data analytics in risk management, which highlights the diverse use of different types of structured and unstructured data in proactive risk management. In the last chapter of the book (Manufacturing-Supply

Chain Risk Under Tariffs Impact in a Local Market), Omar Alhawari and Gürsel Süer address a manufacturing-supply chain network, and study the impact of imposed tariffs on imported raw materials on manufacturer and suppliers. They conduct a case study and discuss the impact of the trade tariffs on the manufacturing system design considering the number of manufacturing cells open to meet the demand. They discuss how a low market demand enforces manufacturers to change their system design to absorb the impact of high imposed tariffs.

I would like to thank all the authors who have contributed to this book. Although the COVID-19 pandemic caused a significant delay in editorial process and reviewing chapters, I appreciate the authors' patience, and also their great effort in revising their chapters, in particular, those who were requested for multiple revisions. My special thanks go to the co-editor, Dr. Henry Xu of the University of Queensland for helping out with reviewing several chapters.

Also, I would like to express my gratitude to Matthew Amboy of Springer New York and Jialin Yan, Associate Editor, for their help and support on this project, and also thanks to the project team in Springer including the project coordinator, Ramya Prakash.

I would also like to thank my wife Miya and sons Nima, Yuma, and Toma for allowing me to devote the time necessary to complete this book during the hard times of the pandemic, especially during our temporary stay in Toronto, Canada. I dedicate this book to them.

Toronto, Canada Yacob Khojasteh
May 2022

Contents

Part I Review on Supply Chain Risk Management

Risks in Supply Chain Management 3
Jestin Johny and Amulya Gurtu

**Supply Chain Risk Management: An Enterprise View
and a Survey of Methods** 27
Mikaella Polyviou, George Ramos, and Eugene Schneller

Part II Supply Chain Risk Strategies and Developments

Developing a Supply Chain Stress Test 61
Lan Luo and Charles L. Munson

**Retail Supply Chain Risk and Disruption: A Behavioral Agency
Approach** 81
Raul Beal Partyka, Fernando Gonçalves Picasso, and Ely Laureano Paiva

**Mitigation of Supply Chain Vulnerability Through Collaborative
Planning, Forecasting, and Replenishment (CPFR)** 95
Leonardo de Carvalho Gomes

**The Boom and Bust of Medical Supplies During the COVID-19
Pandemic** 121
Tung Nhu Nguyen

Part III Supply Chain Sustainability and Resilience

Sustainability Practices for Enhancing Supply Chain Resilience 143
Alejandro Ortiz-Perez, Elena Mellado-Garcia, and Natalia Ortiz-de-
Mandojana

A Theoretical Framework for Supply Chain Resilience Planning 159
Jennifer F. Helgeson and Alfredo Roa-Henriquez

Toward a Resilient Supply Chain 191
Sandeep Ramachandran and Ganesh Balasubramanian

Integrated Optimization of Resilient Supply Chain Network Design and Operations Under Disruption Risks 205
Zhimin Guan, Jin Tao, and Minghe Sun

Balancing Sustainability Risks and Low Cost in Global Sourcing 239
Gbemileke A. Ogunranti and Avijit Banerjee

Part IV Supply Chain Analysis and Risk Management Applications

A Bi-objective, Risk-Aversion Optimization Model and Its Application in a Biofuel Supply Chain 275
Krystel K. Castillo-Villar and Yajaira Cardona-Valdes

Conceptualizing and Modeling Supply Chains in the Hazard Context 293
Douglas S. Thomas and Jennifer F. Helgeson

Developing Predictive Risk Analytic Processes in a Rescue Department 311
Mika Immonen, Jouni Koivuniemi, Heidi Huuskonen, and Jukka Hallikas

A Manufacturing-Supply Chain Risk Under Tariffs Impact in a Local Market 331
Omar Alhawari and Gürsel Süer

Part I
Review on Supply Chain Risk Management

Risks in Supply Chain Management

Jestin Johny and Amulya Gurtu

Abstract Globalization created complex supply chain networks that rely heavily on superior performance and efficient operational systems. Global supply chains require integrated information flow across boundaries for the smooth movement of goods and services. The risks in SCM are due to vulnerabilities and uncertainties at the operational levels in the global supply chain, posing a challenge to the rapidly evolving logistics industry. This chapter focuses on the existing literature on supply chain risk management to identify strategies to minimize these risks due to the competitive business environment by implementing risk mitigation, identification, and assessment in the supply chain networks. The main contribution of this chapter is the discussion on the existing literature on risks associated with supply chain integration and perceptions toward different types of supply chain risks.

Abbreviations

CAGR	Compounded annual growth rate
EDI	Electronic data interchange
FDI	Foreign direct investment
GPS	Global positioning system
GSCM	Green supply chain management
JIT	Just in time
OECD	Organization for economic cooperation and development
OEM	Original equipment manufacturer
SC	Supply chain
SCI	Supply chain integration

J. Johny
N.L. Dalmia Institute of Management Studies and Research, Mumbai, Maharashtra, India

A. Gurtu (✉)
University of Wisconsin-Green Bay, Cofrin School of Business, Green Bay, WI, USA
e-mail: gurtua@uwgb.edu

SCM	Supply chain management
SCRM	Supply chain risk management
SME	Small medium enterprises
SSCM	Sustainable supply chain management

1 Introduction

The trajectory of globalization and surge in business activities gave rise to developing the structure for efficient supply chains and facilitated commercializing the term *"Supply Chain Management."* The field of supply chain management has generated curiosity among academicians and practitioners in the last three decades. Nevertheless, the use of the term supply chains could be traced back to the mid-twentieth century (Gurtu et al., 2017, p. 2). For example, supply chains of animal fur from North America to Europe existed in the seventeenth century (Ray, 2005), and supply chains of spices from India to Europe via Africa existed many centuries before that (Gurtu et al., 2017, p. 1).

The early definition of a supply chain was "the connected series of activities which is concerned with planning, coordinating and controlling material, parts and finished goods from supplier to customer" (Stevens, 1989, p. 3). Further, Mentzer et al. (2001) defined a supply chain "as a set of three or more entities directly involved in the upstream and downstream flows of products, services, finances, and/or information from a source to a customer." Based on the above classification of supply chains, Mentzer et al. (2001, p. 18) supplemented supply chain management (SCM) as "the systemic, strategic coordination of the traditional business functions and the tactics across these business functions within a particular company and across businesses within the supply chain, to improve the long-term performance of the individual companies and the supply chain as a whole." This definition is widely cited in many papers on supply chain management (Ahi & Searcy, 2013; Burgess et al., 2006; Gaiardelli et al., 2007; Gundlach et al., 2006; Svensson, 2002) and aligns with other contributions. For example, the Council of Supply Chain Management Professionals (CSCMP) defined SCM as "the planning and management of all activities involved in sourcing and procurement, conversion, and all logistics activities. Importantly, it also includes coordination and collaboration with channel partners, which can be suppliers, intermediaries, third-party service providers, and customers. Thus, supply chain management integrates supply and demand management within and across companies."

Global supply chains are complex and require greater synchronization in information flow within and across national/international boundaries to move goods and services (Mentzer et al., 2001). For this purpose, organizations source products and services from the best and the most economical source from various locations across the globe (AlHashim, 1980). However, risks and vulnerabilities are associated with global supply chains and their management. "The globalization of supply chains has increased the significance of logistics" (Gurtu, 2019; Gurtu & Johny, 2021). One of the risks and vulnerabilities comes from logistics. Uncertainties in logistics operations exist at strategic, tactical, and operational levels (Schmidt & Wilhelm, 2010).

The global logistics industry is evolving rapidly, and the latest technologies in supply chains have affected organized and unorganized logistics vendors. For example, DHL Express introduced a new TC55 technology that works on the Android platform by using the navigation skills of GPS. The global logistics market is projected to rise at a 7.4% CAGR over the forecast period from 2018 to 2026 (Transparency Market Research, 2020). The global transportation industry was $1.6 trillion in 2015, and it contributed to about 8% of the GDP in the USA (Gurtu, 2019).

This chapter provides an overview of the current literature on supply chain risk management and various types of risks due to supply chain integration. The highly competitive business environment, coupled with a desire to improve the quality of service to customers, lower input costs, and improve profitability, among other things, have created challenges and opportunities for improving the efficiencies in supply chains. The chapter is divided into eight sections. Section 2 describes global supply chain risks and contingency strategies to minimize those risks. Section 3 discusses the concepts and definitions of Supply Chain Risk Management (SCRM) discussed by various academicians and experts in supply chain management. Section 4 explains why managing risk in supply chain integration at the upstream and downstream flow of information in the supply chain network is important. Section 5 narrates that managing risks in a supply chain integration is vital due to the complexity and economic challenges posed by uncertainty in the global environment. Section 6 elucidates the characteristics in various disciplines of supply chain followed by supply chain vulnerability and risk mitigation strategies. Section 7 discusses managerial implications and Sect. 8 concludes the discussion with future risk mitigation strategies.

2 Supply Chain Risk Management

Profit maximization is one of the main objectives of corporate organizations. Equilibrium between output (efficiency) and viability (effectiveness) (Mentzer & Firman, 1994) for the passage of goods and raw materials between nations in a timely and seamless manner (Bowersox & Calantone, 2018) is required to arrive at the ultimate objective of profit. Global supply chains involve economies, cultures, politics, infrastructure, and business environments (Schmidt & Wilhelm, 2010). Managing and controlling global value chains generate conflicts between the centralized team and the local administration of the entire structure (Wildt, 1982).

Global supply chains have greater uncertainties as compared to local or regional supply chains. The 9/11 attacks in the USA, the pandemics crisis like Ebola, and SARS, the nuclear disaster of Fukushima, natural disasters like a tsunami in Asia, and volcano eruption in Italy were risks and have caused significant damage to global supply chains. However, the latest pandemic, COVID-19, has raised the level of risk in global supply chains to a different level. The pandemic has caused disruptions in various layers of supply chain management, like distribution, packaging, and sourcing of raw materials (Aldaco et al., 2020; Choudhury, 2020). The recent pandemic disrupted the trade flows among countries, particularly between border-sharing countries, resulting in lower export earnings (Gurtu et al., 2022a). For example, the NBA announced its season's cancelations on March 12, 2020, after a player was found infected with COVID-19 (Aschburner, 2020). The pandemic is uncertain, and its existence will depend on various circumstances leading to a few months or more (Smit et al., 2020). The bath tissue industry was severely affected in the USA and the EU due to COVID-19, compounded by a surge in demand in residential consumption and a decline in commercial consumption (Gurtu et al., 2022b). Therefore, organizations must have risk mitigation strategies in the occurrence of such events. This preparedness requires leadership to deal with global marketing and manufacturing strategies (Bowersox & Calantone, 2018; Wojakowski, 1971). Hauser (2003) suggests a business framework to assess and manage risk in an organization. The framework suggested by Hauer allows organizations to analyze, prioritize, and measure the economic impact of risk on various business initiatives. The business case framework allows for identifying vulnerabilities, the organization desired risk profile, and measuring performance to achieve value propositions aligned with the company's business goals. Implementation of these frameworks requires a leadership focus as specific structural and procedural changes are required. Strategy implementation requires discipline, commitment from employees/employers, creativity, and superior execution skills (Freedman, 2003).

3 SCRM Concept and Definitions

Natural disasters and human-made accidents have increased in high, medium, and low-income countries during the past decades (Coleman, 2006). Natural disasters, terrorism, and other unpredictable events increase the risk uncertainties global supply chains face (Brown et al., 2006; Chopra et al., 2007; Stewart, 1995). Table 1 provides existing literature and findings on risk and Supply chain integration. Uncertainty can be defined as the inability to develop a planned estimate or come up with possible solutions related to choose or event. For example, at the beginning of 2018, KFC's apologized for running out of chicken in the UK. Three weeks after the announcement, nearly three percent of KFC's restaurants in the U.K. remained closed. The reason for such a disaster was KFC's new approach to the supply chain system. Previously, KFC used to work on a single distribution platform from farms to outlets, and that is how big supermarkets have been working from warehouses in Daventry (UK) for many years. The chicken did not get to KFC outlets because an entirely new system and new processes were being implemented. Experience shows that new software and technologies lead to teething problems. Under the new KFC approach for the entire market of the UK, DHL has taken over the delivery infrastructure, and QSL supplied the software and processes for the

Table 1 Existing literature and findings on risk and supply chain integration

Author	Journal	Method	Main findings
Jüttner (2005)	IJLM	Survey and Focus groups	Traditional approaches to risk management derived from a single firm perspective are not ideally suited to the requirements in s supply chain context
Kleindorfer and Saad (2009)	POM	Empirical Quantitative	Formulated ten principles for risk management deriving from supply chain and industrial risk management literature
Braunscheidel and Suresh (2009)	JOM	Survey	Market orientation impact internal and external integration, along with external flexibility Learning orientation influences internal integration but does not influence external integration Internal Integration leads to External Integration Internal Integration and External Integration I do not lead to external flexibility External Integration, external flexibility, and Internal Integration positively impact supply chain agility which is a risk management initiative
Chen et al. (2013)	IJPR	Survey	Process and demand risks negatively impact supply chain agility which is a risk management initiative Supply chain collaboration reduces all the risks associated with supply chains
Lavastre et al. (2014)	IJPR	Mixed-Method	Firms need to consider SCRM at a strategic level, requiring long-term information exchanges with partners In SCRM, the length of a relationship is very important as methods employ differ as per the duration of a relationship Successful implementation methods based on partner collaborations are most effective
Ellinger et al. (2015)	IJLRA	Survey	Supplier, customer, and internal integration mediates the relationship between SCRM and learning orientation
Gualandris and Kalchschmidt (2015)	IJLM	Field Interviews and Survey	Balanced use of integration practices with SCRM approaches, i.e., revenue sharing and dual sourcing can enhance competitive advantage
Fan et al. (2015)	IJPE	Survey	Risk information sharing and risk-sharing mechanisms are positively associated with financial performance Relationship length and supplier trust positively moderate risk information sharing financial performance evaluation while shared SCRM understanding does not Shared SCRM understanding moderates the relationship between risk-sharing mechanism and financial performance while relationship length and supplier does not

(continued)

Table 1 (continued)

Author	Journal	Method	Main findings
Kauppi et al. (2016)	IJPE	Survey	Traditional risk management practices and external SCI practices complement each other in facing disruption risk and improving operational performance
Brusset and Teller (2017)	IJPE	Survey	Integration Capabilities and flexibility capabilities are positively associated with resilience External capabilities do not lead to resilience Customer risk does not have moderating effects between external capabilities, integrative capabilities, and resilience
Revilla and Saenz (2017)	IJOPM	Survey	Firms having inter-organizational collaboration and integral relationships face the lowest level of supply chain disruptions At the supply chain level, having control over internal operations alone is not enough to handle disruptions
Zhu et al. (2017)	IJLM	Systematic Literature review	Supply chain risks transmit through supply chain members and should be managed for the end-to-end supply chain as a whole

IJLM International Journal of Logistics Management, *POM* production and operations management, *JOM* Journal of Operations Management, *IJPR* International Journal of Production Research, *IJPE* International Journal of Production Economics, *IJLRA* International Journal of Logistics Research Applications, *IJOPM* International Journal of Operations and Production Management

information systems. A small administrative error affected KFC's entire supply chain system, which could have been avoided if the approach had been rolled out in a phased manner, for example, region-wise. Modern supply chains are complex and critical, as it involves different organizations. Each organization contributes unique expertise and resources (Wilding, 2018). This illustrates the inability to manage risks and recover from breakdowns and disruptions. Analyzing the risks can lead to new insights for various pragmatic strategies that mitigate various sources of risk (Sheffi, 2005; Sodhi & Tang, 2012).

The business environment is becoming highly competitive because of uncertainties attached due to variations in supply and demand (Christopher & Lee, 2004). In the era of Industry 4.0 and the circular economy, small and medium enterprises (SMEs) are under pressure to meet environmental norms in order to be a part of the global value chain framework. However, SMEs lack the knowledge and expertise in this area and do not have the resources to acquire it. Therefore, the opinion of consultants and original equipment manufacturers (OEMs) should be taken into consideration. Nowadays, organizations are obliged to adopt new ways of managing a business due to the influence of new technologies, financial instability, and outsourcing, among others (Stefanovic et al., 2009). Supply chain risk can be termed an experience to an event that causes disruption and further affects the supply chain network.

Risk management is becoming an integral part of a holistic SCM design (Christopher & Lee, 2004). Risk itself can be termed as disruption, vulnerability, uncertainty, disaster, peril, and hazard. According to Knight (1964), ambiguity is infinite as it lacks complete certainty and has more than one possibility. In contrast, a risk results from some probable uncertainty with some loss or unwanted outcomes (Hubbard, 2009, 2014). Further, Williams et al. (2008) say that a supply chain is a subcomponent of an overall risk management strategy within the organization. For the definition of supply chain risk, uncertainty, and vulnerability, refer to Appendix.

4 Why Risk Management in Supply Chains

A supply chain is a complex network of upstream (supplier) and downstream (customer) flow of information, service, products, materials, and finance (Jüttner, 2005; Mentzer et al., 2001). Therefore, Supply chain management is the coordination of these interdependent activities. Risk is a disruption between the supply chain entities, as stated above. Hence, the approach toward risk must have a broader scope that provides critical insights for performing across a supply chain (Jüttner, 2005).

Supply Chain Integration (SCI) refers to the extent of strategic alignment and interconnection of a firm and its supply chain partners consisting of internal (cross-functional) and external (customer and supplier) integration (Flynn et al., 2010). For example, Tesco stated that the company focused on better services to customers, re-engaging their colleagues, completely resetting the relationship with the suppliers. As a result, the company has been able to add value to its shareholders. This allowed more popular products to get onto shelves and achieve a strong operational and financial position to deal with the challenges like the Covid-19 situation (Green, 2020). Further, supplier integration refers to information sharing and coordination by providing insights into the supplier's capability, process, planning, and product design for effective operations management (Schoenherr & Swink, 2012). The integration of suppliers and customers lowers supply chain risk when the network shares information throughout the chain (Waters, 2011). However, Laurence Boone, chief economist at the Organization for Economic Cooperation and Development (OECD), said that the outbreak of COVID-19 could lead firms to move away from extremely integrated supply chains in order to mitigate risk in the future (Hart, 2020). The requirement of quality information is needed for anticipating environmental uncertainty. Supplier and customer integration acts as the input or source of information processing to analyze and examine the information gathered from suppliers and customers.

Fig. 1 A five-step process for global supply chain risk management and mitigation (adapted from Manuj & Mentzer, 2008)

5 Risks in Supply Chain Integration

Globalization and consolidation of firms within and outside the industry are responsible for increased uncertainty and complexity in the global supply chain (Abrahamsson et al., 2010). Uncertainties are noticeable due to economic challenges like fluctuations in oil prices and currency rates. Further, nations' competitive advantages affect supply chain systems at the macro and micro levels. However, the uncertainty of lead times and supplier reliability also adds to the worry of supply chain performance (Schmidt & Wilhelm, 2010). As stated by Hocking et al. (1982), logistics activities have the chance of being affected by lead-time uncertainty. Supply chain risk is a complex phenomenon divided into sources and types of risk (Svensson, 2000, 2002). According to Svensson's theory, risk can be atomistic or holistic. Atomistic sources of risk signify selected and limited parts of a supply chain suitable for low-value, non-complex, and commonly available parts and materials. The holistic view of the risk indicates that end-to-end supply chain analysis is required to assess the overall risks. Further, Manuj and Mentzer (2008) have identified five-step risk management and mitigation (Fig. 1). In the first step of the risk management framework, the objective is to identify a risk profile that broadly contains risk as atomistic/holistic, quantitative, and qualitative that affects global operations. Risk assessment and evaluation describes which risks identified in the first step are critical for global supply chain operations. After assessing and evaluating, the next step is to select appropriate strategies to manage the risk, which helps in reducing the probabilities of losses associated with risk events. Implementation of strategies for risk requires specific structural and procedural changes with current globalization trends. Strategy Implementation requires discipline, commitment, creativity, leadership, and superior execution skills (Freedman, 2003). The last step describes mitigation planning that provides a firm with a more mature decision-making process in facing losses caused by unexpected events.

The pace of evolving complex business environments and complicated operational strategies of firms contribute to the risk and vulnerability of supply chains. For example, the fire at Philips plant in 2000 that affected both Nokia and Ericsson disrupted their supply chains. Likewise, the Fukushima Daiichi nuclear disaster in Okuma affected Japan in 2011 and impacted global supply chains. Thus, risks in increasingly complex supply chain networks have brought risk management to the forefront of research and managerial focus. The supply chain partners can identify and manage risk through coordinated approaches (Jüttner et al., 2003). As a result, they play a crucial role in dealing with the dynamic and uncertain business environment and are widely adopted by firms to address increasing risks (Lavastre et al., 2014; Manuj et al., 2014).

The evolving business environment and rising competition are searching and finding efficient supply chain partners within a supply chain network. The pace of innovations in the digital marketplace permits buyers and sellers to transact in a multidimensional marketplace, connecting to multiple trading partners. With increasing development in the digital revolution, many industries are investing in e-collaboration technologies to advance their supply chain capabilities (Foley & Kontzer, 2004). Intel, Wal-Mart, and Kraft Foods are a few examples of organizations that have taken such initiatives. Firms have launched digital platforms for business transactions with their partners/suppliers. SCM e-collaboration technologies enable us to combine and collaborate irrespective of physical proximity while increasing the speed of response, reducing wasteful processes, and e-procurement in the form of real-time demand forecasting. For example, global trade depends on the physical verification of documents, and mediators are involved in the smooth functioning of trade. Hence, Blockchain technology is expected to transform the global supply chain platform by eliminating brokers through a digital platform. In addition, blockchain facilitates effective measurement of the outcome and performance of the SCM process (Gurtu & Johny, 2019). There are organizational and strategy implementation risks, and technology and relational risk in supply chain management. There are two sub-sections. Section 5.1 describes risk in the organization where the top management and different functional departments are involved in achieving strategic fit in the uncertain business environment. Section 5.2 explains risk in technology adoption and how supply chain partners can respond quickly to improve JIT and increase efficiency in the supply chain network.

5.1 Organizational and Strategy Implementation Risk

Risk in the organization refers to the extent of top management involvement and commitment to resources and financial capital. The cost of having an SCM digital platform demands high capital investment. They include connection, hardware, software, set-up, and maintenance costs (Commerce Net, 1998; Iacovou et al., 1995; Nath et al., 1998; Senn, 1998). The rise in IT investment costs has forced supply chain partners to opt for cheaper applications to secure a business system. Also, the training required to get hands-on experience further pushes capital investments. The lack of global standards and guidelines on implementing sustainable technologies is a challenge for SMEs in developing economies (Shin et al., 2019). Leitão et al. (2016) and Bedekar (2016) have highlighted the IT infrastructure-based issues in SMEs, such as weak network connectivity. Feng et al. (2018) highlighted the lack of a trained workforce in management and staff and stressed the need for training about the technologies to improve the skills of management and staff. Moreover, a lack of IT audit systems results in weak software development procedures, leading to poor operational linkages that hinder information sharing.

Implementation of risk also increases due to a lack of technical knowledge skills and training (Premkumar, 2003).

5.2 Technology and Relational Risk

Technological risk emerges from poor integration and security issues that impact the confidentiality, integrity, and authentication of supply chain transactions (Bhimani, 1996). These risks impact supplier operations as they are required to adopt different technological solutions provided by buyers. Technological integration issues stem from supply chain activities and functions that are not networked together and are not likely to foster SCM performance. These risks are derived from a lack of experience about security, the ability to audit e-collaboration systems, task uncertainties, and environmental uncertainties. In addition, the spread of SCM e-collaboration technologies has left most supply chain partners uncertain of their business operations and unaware of the full potential of the technical solutions (Ghosh, 1998).

Further, relational risk refers to the choices, anxiety, and low self-esteem toward approving or disapproving, leading to uncertainties (Cunningham, 1967). Insufficient cooperation and negative attitudes among buyer–supplier relationships may result from exercising coercive power (Blau, 1964). Moreover, power is the capability of a firm to exert influence on another firm to act in a prescribed manner. Power plays a vital role in SCM e-collaboration strategies. For example, previous research in the automotive industry suggests that Ford applied power when its EDI network was first introduced. Its main objective was to gain a competitive advantage by locking its suppliers into its system and its competitors out. EDI allows companies to exchange data electronically rather than on paper. It is an essential component of automation in business processes.

SCM performance results from SC responsiveness, which refers to efficiencies along the supply chain in reducing response lag time and meeting demands for different supply chain activities. The degree to which supply chain partners can respond quickly to demand and environmental changes in the marketplace. Previous studies suggest that effective communication and trust support intra-inter relationships in SCM (Daft & Huber, 1986; Mohr & Nevin, 1990; Morgan & Hunt, 1994).

6 Supply Chain Risk Characteristics

There are numerous definitions of supply chain risk from multiple disciplines in the academic literature. Finance, insurance, emergency management, health, safety, and environment streams have further defined supply chain risk. The supply chain risk is related to the organization's objectives, which need to be proficient enough to achieve it. The achievement of these objectives depends on the elucidation toward unpredicted and ambiguous developments in supply chains. The next sub-sections describe the risks and measures to minimize the impact of the risks in the supply chain network. Section 6.1 explains risk in the firm's capital resources, which may lower the firm's image in the long run. Section 6.2 elucidates the disruptive risks due to changes in the business environment, higher customer expectations, and the collaborative efforts in the global supply chain network to reduce risk impact. Section 6.3 labels vulnerability as the supply chain's exposure to uncertain events and the ability to overcome such unpredictable events. Section 6.4 describes globalization and outsourcing as the critical factors for risk drivers and strategies to mitigate these risk drivers in the supply chain network.

6.1 Financial Risk Objective

The objective of this area is to plan, monitor, and control the company's capital resources. It is necessary to predict and control uncertain developments, which may further lead to the depreciation of brand value and objectives. Risk comprises both gains and losses around the expected value of financial returns. These risks define the loss due to fluctuations in market price, debt payments, and exchange rates, among others—risk management models in finance attempt to predict the consequences of the movements in the operational functions. Operational risk is defined by the Basel II Committee as "the risk of loss resulting from inadequate or failed internal processes, people, and systems" (BIS, 2006). Operational risks better reflect the complexity, uncertainty, and diversity of valid risk sources for supply networks. SCM significantly influences business goals and provides a competitive advantage when designed appropriately (Wagner & Bode, 2008). The goals in SCM can be achieved in two ways: efficiently and effectively (Håkansson & Prenkert, 2005; Möller & Törrönen, 2003). Effectiveness means that achieving a predefined goal can be guaranteed even if conditions are adverse, and efficiency refers to minimal spending of resources to reach this objective. The primary function of a supply chain is to satisfy customer demand and resource availability. Efficiently designed supply chains provide the possibility of higher competitive advantages. Supply chain efficiency refers to a cost and waste minimal execution of supply chain activities (Borgström, 2005). The quest for efficiency and effectiveness are conflicting and need to be carefully balanced (Kull & Closs, 2008).

6.2 Disruptive Risk

The essential feature of risk perception in the research inquiry is the availability of probability distributions. The construction or identification of an optimal or satisfying decision is well supported by decision theory for decision-makers. The effective and efficient supply chain practice in today's globalized world depends on the collaboration between geographically dispersed organizations (Kovács & Paganelli, 2003). Rosenhead et al. (2017) were the first authors to classify a decision process according to the available information into three categories, i.e., certainty, risk, and uncertainty. Under *certainty*, all parameters are deterministic and known. Therefore, the relation between information and decision is unambiguous. Reasons for the need to decide under these circumstances vary from lacking time and resources to collect, process, and evaluate the information for knowing the structural complexity of systems that act as a hurdle in predicting the consequences of a decision. To discern between these different situations, two categories are introduced: decision-making under *risk* relies on the probability distribution, and decision-making under *uncertainty* due to lack of information about the likelihood of parameter changes. Therefore, supply chain risks address both decision-making under risk and uncertainty.

The international standard, ISO-14971 defines and measures a risk R as the product of probability and harm of an event e: $R = P_e S_e$, where S_e and P_e refer to the severity and probability of e, respectively. Furthermost, supply chain risk definitions begin with the assumption that events are a significant factor in determining risk (Waters, 2011). Supply chain risk is understood as a triggering event. (Wagner & Bode, 2008) state that disruption is characterized by its probability, severity, and effects. Many supply chain risk analysis methods relate to the source or root cause, while a few authors relate risk to ultimate consequences. This is to know the likely causes and take measures to reduce the chances of their occurrence in the future. However, the increased complexity of the supply network, changes in the environmental conditions, and market signals drive inadequate mitigation (Jüttner et al., 2003).

6.3 Supply Chain Vulnerability

The notion of the extent to which a supply chain is prone to a specific or vague event is called supply chain *vulnerability*. A concept closely associated to supply chain vulnerability is *supply chain resilience* describing the ability of a supply chain to overcome disruption. Generally, vulnerability is defined as a concept that describes the character and circumstances of a community, system, or asset that makes it susceptible to the damaging effects of a hazard (UN Office for Disaster Risk Reduction, 2004). Defining supply chain vulnerability as a risk requires understanding what is meant by supply chain risk (Peck, 2007).

The understanding of *resilience* relates to the ability of the underlying system to adjust or maintain essential functions under stressful and harsh conditions. The first definitions of resilience related to supply chain management were developed in 2004 at Cranfield University and MIT. Consequently, definitions on supply resilience either refer to the ability to overcome supply chain disruptions (Barroso et al., 2009; Pettit et al., 2010; Svensson, 2000, 2004) or the ability to reduce supply chain risk (Falasca et al., 2008; Jüttner & Maklan, 2011; Peck, 2005).

Fig. 2 Supply chain risk management construct

6.4 Risk Drivers and Mitigation Strategies

According to March and Shapira (1987), "risk" can be defined as "the variation in the distribution of possible supply chain outcomes, their likelihood, and their subjective values." However, decision-making in risk uncertainty leads to adverse implications and chaos in the supply chain network. Braithwaite and Hall (1999) emphasized that the relationship between corporate strategy, risk, and the implications for supply chain management are poorly understood and need further explanation.

The four basic supply chain risk management constructed from the theoretical discussion are summarized in Fig. 2. Most academics and practitioners recommend that the influences on contemporary supply chain management in the last decade, like globalization or outsourcing, have intensified the risk exposure of supply chain management (Engardio, 2001; McGillivray, 2000). Svensson (2002) uses '*calculated risk*' to increase a company's competitive advantage, profitability, or reduce costs. Risk mitigating strategies are the strategic moves establishments consciously undertake to mitigate the uncertainties identified from the various risk sources (Miller, 1992). The term supply chain risk vulnerability is "the propensity of risk sources and risk drivers to outweigh risk mitigating strategies, thus causing adverse supply chain consequences." From a single firm perspective, the adverse consequences affect a firm's goal accomplishment (Svensson, 2002). The sources of risk in supply chains can be categorized into three categories, environmental risks, network-related risks, and organizational risks. Environmental risk sources comprise any uncertainties arising from supply chain environment interaction. Organizational risk sources lie within the boundaries of the supply chain parties and range from labor or production uncertainties. Network related risk sources as the interactions between organizations within the supply chains. Further, supply chain risk consequences depends on specific supply chain context like, risks associated with stockouts, high levels of inventory and political ramifications. Due to the risk consequences, organizations are under constants pressure of operational efficiency due to competitive business environment. Risk mitigating strategies refers to managing risks through flexibility, cooperation, and control of supply chain disruptions.

7 Managerial and Theoretical Implications

Supply chain decisions influence other supply chain members through direct and indirect effects (Novack et al., 1992). Therefore, decisions on risk management should reflect strategies and processes across the global supply chain. Global supply chain risk decisions may be classified into horizontal, vertical, and diagonal effects. For example, a firm will choose multiple suppliers for critical components if it assumes an equivocation strategy to safeguard itself against a disruptive supply chain environment (horizontal effect). On the other hand, suppose a firm chooses to adopt a delay strategy, like labeling and assembly at fulfillment centers close to customers. In that case, it will modify the expectations from suppliers and affect how downstream supply chain members align their systems to the delay strategy (a vertical effect).

Selecting the right risk strategy is crucial for the supply chain network sustainability in the long-term perspective. The managerial perceptions of risks (Sodhi et al., 2012; Zsidisin, 2003) are critical for SCRM and studied by researchers. Choosing the appropriate risk management strategy in risk-averse, risk-neutral, risk-sharing, or risk-taking (Vanany et al., 2009) will directly impact the mitigation. Further, sustainability factors will considerably influence SC design in the future. Non-compliance with sustainability factors could lead to SC risk and disruptions. Therefore, there is a need to mitigate risk and implement sustainability practices. An organization's focus should be on economic sustainability to remain in the business for the long run. Nevertheless, the organization's upper management and SC managers must be aligned with the organizational philosophy regarding competing needs of economic viability and environmental and social sustainability (Gurtu et al., 2017).

SCM collaboration and outsourcing by risk-sharing or contracts amongst SC partners can help to improve the network efficiency (Urciuoli, 2009), which can also be termed *knowledge sharing*. The development of supplier partnerships and strategic alliances is becoming an essential element for long-term profitability and a robust risk mitigation strategy. Contingency planning strategies need to be industry or SC-specific (Jüttner et al., 2003). However, apart from collaboration, risk traceability strategies can be improved if the information is readily available quickly and accurately. Communication and information-sharing technologies are expected to make a significant impact in terms of traceability in SCM. For instance, it is increasingly difficult to find a product made genuinely in a single country, i.e., all inputs for a finished product are from a single country. Therefore, national and regional policies alone are no longer enough to address today's global supply chains (Gurtu et al., 2017).

It is not easy to establish whether quantitative models provide a better understanding and theory than qualitative work. However, research should have a practical use. Various authors have suggested the requirement for better risk management. Some of the proponents have suggested the following for better research in SCRM; empirically grounded research (Jüttner et al., 2003), quantitative tools like mathematical programming models, simulation models (Rao & Goldsby, 2009), analytical/network hierarchy process (Vanany et al., 2009), complexity and graph theory (Colicchia et al., 2012), development of well-grounded models by considering other interdisciplinary research approaches (Khan & Burnes, 2007).

Managers should appreciate the importance of integrative practices for mitigating supply chain risks. Key supply chain partners are sources of external environment information, a critical input for a firm's process, especially in risk-prone situations. At the same time, cross-functional integration among different departments acts as an information processing capability for absorbing, processing, and timely implementation of information for responding to changes in the external environment. Poor communication with key supply chain partners would have a cascading effect on firms' information-sharing process. In contrast, poor internal integration within the firm would affect the processing and timely utilization of information gathered from external integration. Hence, supply chain managers should develop integrative practices and critical supply chain partners to manage risk and enhance operational performance.

8 Conclusions

The awareness of supply chain risk management has been raised by disruptive events impacting the business environment. These incidents have raised the importance of robust supply chain networks for the entire industry. However, understanding the meaning of supply chain risk, which information to share and monitor in light of these disruptive events is varied. Global supply chains have become complex with newer risks, where demand for innovative and sustainable ways to manage complexity in the global supply chain network is needed. Further, global sourcing primarily occurs from emerging markets and developing economies. In these contexts, substantial leverage effects for sustainability in supply chains can be expected by reducing adverse societal impacts and minimizing related risks (Kelling et al., 2020). Flexibility, organizational learning, and information systems are recognized as critical enablers in the process of risk management and mitigation. Therefore, there is a need to understand how these enablers influence risk and effectively make sound supply chain decisions.

The top-level management should undertake risk management procedures to improve external and internal levels of cooperation. Firms also need to address vulnerability issues related to hackers, computer viruses, spam, and telecommunication failures. The challenge is not limited to technology, but the trust factor, data sharing, and security of firms. Risk in SCM refers to the potential occurrence of events associated with inbound and outbound activities that can have significant detriments to the purchasing firm (Zsidisin, 2003). Several factors have emerged which have increased supply chain risk. These are (a) technological changes and innovations, (b) globalization, (c) outsourcing, (d) reduction in the supplier base, and (e) currency fluctuations. All these risk drivers are changes to the modern structure of the supply chain, which directly impact the supply chain network. For example, Sheffi (2001) suggests holding strategic emergency stocks to be used only in the case of extreme disruptions and adopting a dual-sourcing strategy where offshore suppliers are used for the bulk of the procurement volume and local suppliers in the case of disruption.

Firms require a robust supply chain strategy to sustain their operations when significant disruptions occur. Tang (2007) said that some organizations' supply chains are efficient and resilient because they changed their strategy to meet customer demand during shortages and crises. For example, when the Indonesian Rupiah was devalued by more than 50% in 1997, many Indonesian suppliers could not pay for imported components. This event was a shock wave for many US customers who had outsourced their operations to Indonesia. Hence, robust supply chain strategies are necessary for every firm to survive and be sustainable in the long run.

Appendix: Some Definitions of Supply Chain Risk, Uncertainty, and Vulnerability

Theme	Author(s) (year)	Journal	Definition
Vulnerability	Craighead et al. (2007)	Decision Science	Supply Chain Vulnerability $= S_e P_e$ Where S_e is the severity and P_e is the probability of an event
	Jenelius and Mattsson (2012)	Transportation Research Part A	
	Sheffi and Rice Jr (2005)	MIT Sloan Management Review	
	Peck (2005)	International Journal of Physical Distribution and Logistics Management	"Something is at risk; vulnerable: likely to be lost or damaged"
	Svensson (2000)	International Journal of Physical Distribution and Logistics Management	Vulnerability as a construct consisting of two components, namely inductive approach, and deductive approach
	Svensson (2004)	Journal of Business and Industrial Marketing	"Disturbance" and "the negative consequence of disturbance"
Risk	Barroso et al. (2009)	IEEE	Incapacity of the supply chain to react to disruptions at a given

(continued)

Theme	Author(s) (year)	Journal	Definition
			moment and meet supply chain objectives
	Christopher and Peck (2004)	The International Journal of Logistics Management	"Exposure to serious disturbance arising from risks within the supply chain as well as risks external to the supply chain"
	March and Shapira (1987)	Management Science	The variation in the distribution of possible supply chain outcomes, their likelihood, and their subjective values
	Wynne and Conrad (1980)	Conrad, J. (Ed), Society, Technology and Risk Assessment	The potential for unwanted negative consequences that arise from an event or activity
	Jüttner et al. (2010)	International Journal of Logistics Research and Applications	The possibility and effect of mismatch between supply and demand
	Peck (2007)	International Journal of Logistics Research and Applications	Anything that disrupts or impedes the information, material, or product flows from original suppliers to the delivery of the final product to the ultimate end user
	Zsidisin (2003)	Journal of Purchasing and Supply Management	The occurrence of an event that makes a supply chain unable to satisfy customers' demands, threats, and safety
Uncertainty	Milliken (1987)	Academy of Management Review	Perceived inability to predict something accurately
	van der Vorst and Beulens (2002)	International Journal of Physical Distribution and Logistics Management	A situation where the decision-maker lacks information about the network and environment and hence cannot predict the event's impact on supply chain behavior

References

Abrahamsson, M., Aldin, N., & Stahre, F. (2010). Logistics platforms for improved strategic flexibility. *International Journal of Logistics Research and Applications, 6*(3), 85–106. https://doi.org/10.1080/13675560310001 23061

Ahi, P., & Searcy, C. (2013). A comparative literature analysis of definitions for green and sustainable supply chain management. *Journal of Cleaner Production, 52*, 329–341. https://doi.org/10.1016/j.jclepro.2013.02.018

Aldaco, R., Hoehn, D., Laso, J., Margallo, M., Ruiz-Salmon, J., Cristobal, J., Kahhat, R., Villanueva-Rey, P., Bala, A., Batlle-Bayer, L., Fullana-I-Palmer, P., Irabien, A., & Vazquez-Rowe, I. (2020). Food waste management during the COVID-19 outbreak: a holistic climate, economic and nutritional approach. *Science of the Total Environment, 742*, 140524. https://doi.org/10.1016/j.scitotenv.2020.140524

AlHashim, D. D. (1980). Internal performance evaluation in American multinational enterprises. *Management International Review, 20*(3), 33–39. Retrieved from http://www.jstor.org/stable/40227535

Aschburner, S. (2020). Coronavirus pandemic causes NBA to suspend season after player tests positive. Retrieved on March 14, 2021 from https://www.nba.com/news/coronavirus-pandemic-causes-nba-suspend-season

Barroso, A. P., Machado, V. H., & Machado, V. C. (2009). Identifying vulnerabilities in the supply chain. In *2009 IEEE international conference on industrial engineering and engineering management* (pp. 1444–1448). https://doi.org/10.1109/IEEM.2009.5373062

Bedekar, A. (2016). Opportunities & challenges for IoT in India. Retrieved on March 30, 2020, from https://www.siliconindia.com/magazine-articles-in/Opportunities-&-Challenges-for-IoT-in-India-QPHA241372671.html

Bhimani, A. (1996). Securing the commercial Internet. *Communications of the ACM, 39*(6), 29–35. https://doi.org/10.1145/228503.228509

BIS. (2006). *Basel II: International convergence of capital measurement and capital standards: A revised framework—comprehensive version.* @bis_org. Retrieved from https://www.bis.org/publ/bcbs128.htm

Blau, P. M. (Ed.). (1964). *Exchange and power in social life.* Wiley.

Borgström, B. (2005). *Exploring efficiency and effectiveness in the supply chain: A conceptual analysis.* Paper presented at the Proceedings from the twenty-first IMP conference, Rotterdam, Netherlands. http://urn.kb.se/resolve?urn=urn:nbn:se:hj:diva-10463

Bowersox, D. J., & Calantone, R. J. (2018). Executive insights: Global logistics. *Journal of International Marketing, 6*(4), 83–93. https://doi.org/10.1177/1069031x9800600410

Braithwaite, A., & Hall, D. (1999). Risky business? Critical decisions in supply chain management: part 1 & 2. *Supply Chain Practice*, Part 1: 40–57; Part 42: 44–58.

Braunscheidel, M. J., & Suresh, N. C. (2009). The organizational antecedents of a firm's supply chain agility for risk mitigation and response. *Journal of Operations Management, 27*(2), 119–140. https://doi.org/10.1016/j.jom.2008.09.006

Brown, G., Carlyle, M., Salmerón, J., & Wood, K. (2006). Defending critical infrastructure. *Interfaces, 36*(6), 530–544. https://doi.org/10.1287/inte.1060.0252

Brusset, X., & Teller, C. (2017). Supply chain capabilities, risks, and resilience. *International Journal of Production Economics, 184*, 59–68. https://doi.org/10.1016/j.ijpe.2016.09.008

Burgess, K., Co-Editors: Benn Lawson, P. D. C., Singh, P. J., & Koroglu, R. (2006). Supply chain management: a structured literature review and implications for future research. *International Journal of Operations & Production Management, 26*(7), 703-729. https://doi.org/10.1108/01443570610672202

Chen, J., Sohal, A. S., & Prajogo, D. I. (2013). Supply chain operational risk mitigation: a collaborative approach. *International Journal of Production Research, 51*(7), 2186–2199. https://doi.org/10.1080/00207543.2012.727490

Chopra, S., Reinhardt, G., & Mohan, U. (2007). The importance of decoupling recurrent and disruption risks in a supply chain. *Naval Research Logistics, 54*(5), 544–555. https://doi.org/10.1002/nav.20228

Choudhury, N. R. (2020). Food sector faces multipronged consequences of COVID-19 outbreak. *Global Trade Magazine.* Retrieved on June 28, 2021, from https://www.globaltrademag.com/food-sector-faces-multipronged-consequences-of-covid-19-outbreak/

Christopher, M., & Lee, H. (2004). Mitigating supply chain risk through improved confidence. *International Journal of Physical Distribution & Logistics Management, 34*(5), 388–396. https://doi.org/10.1108/09600030410545436

Christopher, M., & Peck, H. (2004). Building the resilient supply chain. *The International Journal of Logistics Management, 15*(2), 1–14. https://doi.org/10.1108/09574090410700275

Coleman, L. (2006). Frequency of man-made disasters in the 20th century. *Journal of Contingencies and Crisis Management, 14*(1), 3–11. https://doi.org/10.1111/j.1468-5973.2006.00476.x

Colicchia, C., Wilding, R., & Strozzi, F. (2012). Supply chain risk management: a new methodology for a systematic literature review. *Supply Chain Management: An International Journal, 17*(4), 403–418. https://doi.org/10.1108/13598541211246558

Commerce Net. (1998). Overview of the Barriers & Inhibitors, research project. *Commerce Net Research Bulletin, 98*(April 23, 2021), 08. Retrieved from http://www.commerce.net/research/pw/bulletin/98-08-b.html

Craighead, C. W., Blackhurst, J., Rungtusanatham, M. J., & Handfield, R. B. (2007). The severity of supply chain disruptions: Design characteristics and mitigation capabilities. *Decision Sciences, 38*(1), 131–156. https://doi.org/10.1111/j.1540-5915.2007.00151.x

Cunningham, S. M. (1967). The major dimensions of perceived risks. In D. F. Cox (Ed.), *Risk taking and information handling in consumer behavior* (pp. 82–108). Division of Research, Graduate School of Business Administration, Harvard University.

Daft, R., & Huber, G. (1986). *How organizations learn: A communication framework*. Defense Technical Information Center. Retrieved from https://books.google.com/books?id=7AW1twAACAAJ

Ellinger, A. E., Chen, H., Tian, Y., & Armstrong, C. (2015). Learning orientation, integration, and supply chain risk management in Chinese manufacturing firms. *International Journal of Logistics Research and Applications, 18*(6), 476–493. https://doi.org/10.1080/13675567.2015.1005008

Engardio, P. (2001). Why the supply chain broke down. *Business Week(3724)*, 41–41. Retrieved from https://ezproxy.uwgb.edu:2443/login?url=https://search.ebscohost.com/login.aspx?direct=true&AuthType=ip,uid&db=buh&AN=4171504&site=ehost-live&scope=site.

Falasca, M., Zobel, C. W., & Cook, D. (2008). *A decision support framework to assess supply chain resilience*. Paper presented at the proceedings of the 5th international ISCRAM conference.

Fan, T., Tao, F., Deng, S., & Li, S. (2015). Impact of RFID technology on supply chain decisions with inventory inaccuracies. *International Journal of Production Economics, 159*, 117–125. https://doi.org/10.1016/j.ijpe.2014.10.004

Feng, L., Zhang, X., & Zhou, K. (2018). Current problems in China's manufacturing and countermeasures for industry 4.0. *EURASIP Journal on Wireless Communications and Networking, 2018*(1). https://doi.org/10.1186/s13638-018-1113-6

Flynn, B. B., Huo, B., & Zhao, X. (2010). The impact of supply chain integration on performance: A contingency and configuration approach. *Journal of Operations Management, 28*(1), 58–71. https://doi.org/10.1016/j.jom.2009.06.001

Foley, J., & Kontzer, T. (2004). Coca-Cola plans to refresh supply chain. *Information Week, February(976)*, 22. Retrieved from https://www.informationweek.com/coca-cola-plans-to-refresh-supply-chain/17700199

Freedman, M. (2003). The genius is in the implementation. *Journal of Business Strategy, 24*(2), 26–31. https://doi.org/10.1108/02756660310508164

Gaiardelli, P., Saccani, N., & Songini, L. (2007). Performance measurement of the after-sales service network—Evidence from the automotive industry. *Computers in Industry, 58*(7), 698–708. https://doi.org/10.1016/j.compind.2007.05.008

Ghosh, S. (1998). Making business sense of the Internet. *Harvard Business Review, 76*(2), 126–135. Retrieved from https://www.ncbi.nlm.nih.gov/pubmed/10177862

Green, W. (2020). Tesco 'completely reset' relationships with suppliers. Retrieved on March 30, 2021, from https://www.cips.org/supply-management/news/2020/april/tesco-completely-reset-relationships-with-suppliers/

Gualandris, J., & Kalchschmidt, M. (2015). Supply risk management and competitive advantage: a misfit model. *The International Journal of Logistics Management, 26*(3), 459–478. https://doi.org/10.1108/IJLM-05-2013-0062

Gundlach, G. T., Hausman, A., Bolumole, Y. A., Eltantawy, R. A., & Frankel, R. (2006). The changing landscape of supply chain management, marketing channels of distribution, logistics and purchasing. *Journal of Business & Industrial Marketing, 21*(7), 428–438. https://doi.org/10.1108/08858620610708911

Gurtu, A. (2019). A pioneering approach to reducing fuel cost and carbon emissions from transportation. *Transportation Journal, 58*(4). https://doi.org/10.5325/transportationj.58.4.0309

Gurtu, A., & Johny, J. (2019). Potential of blockchain technology in supply chain management: a literature review. *International Journal of Physical Distribution & Logistics Management, 49*(9), 881–900. https://doi.org/10.1108/ijpdlm-11-2018-0371

Gurtu, A., Johny, J., & Chowdhary, R. (2022a). Effects of Free Trade Agreements on Trade Activities of Signatory Countries. *The Indian Economic Journal, 70*(2), 1–24. https://doi.org/10.1177/00194662221104750

Gurtu, A., Johny, J., & Buechse, O. (2022b). Paper and packaging industry dynamics during COVID-19 and their strategies for the future. *Strategic Management, 00*, 18–18. https://doi.org/10.5937/StraMan2200017G

Gurtu, A., & Johny, J. (2021). Supply chain risk management: Literature review. *Risks, 9*(1), 16. https://doi.org/10.3390/risks9010016

Gurtu, A., Searcy, C., & Jaber, M. Y. (2017). Sustainable supply chains. In M. Khan, M. Hussain, & M. M. Ajmal (Eds.), *Green supply chain management for sustainable business practice* (pp. 1–26). IGI Global. https://doi.org/10.4018/978-1-5225-0635-5.ch001

Håkansson, H., & Prenkert, F. (2005). Exploring the exchange concept in marketing. In H. Håkansson, D. Harrison, & A. Waluszewski (Eds.), *Rethinking marketing: Developing a new understanding of markets* (pp. 75–97). Wiley. Retrieved from https://www.wiley.com/en-us/Rethinking+Marketing:+Developing+a+New+Understanding+of+Markets-p-9780470021477

Hart, C. (2020). Coronavirus threatens integrated supply chains. Retrieved on April 9, 2021, from https://www.cips.org/supply-management/news/2020/march/coronavirus-threatens-integrated-supply-chains-says-oecd/

Hauser, L. M. (2003). Risk-adjusted supply chain management. *Supply Chain Management Review, 7*(6), 64–71.

Hocking, B. M., Campbell, M. J., & Storey, E. (1982). Infant feeding patterns. *Australian Dental Journal, 27*(5), 300–305. https://doi.org/10.1111/j.1834-7819.1982.tb05251.x

Hubbard, D. W. (2009). *The failure of risk management: Why it's broken and how to fix it* (1st ed.). Wiley. ISBN: 9780470387955.

Hubbard, D. W. (2014). *How to measure anything: Finding the value of intangibles in business* (3rd ed.). Wiley. ISBN:1118836448.

Iacovou, C. L., Benbasat, I., & Dexter, A. S. (1995). Electronic data interchange and small organizations: Adoption and impact of technology. *MIS Quarterly, 19*(4). https://doi.org/10.2307/249629

Jenelius, E., & Mattsson, L.-G. (2012). Road network vulnerability analysis of area-covering disruptions: A grid-based approach with case study. *Transportation Research Part A: Policy and Practice, 46*(5), 746–760. https://doi.org/10.1016/j.tra.2012.02.003

Jüttner, U. (2005). Supply chain risk management. *The International Journal of Logistics Management, 16*(1), 120–141. https://doi.org/10.1108/09574090510617385

Jüttner, U., & Maklan, S. (2011). Supply chain resilience in the global financial crisis: an empirical study. *Supply Chain Management: An International Journal, 16*(4), 246–259. https://doi.org/10.1108/13598541111139062

Jüttner, U., Peck, H., & Christopher, M. (2003). Supply chain risk management: outlining an agenda for future research. *International Journal of Logistics Research and Applications, 6*(4), 197–210. https://doi.org/10.1080/13675560310001627016

Jüttner, U., Peck, H., & Christopher, M. (2010). Supply chain risk management: outlining an agenda for future research. *International Journal of Logistics Research and Applications, 6*(4), 197–210. https://doi.org/10.1080/13675560310001627016

Kauppi, K., Longoni, A., Caniato, F., & Kuula, M. (2016). Managing country disruption risks and improving operational performance: risk management along integrated supply chains.

International Journal of Production Economics, 182, 484–495. https://doi.org/10.1016/j.ijpe.2016.10.006

Kelling, N. K., Sauer, P. C., Gold, S., & Seuring, S. (2020). The role of institutional uncertainty for social sustainability of companies and supply chains. *Journal of Business Ethics, 173*(4), 813–833. https://doi.org/10.1007/s10551-020-04423-6

Khan, O., & Burnes, B. (2007). Risk and supply chain management: creating a research agenda. *The International Journal of Logistics Management, 18*(2), 197–216. https://doi.org/10.1108/09574090710816931

Kleindorfer, P. R., & Saad, G. H. (2009). Managing disruption risks in supply chains. *Production and Operations Management, 14*(1), 53–68. https://doi.org/10.1111/j.1937-5956.2005.tb00009.x

Knight, F. H. (1964). *Risk, uncertainty and profit*. Augustus M. Kelley.

Kovács, G. L., & Paganelli, P. (2003). A planning and management infrastructure for large, complex, distributed projects—beyond ERP and SCM. *Computers in Industry, 51*(2), 165–183. https://doi.org/10.1016/s0166-3615(03)00034-4

Kull, T., & Closs, D. (2008). The risk of second-tier supplier failures in serial supply chains: Implications for order policies and distributor autonomy. *European Journal of Operational Research, 186*(3), 1158–1174. https://doi.org/10.1016/j.ejor.2007.02.028

Lavastre, O., Gunasekaran, A., & Spalanzani, A. (2014). Effect of firm characteristics, supplier relationships and techniques used on Supply Chain Risk Management (SCRM): an empirical investigation on French industrial firms. *International Journal of Production Research, 52*(11), 3381–3403. https://doi.org/10.1080/00207543.2013.878057

Leitão, P., Colombo, A. W., & Karnouskos, S. (2016). Industrial automation based on cyber-physical systems technologies: Prototype implementations and challenges. *Computers in Industry, 81*, 11–25. https://doi.org/10.1016/j.compind.2015.08.004

Manuj, I., Esper, T. L., & Stank, T. P. (2014). Supply chain risk management approaches under different conditions of risk. *Journal of Business Logistics, 35*(3), 241–258. https://doi.org/10.1111/jbl.12051

Manuj, I., & Mentzer, J. T. (2008). Global supply chain risk management strategies. *International Journal of Physical Distribution & Logistics Management, 38*(3), 192–223. https://doi.org/10.1108/09600030810866986

March, J. G., & Shapira, Z. (1987). Managerial perspectives on risk and risk taking. *Management Science, 33*(11), 1404–1418. https://doi.org/10.1128//mnsc.33.11.1404

McGillivray, G. (2000). Commercial risk under JIT. *Canadian Underwriter, 67*(1), 26–30.

Mentzer, J. T., DeWitt, W., Keebler, J. S., Min, S., Nix, N. W., Smith, C. D., & Zacharia, Z. G. (2001). Defining supply chain management. *Journal of Business Logistics, 22*(2), 1–25. https://doi.org/10.1002/j.2158-1592.2001.tb00001.x

Mentzer, J. T., & Firman, J. (1994). Logistics control systems in the 21st century. *Journal of Business Logistics, 15*(1), 215–227. Retrieved from https://ezproxy.uwgb.edu:2443/login?url=https://search.ebscohost.com/login.aspx?direct=true&AuthType=ip,uid&db=bth&AN=9705251044&site=ehost-live&scope=site.

Miller, K. D. (1992). A framework for integrated risk management in international business. *Journal of International Business Studies, 23*(2), 311–331. https://doi.org/10.1057/palgrave.jibs.8490270

Milliken, F. J. (1987). Three types of perceived uncertainty about the environment: State, effect, and response uncertainty. *Academy of Management Review, 12*(1), 133–143.

Mohr, J., & Nevin, J. R. (1990). Communication strategies in marketing channels: a theoretical perspective. *Journal of Marketing, 54*(4). https://doi.org/10.2307/1251758

Möller, K. E. K., & Törrönen, P. (2003). Business suppliers' value creation potential. *Industrial Marketing Management, 32*(2), 109–118. https://doi.org/10.1016/s0019-8501(02)00225-0

Morgan, R. M., & Hunt, S. D. (1994). The commitment-trust theory of relationship marketing. *Journal of Marketing, 58*(3). https://doi.org/10.2307/1252308

Nath, R., Akmanligil, M., Hjelm, K., Sakaguchi, T., & Schultz, M. (1998). Electronic commerce and the internet: issues, problems, and perspectives. *International Journal of Information Management, 18*(2), 91–101. https://doi.org/10.1016/s0268-4012(97)00051-0

Novack, R. A., Rinehart, L. M., & Wells, M. V. (1992). Rethinking concept foundations in logistics management. *Journal of Business Logistics, 13*(2), 233–267. Retrieved from https://ezproxy.uwgb.edu:2443/login?url=https://search.ebscohost.com/login.aspx?direct=true&AuthType=ip,uid&db=buh&AN=9706191136&site=ehost-live&scope=site.

Peck, H. (2005). Drivers of supply chain vulnerability: an integrated framework. *International Journal of Physical Distribution & Logistics Management, 35*(4), 210–232. https://doi.org/10.1108/09600030510599904

Peck, H. (2007). Reconciling supply chain vulnerability, risk and supply chain management. *International Journal of Logistics Research and Applications, 9*(2), 127–142. https://doi.org/10.1080/13675560600673578

Pettit, T. J., Fiksel, J., & Croxton, K. L. (2010). Ensuring supply chain resilience: development of a conceptual framework. *Journal of Business Logistics, 31*(1), 1–21. https://doi.org/10.1002/j.2158-1592.2010.tb00125.x

Premkumar, G. P. (2003). Perspectives of the e-marketplace by multiple stakeholders. *Communications of the ACM, 46*(12), 279–288. https://doi.org/10.1145/953460.953512

Rao, S., & Goldsby, T. J. (2009). Supply chain risks: a review and typology. *The International Journal of Logistics Management, 20*(1), 97–123. https://doi.org/10.1108/09574090910954864

Ray, A. J. (2005). *Indians in the fur trade: their role as trappers, hunters, and middlemen in the lands southwest of Hudson Bay, 1660-1870: with a new introduction*. University of Toronto Press. https://doi.org/10.3138/j.ctt1287w22

Revilla, E., & Saenz, M. J. (2017). The impact of risk management on the frequency of supply chain disruptions. *International Journal of Operations & Production Management, 37*(5), 557–576. https://doi.org/10.1108/ijopm-03-2016-0129

Rosenhead, J., Elton, M., & Gupta, S. K. (2017). Robustness and optimality as criteria for strategic decisions. *Journal of the Operational Research Society, 23*(4), 413–431. https://doi.org/10.1057/jors.1972.72

Schmidt, G., & Wilhelm, W. E. (2010). Strategic, tactical and operational decisions in multi-national logistics networks: A review and discussion of modelling issues. *International Journal of Production Research, 38*(7), 1501–1523. https://doi.org/10.1080/002075400188690

Schoenherr, T., & Swink, M. (2012). Revisiting the arcs of integration: Cross-validations and extensions. *Journal of Operations Management, 30*(1–2), 99–115. https://doi.org/10.1016/j.jom.2011.09.001

Senn, J. A. (1998). Expanding the reach of electronic commerce: The internet EDI alternative. *Information Systems Management, 15*(3), 7–15. https://doi.org/10.1201/1078/43185.15.3.19980601/31129.2

Sheffi, Y. (2001). Supply chain management under the threat of international terrorism. *The International Journal of Logistics Management, 12*(2), 1–11. https://doi.org/10.1108/09574090110806262

Sheffi, Y. (2005). *The resilient enterprise: Overcoming vulnerability for competitive advantage*. MIT Press. ISBN: 9780262283489.

Sheffi, Y., & Rice, J. B., Jr. (2005). A supply chain view of the resilient enterprise. *MIT Sloan Management Review, 47*(1), 41.

Shin, H., Hwang, J., & Kim, H. (2019). Appropriate technology for grassroots innovation in developing countries for sustainable development: The case of Laos. *Journal of Cleaner Production, 232*, 1167–1175. https://doi.org/10.1016/j.jclepro.2019.05.336

Smit, S., Hirt, M., Buehler, K., Lund, S., Greenberg, E., & Govindarajan, A. (2020). *Safeguarding our lives and our livelihoods: The imperative of our time*. Retrieved from https://www.mckinsey.com/business-functions/strategy-and-corporate-finance/our-insights/safeguarding-our-lives-and-our-livelihoods-the-imperative-of-our-time

Sodhi, M. S., Son, B.-G., & Tang, C. S. (2012). Researchers' perspectives on supply chain risk management. *Production and Operations Management, 21*(1), 1–13. https://doi.org/10.1111/j.1937-5956.2011.01251.x

Sodhi, M. S., & Tang, C. S. (2012). *Managing supply chain risk.* Springer. https://doi.org/10.1007/978-1-4614-3238-8. ISBN: 978-1-4614-3237-1, 978-1-4614-3238-8.

Stefanovic, D., Stefanovic, N., & Radenkovic, B. (2009). Supply network modelling and simulation methodology. *Simulation Modelling Practice and Theory, 17*(4), 743–766. https://doi.org/10.1016/j.simpat.2009.01.001

Stevens, G. C. (1989). Integrating the supply chain. *International Journal of Physical Distribution & Materials Management, 19*(8), 3–8. https://doi.org/10.1108/eum0000000000329

Stewart, G. (1995). Supply chain performance benchmarking study reveals keys to supply chain excellence. *Logistics Information Management, 8*(2), 38–44. https://doi.org/10.1108/09576059510085000

Svensson, G. (2000). A conceptual framework for the analysis of vulnerability in supply chains. *International Journal of Physical Distribution & Logistics Management, 30*(9), 731–750. https://doi.org/10.1108/09600030010351444

Svensson, G. (2002). A conceptual framework of vulnerability in firms' inbound and outbound logistics flows. *International Journal of Physical Distribution & Logistics Management, 32*(2), 110–134. https://doi.org/10.1108/09600030210421723

Svensson, G. (2004). Vulnerability in business relationships: the gap between dependence and trust. *Journal of Business & Industrial Marketing, 19*(7), 469–483. https://doi.org/10.1108/08858620410564418

Tang, C. S. (2007). Robust strategies for mitigating supply chain disruptions. *International Journal of Logistics Research and Applications, 9*(1), 33–45. https://doi.org/10.1080/13675560500405584

Transparency Market Research. (2020). Globalization to drive global logistics market, notes transparency market research. Retrieved on June 18, 2021 from https://www.openpr.com/news/1909066/globalization-to-drive-global-logistics-market-notes

UN Office for Disaster Risk Reduction. (2004). *Living with risk: a global review of disaster reduction initiatives* (Vol. 1). UN. Retrieved from https://digitallibrary.un.org/record/524401?ln=enISBN:9211010640

Urciuoli, L. (2009). Supply chain security—mitigation measures and a logistics multi-layered framework. *Journal of Transportation Security, 3*(1), 1–28. https://doi.org/10.1007/s12198-009-0034-3

Vanany, I., Zailani, S., & Pujawan, N. (2009). Supply chain risk management. *International Journal of Information Systems and Supply Chain Management, 2*(1), 16–33. https://doi.org/10.4018/jisscm.2009010102

van der Vorst, J. G. A. J., & Beulens, A. J. M. (2002). Identifying sources of uncertainty to generate supply chain redesign strategies. *International Journal of Physical Distribution & Logistics Management, 32*(6), 409–430. https://doi.org/10.1108/09600030210437951

Wagner, S. M., & Bode, C. (2008). An empirical examination of supply chain performance along several dimensions of risk. *Journal of Business Logistics, 29*(1), 307–325. https://doi.org/10.1002/j.2158-1592.2008.tb00081.x

Waters, C. D. J. (2011). *Supply chain risk management: vulnerability and resilience in logistics* (2nd ed.). Kogan Page. ISBN: 9780749463939.

Wilding, R. (2018). KFC: an important MBA case study. Retrieved on March 30, 2020, from https://www.cips.org/supply-management/opinion/2018/february/kfc-an-important-mba-case-study/

Wildt, J. (1982). [Peritonsillar abscess. Report of a child aged 17 months]. *Ugeskr Laeger, 144*(46), 3431–3432. Retrieved from https://www.ncbi.nlm.nih.gov/pubmed/6963070

Williams, Z., Waller, M., Lueg, J. E., & LeMay, S. A. (2008). Supply chain security: an overview and research agenda. *The International Journal of Logistics Management, 19*(2), 254–281. https://doi.org/10.1108/09574090810895988

Wojakowski, I. (1971). [Pelvic fractures in iron ore miners with reference to accident exposure]. *Wiad Lek, 24*(18), 1729–1732. Retrieved from https://www.ncbi.nlm.nih.gov/pubmed/5117443

Wynne, B., & Conrad, J. (1980). *Society, technology and risk assessment*. Academic.

Zhu, Q., Krikke, H., & Caniëls, M. C. J. (2017). Integrated supply chain risk management: a systematic review. *The International Journal of Logistics Management, 28*(4), 1123–1141. https://doi.org/10.1108/ijlm-09-2016-0206

Zsidisin, G. A. (2003). A grounded definition of supply risk. *Journal of Purchasing and Supply Management, 9*(5–6), 217–224. https://doi.org/10.1016/j.pursup.2003.07.002

Supply Chain Risk Management: An Enterprise View and a Survey of Methods

Mikaella Polyviou, George Ramos, and Eugene Schneller

Abstract As supply chains have become more global and complex, supply chain disruptions have become more frequent (Resilinc. Supply chain disruptions-Resioinc's mid-year report. https://www.resilinc.com/in-the-news/supply-chain-disruptions-resilincs-mid-year-report/, 2021) and severe (Craighead et al. Decision Sciences 38(1):131–156, 2007). It is thus imperative for public and private enterprises to develop and implement strategies to prevent supply chain disruptions from occurring and recover quickly from them when they occur. Enterprises can do so by first establishing an effective supply chain risk management (SCRM) process that identifies, assesses, and proposes strategies to manage and monitor supply chain risks. In this chapter, we review the SCRM process and describe its four stages: risk identification, risk assessment, risk management, and risk monitoring. In doing so, we propose practices and strategies that help enterprises identify, assess, manage (accept, avoid, transfer, or mitigate), and monitor supply chain risks. We also provide examples of how enterprises across industries have implemented SCRM and identify key technologies employed within this process. Finally, we review recent research on behavioral influences in the context of SCRM. The chapter, overall, emphasizes the impact of continued risk for supply chains due to the COVID-19 pandemic. This chapter serves as a resource to academics, students, and practitioners into the SCRM process, actionable strategies employed within each stage of this process, and behavioral factors influencing it.

M. Polyviou (✉) · E. Schneller
Department of Supply Chain Management, W. P. Carey School of Business, Arizona State University, Tempe, AZ, USA
e-mail: Mikaella.Polyviou@asu.edu

G. Ramos
Resilinc, Milpitas, CA, USA

© The Author(s), under exclusive license to Springer Nature Switzerland AG 2022
Y. Khojasteh et al. (eds.), *Supply Chain Risk Mitigation*, International Series in Operations Research & Management Science 332,
https://doi.org/10.1007/978-3-031-09183-4_2

1 Introduction

As supply chains have become more global and complex, enterprises operating in these supply chains have become more vulnerable to supply chain risks (World Economic Forum, 2013). Such vulnerabilities became apparent, for example, as the world faced the COVID-19 pandemic in 2019 through 2022. The pandemic has exposed the dependency of nations, hospitals, and businesses worldwide on Chinese suppliers and manufacturers for masks, gowns, and other protective equipment (Bradsher, 2020). It also exposed the bottlenecks and rigidity in supply chains after consumer demand for staples, such as toilet paper, yogurt, or meat, surged while industrial demand plummeted (Smith, 2020). As a result, the importance and consequences of not managing supply chain risks became clear to nations, public and private enterprises, and consumers.

Supply chain risks refer to "events or conditions that [have the potential to] adversely influence any part of a supply chain leading to operational, tactical, or strategic level failures or irregularities" (Ho et al., 2015, p. 5). They are characterized by the probability of their occurrence and the severity of their impact (Ho et al., 2015; Sheffi & Rice, 2005). Most supply chain risks are not "black swan" events (Taleb, 2007) but are common and predictable; thus, they may be considered "white swans" (Akkermans & Van Wassenhove, 2013, 2018). Supply chain risks that materialize often lead to supply chain disruptions, which refer to interruptions in the materials, services, information, or financial flow from one organization to another in a supply chain (Kim et al., 2015; Polyviou et al., 2018). Indeed, supply chain disruptions occur increasingly (Resilinc., 2021), with severe consequences for enterprises (Hendricks & Singhal, 2003, 2005).

Risks can disrupt not only an enterprise's direct operations but also those of its trading partners, thereby disrupting the enterprise itself as a consequence. Indeed, the Business Continuity Institute revealed that the majority of supply chain disruptions experienced by an enterprise originated outside its boundaries: 48.9% of disruptions occurred at a first-tier supplier, 24.9% at a second-tier supplier, and 12.2% at a tier beyond second-tier suppliers (Business Continuity Institute, 2018).

Meanwhile, an enterprise might also be affected by disruptions occurring downstream in its supply chain. For example, a significant customer going out of business or changing course in product or service offerings can severely disrupt a supplier's operations. In another example, farmers in the USA were forced to dump or dispose of milk and other fresh foods due to a significant drop in demand from restaurants, hotels, schools, and other food service providers at the beginning of the COVID-19 pandemic in the USA (Yaffe-Bellany & Corkery, 2020). As such, an enterprise does not operate in a vacuum. Instead, it depends on a network of suppliers, transportation and logistics providers, dealers, and others to receive and provide goods and services (Sheffi, 2005). Therefore, it needs to look outside its boundaries and work with its trading partners (suppliers, transportation providers, third-party logistics (3PL) providers, or customers) to identify and evaluate risks that can disrupt supply chain operations.

In this way, the academic literature advocates that enterprises focus on supply chain risk management (SCRM), namely the process of identifying, assessing, managing, and monitoring the risks that can disrupt their operations and supply chain networks (Ho et al., 2015). In this chapter, our objectives are to:

- Describe the SCRM process by outlining its stages
- Present the academic literature on each stage of the SCRM process and propose relevant methods and practices within each stage
- Discuss significant business examples that demonstrate effective SCRM
- Present recent academic literature on the behavioral influences in the context of managing supply chain risks and disruptions
- Present recent developments in SCRM in terms of technologies and software

This chapter contributes to the literature in several ways. First, it provides a comprehensive review of methods employed to identify, assess, and mitigate supply chain risks. It also reviews the capabilities and supply chain strategies that enterprises can develop and implement to mitigate the probability and severity of supply chain disruptions proactively. Second, it makes a strong argument for an enterprise and network view of supply chain risks. The chapter emphasizes that an enterprise needs to look not only across functional silos and work with internal stakeholders (e.g., procurement, logistics, operations, sales, finance) but also beyond its boundaries and work with its trading partners (e.g., suppliers, customers, third-party logistics providers or government) to identify and evaluate risks that can disrupt supply chain operations. Finally, this chapter serves as a background resource to academics, students, and practitioners in the SCRM process, familiarizing all with actionable strategies employed within each stage and enhancing understanding of behavioral factors that can potentially influence this process.

2 Supply Chain Risk Management: An Overview

Supply chain risk management (SCRM) is defined as the process to identify, assess, manage, and monitor risks in the supply chain (Ho et al., 2015), as shown in Fig. 1. The stages in this process are summarized below:

1. ***Supply chain risk identification***: This stage involves discovering all relevant risks that can influence an enterprise's operations and supply chain (Zsidisin & Henke, 2019). These risks might stem from internal and external sources relative to the boundaries of an enterprise (Christopher & Peck, 2004). Internal sources might be equipment breakdown, production delays, or accidents, while external sources may include natural disasters, pandemics, cyber-attacks, production or quality problems at suppliers' plants, or transportation accidents.
2. ***Supply chain risk assessment***: Supply chain risks are typically characterized by the probability of their occurrence and the severity of their impact. This stage, as such, involves estimating these variables for the relevant supply chain risks

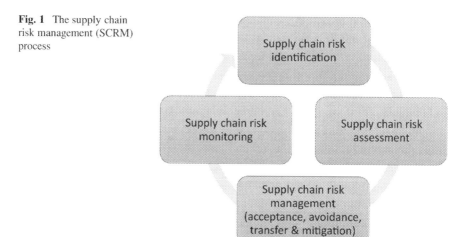

Fig. 1 The supply chain risk management (SCRM) process

identified in the risk-identification stage (Zsidisin & Henke, 2019). This stage also involves prioritizing the supply chain risks according to the enterprise's risk tolerance.

3. **Supply chain risk management:** This stage involves identifying and developing strategies to reduce the probability or severity of the identified supply chain risks. This stage can include risk acceptance (i.e., doing nothing to mitigate the risk) and strategies to avoid, transfer, or mitigate risks (Chapman, 2006).
4. **Supply chain risk monitoring:** This stage involves evaluating the efficacy of the risk treatment strategies developed and implemented in the previous stage. It also includes identifying the opportunities to improve the stages of the SCRM process and updating the process based on the learnings gathered (Zsidisin & Henke, 2019).

We note that Fig. 1 depicts the SCRM process as a cycle, implying that the process is continuous. That is, the supply chain risk monitoring stage informs the supply chain risk identification, assessment, and management stages on an ongoing basis to ensure that the findings and decisions in each stage remain updated and relevant in a continuously changing business environment.

We also note that an enterprise must first be motivated to focus on SCRM and implement a program to identify, assess, mitigate, and actively monitor risks. This motivation is largely driven by the enterprise's orientation toward supply chain risks and disruptions, formally defined as a "general awareness and consciousness of, concerns about, seriousness toward, and recognition of opportunity to learn from supply chain disruptions" (Bode et al., 2011, p. 837). An enterprise, which has a strong orientation toward supply chain disruptions, will consider them a critical issue and take actions that ensure continuity in its supply chain operations (Ambulkar et al., 2015; Bode et al., 2011).

Once an enterprise initiates an SCRM program, it will need to identify where the key vulnerabilities and failure points lie in its supply chain network and which customers need to be prioritized if a supply failure occurs. It also needs to evaluate its

Table 1 Supply chain risk management (SCRM) assessment questions

Assessment questions to ask internally about an enterprise's SCRM capabilities

- What capacity is available, and how quickly can we redirect?
- Do we have emergency management structures and defined roles and responsibilities in place to respond to a crisis?
- How do we procure direct and indirect materials? How are supply chain disruptions accounted for in those procurement processes?
- Who is responsible for SCRM and crisis management at each site?
- What immediate action must we take to minimize loss and liability?
- Do you know your key support groups and their SCRM plans? Are your plans aligned so that you can continue operating?
- Do we need to prioritize customer demand? If so, which customers will be prioritized?
- What are the worst-case financial loss and legal exposure? Do you have a key contact list for individuals required to respond to a crisis?
- How long will it take to resume operations?

Assessment questions to ask suppliers about their SCRM capabilities

- What kinds of business functions are considered critical and have SCRM plans associated with them?
- What kinds of impacts are considered by your risk-mitigation and recovery-planning activities?
- How does senior management support the SCRM program? What management review and corporate governance mechanisms exist?
- Does your SCRM program ensure that all business processes and functions "critical" to your company are identified and documented?
- Does the SCRM documentation cover the components that make/support critical processes to an appropriate level of detail to ensure that single points of failure can be identified?
- Does your SCRM program ensure that business interruption risks are understood and prioritized and their impacts comprehended?
- Have your business groups taken steps to reduce risks?
- How frequently is the risk and impact assessment refreshed so that changes to your business are reflected in the SCRM program?
- Does your SCRM program ensure that the plans in place are well documented and current?
- Do these plans provide effective crisis response and ensure that critical operations continue during a crisis?
- Is the SCRM plan documentation readily available to the people who need it and maintain it?
- What kinds of exercises and drills are performed to ensure the completeness of the plans? Is the enterprise prepared to perform effectively during a crisis?
- Can your senior management confidently answer "Yes" when asked if everything reasonable and prudent has been done to be able to respond to and recover from an emergency?

Adapted from Zsidisin (2007)

current capabilities and those of its suppliers to manage supply chain risks. A CAPS Research study on business continuity management identified two sets of questions that an organization can ask internally and its suppliers when embarking on these initiatives (Zsidisin, 2007). We adjusted these questions to the SCRM context and provided them in Table 1.

3 Approaches to Identify, Assess, Manage, and Monitor Supply Chain Risks

3.1 Supply Chain Risk Identification

The first stage in the SCRM process is supply chain risk identification. This stage involves discovering all relevant risks that can disrupt an enterprise's operations. The objective of an enterprise in this stage should be to develop a risk register, namely a list of identified supply chain risks and a rating of their importance (Sodhi & Tang, 2012).

Risks that can interrupt the flow of materials, services, information, money, or even human resources in a supply chain are numerous. Examples include but are not limited to natural disasters (such as earthquakes, floods, or hurricanes), pandemics (such as the COVID-19 pandemic), geopolitical events (such as political unrest), labor strikes (such as strikes at plants or ports), accidents (such as transportation accidents), supplier-related disruptions (such as factory fires, product quality problems, or production bottlenecks), and security-related events (such as hacking and piracy). Other types of risks may not cause an interruption in the flow. Nevertheless, they may require adjustments to an enterprise's operations or influence its reputation, such as governmental policies and regulations (such as new environmental policies, tariffs, and other trade restrictions) or environmental incidents (such as oil spills). Table 2 presents a sample of the supply chain risk categories identified in the literature.

An enterprise can use different approaches to identify supply chain risks, as shown in Table 3. Every enterprise is responsible for identifying its own risks and typically does so from its own viewpoint (Hallikas et al., 2004). Nonetheless, it must work with key trading partners, such as suppliers, distributors, transportation providers, and customers, to identify and evaluate its dependencies on them and find where vulnerabilities might exist in the supply chain network beyond first-tier supply chain partners (Hallikas et al., 2004). For example, after a severe sub-supplier accident, Ericsson implemented a proactive SCRM approach. It began working with and required its first-tier suppliers to analyze, assess, and manage risks in their supply chains (Norrman & Jansson, 2004). Likewise, General Motors (GM) started working with its first-tier suppliers to assess if any second-tier suppliers were in trouble and proactively mitigate possible disruptions from those sub-tier suppliers (Banker, 2016).

Moreover, some types of enterprises, such as state or federal governments, have a responsibility to fulfill the needs of their constituents even after a supply failure or discontinuity. As such, they have to incorporate the needs and viewpoints of their stakeholders into their SCRM plans. For example, the US Federal Government maintains strategic national stockpiles of medicines and medical devices for use during public health emergencies, such as the COVID-19 pandemic (US Department of Health and Human Services, 2020).

Table 2 Categories of supply chain risks

Author	Concept	Risk categories
Harland et al. (2003)	Supply network risk	1. Strategic risk 2. Operations risk 3. Supply risk 4. Customer risk 5. Asset impairment risk 6. Competitive risk 7. Reputation risk 8. Financial risk 9. Fiscal risk 10. Regulatory risk 11. Legal risk
Christopher and Peck (2004)	Supply chain risk	1. Internal to the firm (Process; Control) 2. External to the firm but internal to the supply chain network (Demand; Supply) 3. External to the network (Environmental)
Manuj and Mentzer (2008)	Supply chain risk	1. Supply risks 2. Demand risks 3. Operational risks 4. Security risks 5. Macro risks 6. Policy risks 7. Competitive risks 8. Resource risks
Tang and Tomlin (2008)	Supply chain risk	1. Supply risks (Supply cost risk; Supply commitment risks) 2. Process risks 3. Demand risks 4. Intellectual property risks 5. Behavioral risks 6. Political/social risks
Wagner and Bode (2008)	Supply chain risk source	1. Demand-side 2. Supply-side 3. Regulatory, legal, and bureaucratic 4. Infrastructure 5. Catastrophic
Rao and Goldsby (2009)	Supply chain risk	1. Environmental risk sources 2. Industry risk sources 3. Organizational risk sources 4. Problem-specific risk sources 5. Decision-maker risk sources
Tummala and Schoenherr (2011)	Supply chain risk	1. Demand risks 2. Delay risks 3. Disruption risks 4. Inventory risks 5. Manufacturing (process) breakdown risks 6. Physical plant (capacity) risks 7. Supply (procurement) risks 8. System risks 9. Sovereign risks 10. Transportation risks

(continued)

Table 2 (continued)

Author	Concept	Risk categories
Pettit et al. (2013)	Supply chain vulnerabilities	1. Turbulence 2. Deliberate threats 3. External pressures 4. Resource limits 5. Sensitivity 6. Connectivity
Ho et al. (2015)	Supply chain risk	1. Macro risks (Natural; Man-made) 2. Micro risks (Demand; Manufacturing; Supply; Infrastructural)

Adapted from Polyviou (2016)

3.2 Supply Chain Risk Assessment

The second stage in the SCRM process is supply chain risk assessment. This stage involves evaluating the probability of occurrence and the severity of impact of the supply chain risks identified in the first stage of supply chain risk identification. A key objective in this stage is for an enterprise to prioritize supply chain risks according to these variables so that it can focus on the high-priority risks. Table 4 provides exemplary methods that can be used to assess supply chain risks. Finally, an enterprise may use different metrics to measure severity, as shown below (Macdonald & Corsi, 2013; Simchi-Levi et al., 2014; US Department of Defense Standard Practice, 2012):

- The number of products affected
- The number of plant locations affected
- The number of customers affected by flow discontinuity
- The extent of damage to or loss of equipment or property
- The extent of damage to the environment
- The financial loss of the enterprise
- The time-to-recover (TTR): The time (e.g., in days or weeks) that a particular node in a supply chain network (e.g., a supplier's factory, a warehouse or distribution center, a transportation center) would need to become fully functional after a supply chain disruption has occurred (Simchi-Levi et al., 2015)

3.2.1 Supply Chain Risk Mapping and Prioritization

An enterprise is unlikely to have all the resources (time, physical, financial, human) to manage every possible risk that could affect its supply chain operations. Therefore, it needs to decide which risks to accept for the short- or long-term, manage actively, monitor actively but not manage, and require its suppliers to monitor. Notably, the enterprise needs to pay attention to the significant, apparent risks akin

Table 3 Methods to identify supply chain risks

Method	Description
Brainstorming	Method in which a team of experts collects a broad set of ideas and ranks them
Checklists	Method in which users refer to a previously developed list of representative supply chain risks that need to be considered
Check-sheets	Method in which users collect past data about events to derive an event's distribution. For example, an enterprise might record late deliveries from suppliers to rate supplier reliability
Delphi method	Multi-round process in which a group of experts anonymously replies to questionnaires about supply chain risks. At the end of each round, the experts receive feedback on the group's responses. The process repeats itself until expert consensus is achieved
Early warning signals	Indicators used to notify users of changing supply chain risks. For example, an enterprise may collect equipment maintenance information, machine reliability information, or product complaints as early indicators to predict supply chain risks
Fault tree analysis	Deductive method that begins with an undesired event (top event) and determines how that event could occur by constructing a logic diagram (i.e., the fault tree)
Failure mode and effects analysis (FMEA)	Method that identifies: (a) the ways in which a product or process can fail (i.e., failure modes) and (b) the consequences of those failures (i.e., effects). Failures can be prioritized according to their frequency, severity, and detectability
Hazard and operability study (HAZOP)	Inductive method that defines possible deviations from the expected or intended performance. "Guide words" are used as a systematic list to identify those deviations. Finally, the criticalities of the deviations are assessed
Interviews and surveys	Use of structured or semi-structured interviews and questionnaires to ask experts to identify supply chain risks
Ishikawa cause and effects diagram (Fishbone diagram)	Method that identifies possible causes for a problem. Causes are typically grouped into the following categories: Methods, Machines, People, Materials, Measurement, and Environment. This method can be used to facilitate brainstorming
Supply chain mapping	Method that involves the mapping of the supply chain network to identify the number and location of suppliers, number and origin of shipments, modes of transport and routes, ports, or 3PL providers. The supply chain map is then used to identify supply chain risks across the different nodes or arcs of the network
Wheel of crises	Method in which possible crises are listed in a wheel. Users turn the wheel and discuss the possible consequences if they face that crisis where the wheel stops. This method can be used to facilitate brainstorming

Based on Harland et al. (2003), Mitroff and Alpaslan (2003), Norrman and Jansson (2004), Peck (2005), Knemeyer et al. (2009), and Tummala and Schoenherr (2011)

Note: Other supply chain risk identification methods can be found in the British Standards Institution (BS 31100:2008) and ISO/ICE (ICE 31010:2019)

Table 4 Methods to assess supply chain risks

Method	Description
Bayesian analysis	Statistical technique that uses distribution data to assess the probability of occurrence of a supply chain risk
Bow-tie analysis	Method that describes and communicates risk scenarios. The bow-tie diagram gives a visual representation of all possible incident scenarios existing around a specific hazard. The bow-tie also shows what an enterprise does to control those scenarios by identifying safety barriers
Probability/severity matrix	Matrix that combines qualitative or semi-quantitative ratings of probability and severity of supply chain risks (see Fig. 2 for an example)
Event tree analysis	Inductive method that begins with a starting event (e.g., component failure) and determines the consequences of that event by constructing a logic diagram (i.e., the event tree). Each path is assigned a probability of occurrence. The user can then calculate the probability of the various possible outcomes
Expert opinion combined with historical data	Method that combines historical data about the occurrence of events (if available) with expert opinion on the events' probability of occurrence and severity of impact
Failure mode and effects analysis (FMEA)	Method that identifies: (a) the ways in which a product or process can fail (i.e., failure modes) and (b) the consequences of those failures (i.e., effects). Failures can be prioritized according to their frequency, severity, and detectability
Hazard and operability study (HAZOP)	Inductive method that defines possible deviations from the expected or intended performance. "Guide words" are used as a systematic list to identify those deviations. Finally, the criticalities of the deviations are assessed
Ishikawa cause and effects diagrams (Fishbone diagram)	Method that identifies possible causes for a problem. Causes are typically grouped into the following categories: Methods, Machines, People, Materials, Measurement, and Environment. This method can be used to facilitate brainstorming
Monte Carlo Simulation	Method used to establish the aggregate variation in a system that results from variations in the system for various inputs. Each input has a defined distribution, and the inputs are related to the output via defined relationships

Based on Harland et al. (2003), Mitroff and Alpaslan (2003), Norrman and Jansson (2004), Zsidisin et al. (2004), Knemeyer et al. (2009), and Tummala and Schoenherr (2011)

Other supply chain risk assessment methods can be found in the British Standards Institution (BS 31100:2008) and ISO (ISO 31010:2019)

Some methods are relevant for supply chain risk identification and risk assessment and thus appear in Tables 3 and 4 of this chapter

to the "elephant in the room" that decision-makers tend to overlook. These risks are labeled as "gray rhinos" and are the high-probability and high-impact events that are generally ignored (Wucker, 2016). Gray rhinos do not occur suddenly but after a

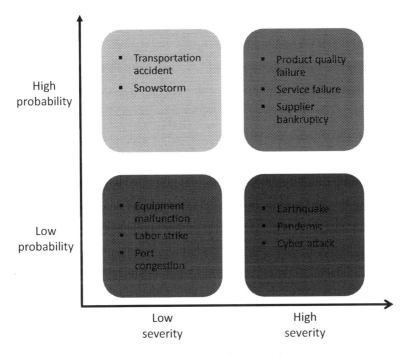

Fig. 2 Risk probability and severity matrix with supply chain risk examples

series of warnings and visible evidence, which decision-makers and organizations tend to overlook until too late (Wucker, 2016).

One method to guide decision-making around prioritizing the management of supply chain risks is to develop a matrix that categorizes them according to low versus high probability and low versus high severity. Figure 2 illustrates an example of such a matrix. Notably, an enterprise needs to define a time interval (e.g., a quarter or a year) by which it will update this categorization. The environment in which it operates changes continuously, and some risks might shift across categories. For instance, we indicated "port congestion" as a low-probability and low-severity event in the example depicted in Fig. 2. Nevertheless, port congestion in the USA in 2021, for instance, is likely considered a high-probably and high-severity event by many enterprises that import goods. In the summer and fall of 2021, the ports of Los Angeles and Long Beach are struggling to handle the overwhelming number of containers arriving—a result of a surge in US consumer demand for imported durable goods post-COVID-19. In this way, containers have been sitting on containerships in the water instead of being processed through the ports, disrupting the supply chain operations of companies such as Nike and Costco (Paris & Smith, 2021).

An enterprise may decide to focus first on high-probability and high-severity risks, as these are very likely to occur and will severely impact operations and disrupt supply continuity when they do occur. It could also focus first on the high-probability and low-severity risks. Even if these risks are not as severe, their frequent

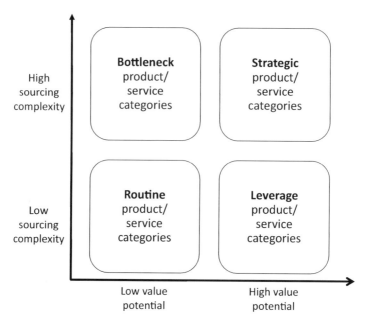

Fig. 3 Portfolio analysis matrix (Kraljic, 1983)

occurrence can accumulate costs and failures with severe long-term implications. Alternatively, it can focus on mapping the risks that exist within its boundaries (such as its plant locations or warehouses) or within its boundaries and at the first-tier suppliers and customers before considering risks in the sub-tiers of its supply chain.

An enterprise could also prioritize among first-tier suppliers and customers. A helpful way to do so is by using portfolio analysis to analyze its goods and services (Kraljic, 1983). With portfolio analysis, an enterprise can categorize goods and services according to their "value potential" (low versus high) and "sourcing complexity" (low versus high) (Kraljic, 1983).

Figure 3 shows a portfolio analysis matrix that can be constructed based on the product's value potential and sourcing complexity. To assess "value potential," metrics might include relative spend, impact to cost, delivery, and reliability, among others (Kraljic, 1983; Lambert, 2008). Metrics to assess "sourcing complexity" might include the number of available suppliers in the market, the complexity of materials requirements, product complexity, logistics complexity, and geographical locations of suppliers, among others (Kraljic, 1983; Lambert, 2008). Essentially, sourcing complexity represents the sourcing constraints that an enterprise will face when searching for alternative supply sources in case of a supply chain disruption. Commodity materials (e.g., those in the routine or leverage categories) typically pose a lower risk to an enterprise, as it can likely find alternative suppliers if a supply chain disruption occurs. Hence, the enterprise may not need to go beyond first-tier suppliers for its SCRM program. Conversely, materials that are sole-sourced or single-sourced (e.g., those possibly in the bottleneck or strategic categories) present

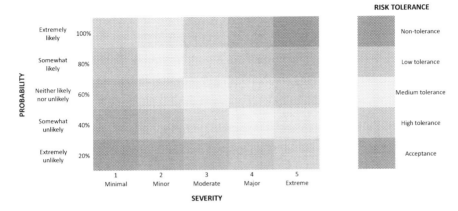

Fig. 4 Supply chain risk prioritization using a risk heat map

a significant risk of supply discontinuity. Therefore, the enterprise needs to closely monitor or work with those suppliers, search for risks beyond first-tier suppliers, and include those in its risk register. It is noteworthy that during COVID-19, items that many had categorized as commodities (e.g., masks and gowns), many of which were manufactured in China, turned out to be critical to the operation of health care organizations and the safety of both workers and patients. Therefore, one must be diagnostic as to the impact of any item as items are classified.

An enterprise can also include its risk tolerance or risk appetite, namely the amount of risk it is prepared to tolerate (be exposed to) at any time (Chapman, 2006), in the probability/severity matrix. Risk tolerance is unique to each enterprise and depends on its culture and objectives as well as the changing environmental conditions (Chapman, 2006).

Figure 4 shows an example of a risk heat map that combines the probability/severity matrix and risk tolerance. By considering its risk tolerance, an enterprise may first focus on the red areas before moving into the orange and yellow areas and decide to accept the supply chain risks in the green areas.

Once an enterprise identifies relevant supply chain risks, it can map them using different methods. For example, GM uses a concentric vulnerability map to map risks (see Fig. 5) (Sheffi, 2005). GM categorizes risks into strategic, financial, operations, and hazard risks. The axes correspond to low versus high probability of occurrence and low versus high impact. The radials show whether these risks originate from GM's internal operations (e.g., at GM's plants) or from the external environment (e.g., natural disasters).

Fig. 5 Concentric vulnerability map: A risk mapping tool (adapted from Sheffi, 2005)

3.2.2 Supplier Risk Assessments

A critical part of this stage in the SCRM process is supplier risk assessment. A supplier risk assessment constitutes a formal evaluation of the financial and operational risks that suppliers may exhibit. As enterprises typically do not have abundant resources to include all their suppliers in a formal supplier risk program, they use various criteria to decide on which ones to include. A recent survey of supply chain professionals by CAPS Research identifies the top criteria enterprises use, as shown in Table 5. Critical categories or spend areas are those categories that an organization considers essential to the business, either because they regard critical materials or materials that feed into multiple production lines. Annual spend represents the amount an enterprise spends with a supplier and is often an indicator of the supplier's importance and, therefore, the amount of risk the supplier poses to the enterprise. Sole-sourcing is when an enterprise sources a particular good or service from one supplier, and only that supplier is available in the supply market (Van Weele, 2010). Single-sourcing is when an enterprise chooses to source a particular good or service

Table 5 Criteria used to include suppliers in a supplier risk program (CAPS Research, 2015)

Criteria used to determine if a supplier should be included in the supplier risk program	% of organizations surveyed
Critical categories/spend areas	82.8
Annual spend	70.3
Supplier product/ service is part of a critical deliverable	68.8
Sole- or single-source supplier	57.8
New supplier (never used before)	34.4
Certain geographic locations	29.7

from one supplier, even when other suppliers may be available in the market (Van Weele, 2010). Both these strategies can be risky as the enterprise depends on this one supplier, and anything the supplier does will influence the enterprise's supply chain operations. A new supplier in the supply base can also present risks because the enterprise has no experience working with this supplier and may be unfamiliar with its processes and general way of doing business. Finally, specific geographic locations may be considered riskier, for example, due to natural disaster risk, geopolitical tensions, trade restrictions, port congestion, or consistency in the quality of procured goods and services. Notably, the criteria in Table 5 could be considered indicators of "value potential" and "sourcing complexity," the two dimensions of Kraljic's portfolio analysis discussed above.

Supplier risk assessments are imperative given that most supply chain disruptions experienced by an enterprise originate in first-tier suppliers (Business Continuity Institute, 2018). Hence, enterprises can conduct formal supplier risk assessments, either internally or using a third party. The same CAPS Research survey mentioned above reported that 86% of the enterprises surveyed conducted supplier risk assessments, 79% scored those assessments, while 29% used third parties to conduct those assessments (CAPS Research, 2015).

3.3 Supply Chain Risk Management

The third stage in the SCRM process involves identifying, evaluating, and implementing strategies to manage the supply chain risks according to the enterprise's risk prioritization. Importantly, SCRM might not always involve mitigating risks, as an enterprise may accept, avoid, or transfer risks.

- *Supply chain risk acceptance*: An enterprise identifies and accepts the supply chain risk. It does not act either because it finds it economical not to do anything or has no alternative and feasible options to transfer or mitigate the risk (Chapman, 2006). Risk acceptance depends on the context in which an enterprise operates. For example, in the health care industry, hospitals can be captive to pharmaceutical manufacturers who hold a patent for a specific drug or medical device for a certain period of years. Hospitals, as such, often accept the risk of sourcing these products. Risk acceptance also depends on the enterprise's risk tolerance or appetite (Chapman, 2006). As mentioned earlier in this chapter, risk tolerance is unique to each enterprise and depends on the enterprise's culture, objectives, industry sector, as well as environmental and business conditions (Chapman, 2006).

- *Supply chain risk avoidance*: An enterprise identifies the supply chain risk and considers it unacceptable. Because the enterprise cannot alter the risk, it chooses to eliminate it before the risk triggers a supply chain disruption (Ritchie & Brindley, 2007). Possible risk avoidance strategies include stopping the sale of a product, exiting a geographical market, or switching a supplier (Manuj & Mentzer, 2008).
- *Supply chain risk transfer*: An enterprise identifies the supply chain risk but transfers responsibility to another party. Possible risk transfer strategies include business interruption or supply chain disruption insurance (Cummings, 2020; Fan & Stevenson, 2018), outsourcing, financial risk transfer mechanisms, or risk-transfer contracts (Olson & Wu, 2010). Notably, risk transfer strategies may not eliminate an enterprise's exposure to the risk or the risk's impact. For instance, Hurricane Maria exposed the vulnerabilities of US hospitals, which relied on group purchasing organizations (GPOs) for a large amount of critical supplies with the notion that GPOs have a diversified supply base. In reality, GPOs were exposed to the same sub-tier suppliers.
- *Supply chain risk mitigation.* An enterprise identifies risk and actively manages it through actions that seek to reduce the probability of the risk's occurrence or the severity of its impact.

In this section, we focus on supply chain risk mitigation approaches. We refer the reader to Chapman (2006) for a more comprehensive review of the abovementioned approaches to supply chain risk avoidance and transfer. Indeed, the majority of the SCRM literature concentrates on supply chain risk mitigation. Table 6 offers an exemplary but not exhaustive list of such mitigation approaches.

We note that there are supply chain risks that an enterprise cannot anticipate. These risks can be highly improbable with highly severe consequences, typically regarded, as mentioned above, as "black swans" (Taleb, 2007), or they are inconceivable by management and organizational systems, typically regarded as "unknown-unknowns" (Ramasesh & Browning, 2014). The enterprise will be unable to develop specific risk mitigation strategies for these types of risks. Therefore, it must build resilience into its supply chain through (a) robustness strategies that help it avoid a supply chain disruption or resist its impact (such as anticipation and visibility capabilities) and (b) recovery strategies that help the enterprise recover from a supply chain disruption quickly (such as agile supply chain redesign) (Pettit et al., 2013; Wieland & Wallenburg, 2013).

Finally, even with a robust set of supply chain risk mitigation strategies available, an enterprise may be unable to implement certain strategies depending on various factors. In this chapter, we discuss two central factors influencing the ability of an enterprise to implement such strategies: the properties of the industry in which it operates and the attributes of its supply chain, specifically supply chain complexity.

Table 6 Exemplary supply chain strategies to mitigate the probability or impact of supply chain risks

Supply chain risk mitigation approach		References
Demand management approaches	Reduction of the forecast horizon	Sodhi and Tang (2012)
	Centralized (decentralized) capacity for unpredictable (predictable) demand, i.e., risk pooling	Chopra and Sodhi (2004, 2014)
	Shift demand across different products	Tang and Tomlin (2008)
	Flexible pricing	Tang and Tomlin (2008)
Supply management approaches	Flexible supplier contracts	Tang and Tomlin (2008)
	Risk-sharing contracts (buybacks, real option-based contracts)	Tang (2006)
	Multi-sourcing	Chopra and Sodhi (2004); Sheffi and Rice (2005); Pettit et al. (2013)
	Favor redundant supplies for high-volume products, low redundancy for low-volume products	Chopra and Sodhi (2004)
	Centralize redundancy for low-volume products to few, key, flexible suppliers	Chopra and Sodhi (2004)
	Supplier selection criteria	Ravindran et al. (2010)
	Supplier performance evaluation criteria	Blome and Schoenherr (2011)
Manufacturing management approaches	Modular product designs	Pettit et al. (2013)
	Multiple plants with interoperability	Sheffi and Rice (2005); Tang and Tomlin (2008)
	Manufacturing postponement	Pettit et al. (2013)
Transportation management approaches	Multiple transportation providers	Pettit et al. (2013)
	Multiple transportation modes	Pettit et al. (2013)
	Re-routing of requirements	Pettit et al. (2013)
Inventory management approaches	Logistics postponement	Zinn and Bowersox (1988)
	Strategic inventory	Pettit et al. (2013)
	Decentralized (centralized) inventory for predictable, lower-value products (for less predictable, higher-value products)	Chopra and Sodhi (2014)
Collaborative approaches with trading partners	Alliances with suppliers, group purchasing enterprises (GPOs), transportation providers, or distributors to identify, assess, share, or mitigate risks	Sheffi and Rice (2005); Pettit et al. (2013)
	Information sharing	Pettit et al. (2013)
	Collaborative Planning, Forecasting, and Replenishment (CPFR)	Pettit et al. (2013)
	Vendor-managed inventory (VMI)	
Product management approaches	Part commonality	Chopra and Sodhi (2004); Pettit et al. (2013)
	Product variability reduction	Pettit et al. (2013)

(continued)

Table 6 (continued)

Supply chain risk mitigation approach		References
	Favor responsiveness to cost for short life-cycle products, or low-volume, unpredictable products	Chopra and Sodhi (2004)
Capacity management approaches	Slack in capacity utilization	Sheffi and Rice (2005); Pettit et al. (2013)
	High redundancy for high-volume products and low redundancy for low-volume products	Chopra and Sodhi (2014)
	Distributed capacity	Pettit et al. (2013)
Financial management approaches	Hedging	Pettit et al. (2013)
	Portfolio diversification	Pettit et al. (2013)
Information management approaches	Information gathering about the business environment, competitors, suppliers, supply markets	Pettit et al. (2013)
	Monitoring and sharing information about early warning signals	Pettit et al. (2013)
General approaches	Business continuity plans internally and with suppliers	Zsidisin et al. (2005)

3.3.1 Industry Constraints and Supply Chain Risk Mitigation

The tolerance for risk and the implementation of supply chain risk mitigation strategies will depend on the industry in which an enterprise operates. For example, some industries, such as aerospace manufacturing or automotive manufacturing, were early adopters of just-in-time (JIT) manufacturing and delivery, leading to reduced inventories. Companies operating in these sectors recognized the need to employ other strategies to mitigate the risk of low buffers in the supply chain. Toyota, for example, seeks to standardize the parts it sources from Japanese suppliers so that the suppliers can share components that can be manufactured in several locations; asks suppliers of specialized parts, which cannot be duplicated across plants, to hold more inventory; and seeks to make parts procurement across geographic regions independent so that a natural disaster in Japan would not affect Toyota's production in other countries (Kim, 2011).

Furthermore, the degree of outsourcing differs across industries, which also influences an enterprise's dependency on its supply base. For example, industries, such as health care, have not only outsourced most of their procurement spend, but they frequently depend on sole or single suppliers, especially when the supplier holds the patent for the manufacturing of a drug or medical device. While this resource dependency (Pfeffer & Salancik, 1978) further increases supply chain risks, enterprises in this industry often accept the risk as a given.

3.3.2 Supply Network Complexity and Supply Chain Risk Mitigation

One stream of work explored how increasing complexity in the supply chain might increase the frequency and impact of supply chain risks. Broadly, supply chain complexity refers to "the level of detail complexity and dynamic complexity exhibited by the products, processes and relationships that make up a supply chain" (Bozarth et al., 2009, p. 80). Others viewed complexity as the combination of "the total number of nodes ... and the total number of forward..., backward..., and within-tier materials flows... within a given supply chain" (Craighead et al., 2007, p. 140).

Prior research largely demonstrated that supply chain complexity is detrimental when it comes to supply chain disruptions. For example, Choi and Krause (2006) argued that higher complexity in the supply base means an enterprise has to deal with many suppliers and, thus, monitor and coordinate more interfaces with those suppliers. Hence, Bode and Wagner (2015), argued that it becomes more difficult for the enterprise to continue having a sufficiently broad view and control over its suppliers, making it more susceptible to experiencing supply chain disruptions more frequently.

Furthermore, as complexity in the supply chain increases, the severity of supply chain disruptions can also increase. For example, Craighead et al. (2007) showed that if a trigger disrupts a part of the supply chain that is more complex, it is expected to affect more nodes or arcs in that network and, thus, increase the impact of the subsequent disruption. Also, Bode and Macdonald (2017) found that when complexity in a supply chain increases, managers are challenged to recognize that a supply chain disruption has happened as well as and diagnose it. This added difficulty can slow down an enterprise's reaction to the supply chain disruption and subsequently exacerbate the disruption's impact.

Recent research, however, has provided evidence that supply chain complexity may be both a detriment and a blessing for supply chains. Wiedmer et al. (2021) examined how various dimensions of supply network complexity (supply, logistics, and product) influence the ability of US automotive supply chains to resist and recover from supply chain disruptions triggered by the 2011 Japan Earthquake and Tsunami. Supply complexity (i.e., the number of suppliers) worsens disruption impact and improves a firm's recovery from the disruption. Logistics complexity (i.e., the number of ocean carriers) does not significantly affect disruption impact but enhances a firm's recovery. Lastly, product complexity (i.e., the number of components in a product) worsens disruption impact but does not significantly affect recovery. Wiedmer et al. (2021), as such, concluded that academics should differentiate between the various types of supply network complexity and the phase of the disruption in which an organization is (i.e., disruption-impact versus disruption-recovery phase).

3.4 Supply Chain Risk Monitoring

As enterprises change and evolve, so does supply chain risk. Hence, they need to regularly scan their internal operations, supply chain network, and external environment to identify new sources of risks or how the already identified risks may have changed. They also need to evaluate whether the established supply chain risk mitigation strategies are effective. This stage is formally labeled as supply chain risk monitoring. Activities in this stage include:

- Updating the risk register if necessary
- Appraising the effectiveness of the supply chain risk mitigation actions. Metrics that can be used to assess effectiveness include time to diagnose a supply chain disruption and implement recovery strategies, time to recover, time to set up alternative sources, operational metrics (such as on-time delivery, lead time), or the number of risk events affecting the supply chain. For specific types of supply chain disruptions, such as product recalls, other metrics can be used, such as the number of downstream partners notified about the recall, the number of responders to the recall notification, the percentage of recalled products, or the time to remove a product from the market
- Evaluating the effects of the risk treatment on the performance of the enterprise (such as product or service quality, on-time delivery, and lead time)
- Identifying opportunities for improvement
- Considering changes in regulations, processes, performance assessment, and the supply chain to update the SCRM plan
- Monitoring how supply chain partners are performing relative to their commitments

Academic research has paid little attention to supply chain risk monitoring (Fan & Stevenson, 2018; Ho et al., 2015). Researchers advised that firms develop data management information systems to monitor risks (Tummala & Schoenherr, 2011) and establish processes to identify and monitor early warning signals (Craighead et al., 2007) to identify new risks or observe trends proactively. Sheffi (2005) has identified the importance of studying "near misses," namely incidents that, if actually occurred, or occurred more frequently or for more extended periods, would have a significant impact and provide the occasion to opine on ways one might act. This takes a significant commitment to vigilance and deterrence. However, there is an important temporal aspect to this—as no risk, as discussed later in this chapter, of the breadth and depth of COVID-19 has occurred in many years. Denial or a lack of incentives, including costs associated with long-term vigilance by management and executive boards, may interfere with a methodical, disciplined, and prudent approach to existing SCRM routines that focus on short-term and more manageable risks (e.g., hurricanes and factory fires). In practice, enterprises tend to incorporate supply chain risk monitoring into existing enterprise routines (Fan & Stevenson, 2018). When it comes to monitoring supplier financial or operational risk, enterprises tend to incorporate it into their regular supplier assessment activities

Table 7 Assessments of highest-risk suppliers (CAPS Research, 2015)

Regularity of supplier risk assessments	% of companies surveyed
Quarterly	30.2
Yearly	23.8
Six months	20.6
Monthly	6.3
Other	12.7

(Blome & Schoenherr, 2011). For instance, CAPS Research reported that most organizations re-assess the highest-risk suppliers quarterly, followed by yearly and bi-annually, as shown in Table 7.

3.4.1 Technologies for Supply Chain Risk Monitoring

Enterprises increasingly utilize specialized software to monitor the risks in their supply chain. Interviews we conducted with supply chain managers from various industries show that best-practice organizations typically employ third-party tools to monitor risks in their supply chains on a real-time basis. For example, Credit Risk Monitor and Dun & Bradstreet's Supplier Risk Manager are popular tools to monitor supplier financial risk. Other tools, such as supply chain network mapping software by Resilinc or Risk Methods, enable enterprises to map their supply chain networks, monitor events that can affect critical nodes or arcs in the network in real-time, and promptly act on threats for supply continuity. Other examples of such software are shown in Fig. 6. These types of software often use technologies including artificial intelligence and machine learning, as shown in Fig. 6.

A recent development in supply chain management has been the use of control towers. In the supply chain context, control towers are cloud-based, digital networks that provide executives visibility into their trading partners and the supply chain. For example, intelligent control towers can help mitigate supply chain risks by providing visibility into events occurring in the supply chain network, identifying how such events can influence lead times based on the enterprise's service-level agreements, offering suggestions for mitigating the risks using artificial intelligence, and even executing these suggestions without human intervention (One Network, 2020).

4 Behavioral Influences in Supply Chain Risk Management

Recent research has begun extending inquiry around SCRM beyond the traditional supply chain tactics to mitigate risks and exploring behavioral supply management issues. Carter et al. (2007, p. 634) define behavioral supply management as "the study of how judgment in supply management decision-making deviates from the assumptions of homo economicus." Recent research examined behavioral issues

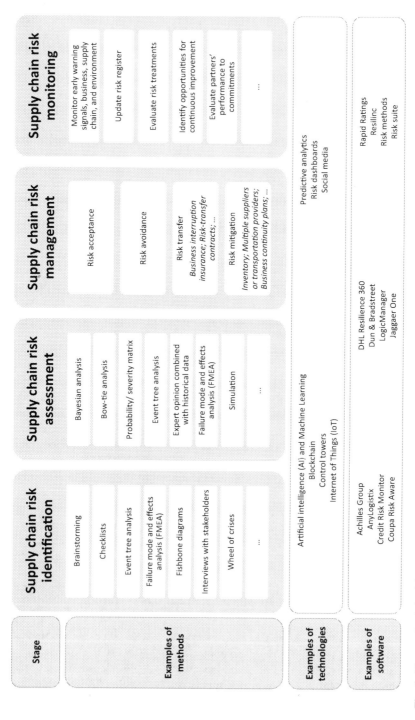

Fig. 6 The supply chain risk management (SCRM) process with examples of methods, technologies, and software

surrounding the identification, assessment, and mitigation of supply chain risks and reactions following a supply chain disruption. Ellis et al. (2010) were among the first to propose and demonstrate empirically that managerial risk perceptions matter when it comes to risk mitigation strategies. The authors showed that supply market characteristics (such as technological uncertainty and availability of suppliers) and product characteristics (such as the degree of importance and customization of a product) influence managerial perceptions of the probability and severity of supply chain disruptions, and in turn, their search for alternative suppliers.

Eckerd and colleagues focused on psychological contract breaches, which occur when "an individual perceives insufficient fulfillment of obligations from an exchange partner" (Eckerd et al., 2013, p. 568). A breach, as such, is a perception that the terms in a psychological contract have been violated or simply not met (Suazo, 2011). Eckerd et al. (2013) found that breach attribution and severity perceptions influence a buyer's ordering behavior. These attributes also trigger an adverse affective reaction, termed psychological contract violation, which influences a buyer's perception of how fair a supplier is. Similarly, Mir et al. (2017) found that breach attribution and severity elicit negative affective reactions, which, in turn, influence supplier switching. Lastly, Eckerd et al. (2016) extended this previous work to examine the role of national culture (China vs. the USA). They found that breach attribution influences post-breach ordering behavior and that breach severity and national culture influence trust towards the supplier.

Other studies explored how an enterprise's communication about risk and resilience influences sourcing decisions. For example, DuHadway et al. (2018) showed that individuals make riskier sourcing decisions after their enterprises communicate progress in reducing supply chain risk levels. Likewise, Mena et al. (2020) found that managerial perceptions of enterprise resilience via systemic communication of resilience initiatives, such as training or corporate announcements by the company's executives or via personal exposure motivate managers to select riskier suppliers. This effect was strengthened when the risk propensity of a decision-maker was higher.

Other researchers focused on the effects of supply chain disruptions on affective reactions and supply management decisions post disruption. For example, Reimann et al. (2017) looked at supplier-included disruptions and examined the conditions that induce buyers to engage in constructive interaction with their suppliers or create conflict between buyers and suppliers. Finally, two studies by Polyviou and colleagues examined the role of emotions in response to supply disruptions and post-disruption sourcing decisions. Polyviou et al. (2018) demonstrated that sourcing managers experience more anger when they consider a supply disruption as controllable by a disrupting supplier rather than nature; as a result, they are less likely to keep sourcing from that supplier afterward. Polyviou et al. (2022) found that sourcing managers experience more guilt following a supply disruption they consider controllable by a disrupting supplier rather than nature, and they had recommended that supplier to their organization before the disruption. In turn, they tend to prefer riskier yet more advantageous suppliers when making new supplier selections after the disruption. In other words, Polyviou et al. (2022)

showed there is a path dependency between prior and new supplier selection decisions when the previously selected supplier later becomes a disrupting supplier. Finally, Chen et al. (2019) found evidence for the "positive supplier performance penalty effect" (p. 1224) such that sourcing managers are more likely to terminate a supplier with stellar performance when that supplier commits an error.

In summary, this literature stream has demonstrated that responses to supply chain risks and disruptions are influenced by the characteristics of the individuals making the decisions, the firm experiencing the risk or disruption, and the environment. Therefore, researchers must not only focus on proposing traditional supply chain strategies to mitigate risks but also on further understanding the cognitions, emotions, and other non-objective factors that can influence decision-making in this context.

5 The Governance Structure of Supply Chain Risk Management: Where Does Risk Responsibility Lie?

Little academic research focuses on the governance structure around SCRM. Although SCRM is traditionally considered everyone's job (Sheffi & Rice, 2005), an enterprise must establish a governance structure around it. The Supply Chain Risk Leadership Council (SCRLC) (2011) offers generic guidelines about a governance structure for risk management. According to SCRLC (2011), an enterprise developing an SCRM program should start with a cross-functional team of decision-makers, including quality, engineering, operations, supply management, logistics, finance, legal, or marketing managers. This team will identify, own, and manage risks at the level they exist, and determine the SCRM program's scope. An enterprise, however, needs to go beyond these guidelines and establish a formal governance structure around SCRM. A governance structure has several benefits, as follows:

- Establishes a formal SCRM process
- Determines formal accountability and ownership of supply chain risks
- Sets the system within which cross-functional managers will come together to identify, evaluate, and alleviate supply chain risks
- Establishes formal channels of communication among the various stakeholders
- Determines the frequency of supply chain risk monitoring and continuous update of the risk register
- Establishes a performance measurement system to assess the effecacy of the SCRM practices implemented not only by the focal enterprise but also by its key trading partners
- Provides the incentives for continuous improvement in SCRM
- Encourages change in the behavior within and beyond the enterprise to truly make SCRM everyone's job

6 Supply Chain Risk Management in the Post-COVID-19 Era

The COVID-19 pandemic brought to the forefront the importance of many of the issues discussed above across many industries. Indeed, it presented a continuing "live study" to understand supply chain disruptions and approaches to alleviate them (Chopra et al., 2021) and brought the importance of large-scale/long-term SCRM efforts to the forefront. The pandemic disrupted supply chains around the globe, impacting production and logistics operations and reducing opportunities to meet the surge in demand for consumer products. Less than 6 months into the pandemic, 94% of Fortune 1000 companies incurred supply chain disruptions triggered by COVID-19, while 75% were negatively affected (Timmermans et al., 2020). Vulnerabilities of supply chains to an external shock such as a pandemic—in concert with little attention by management to plans for regaining resilience—left supply chains across the globe in shambles (Timmermans et al., 2020).

In the USA, the newly elected President Biden issued executive orders focused on managing supply chain risks (The White House, 2021a, 2021b). The EU's "Action Plan on Critical Raw Materials" (European Commission, 2020) presented cross-national strategies to develop a secure and sustainable supply of raw materials for a resilient European economy. The pandemic highlighted the importance of supply chains and SCRM as keys to sustainability and what many have described as a need to "future proof" the supply chain.

The COVID-19 pandemic led to reflection on the unanticipated effects of supply chain practices. For example, the health sector in the USA had embraced JIT inventory and an overemphasis on cost, and it relied on a number of intermediaries (GPOs and distributors) to carry out the sourcing and contracting on behalf of hospitals. However, hospitals across the globe were strained as they had continuously escalated their contracting with a few suppliers, many of whom were located far from their shores. Such single sourcing (see Table 5) and contracting were frequently done without adequate credentialing of suppliers and their suppliers' upstream suppliers, with little network mapping software in place. Indeed, a domino effect could be observed as shortages cascaded. Few health care providers subscribed to the services that detect suppliers' risks.

Disruptions to the health sector supply chain were not a new occurrence. In 2017, the aforementioned Hurricane Maria closed down important suppliers in Puerto Rico—many of which were the sole source for medical goods and operated in an environment that had little investment in pre-disaster mitigation, adaption, and planning as well as little building and maintaining of relationships with diverse partners, businesses and stakeholders—all of which hampered the recovery effort (Kim & Bui, 2019; Sacks et al., 2018). The result of poor disaster management was that healthcare organizations in the USA were forced to postpone elective surgeries and ration critical products (Sacks et al., 2018).

COVID-19 exposed many of the risks associated with global sourcing. In the health care and other sectors, many products were produced in Asia. As those nations secured their domestically manufactured products for their populations, the surges in demand for products in other nations, such as the USA, were not met. Moreover, as shipping became difficult and tariffs were imposed, supply chain disruptions were exacerbated. With a rapidly accelerating demand, the hypergrowth in semiconductor chip demand has significantly impacted the availability of chip-powered equipment. While much focus has been on the auto industry, in the biotech sector, devices such as ventilators and defibrillators, imaging machines, monitors (for glucose and blood pressure), and implantable devices are all dependent on the availability of semiconductor chips, leading companies to consider strategies to mitigate both current and long-term risks (Murray & Bradley, 2021).

One of the conclusions to be drawn from the COVID-19 experience is the inadequacy of ongoing management strategies for dealing with significant periods of uncertainty. As discussed above, governance is an essential aspect of any SCRM program. Health care systems across the USA and other countries had relied on government-sponsored and managed pooled resources. And while pools can be an important part of an SCRM program, many of these stockpiles failed to provide significant quantities of needed products and, when provided, products had not been properly maintained or rotated as expiration dates occurred (Handfield et al., 2020). Common pool resources require a strong governance and management structure.

How should managers be thinking? COVID-19 presented an unusual situation with (a) a sizeable demand increase, (b) a significant dip in supply, and (c) great uncertainty as to the duration of the pandemic. It is an example of an extreme supply chain risk requiring new ways of addressing risk and thinking about new models for supply chain resilience (Sheffi, 2020). Stimulated by this tri-factor situation, researchers have begun to differentiate between strategies and governance systems required for "micro," "meso," and "macro" supply chain disruptions (Azadegan & Dooley, 2021) and strategies associated with different kinds of disruptions (Chopra et al., 2021). SCRM strategies include pools of supplies and capabilities owned by an individual company for use at a time of minimal disruption, tiered, within-industry collaboration of supplies for somewhat longer disruptions, and larger pools of well-managed common goods, generally sourced and managed by the government or their well-managed surrogate, to assure resilience. Noteworthy is the "macro" effort identified by Azadegan and Dooley (2021), which recognizes the role of government and association-sponsored pools, which largely failed in their ability to meet the supply chain disruptions associated with COVID-19 (Handfield et al., 2020) in developing common repositories and strategies.

Over the months of the COVID-19 pandemic, there has been much discussion about "future-proofing" the supply chain and putting forth a "new normal." Needed is a thorough understanding of the markets in which an organization operates and the markets in which trading partners operate. Absent early in the pandemic, in the health care and other sectors, was adequate visibility into the market, understanding of extant and potential resources, and a governance system to manage existing resources (Butt, 2021; Finkenstadt & Handfield, 2021). Clearly, COVID-19 was a critical wake-up call as there has been a rapid evolution in developing and implementing technologies to achieve end-to-end visibility (Sharma et al., 2020).

This discussion of COVID-19 reveals the need for different SCRM strategies to prepare for and mitigate supply chain disruptions of different duration and impacts on an entire industry. The health sector, highlighted in our discussion, is just one of many sectors impacted. Lingering questions include the benefits and costs of reshoring, the likelihood of competitors working collaboratively, the role of government and public–private entities as a buffer, and the financing of resilience initiatives. The World Health Organization had long warned that there were substantial risks for a pandemic of the nature of COVID-19, the West Pacific Region was a hotspot for outbreaks, and there was a need to "establish mechanisms to ensure the timely supply and availability of PPE, vaccines, drugs and other materials to ensure the safety and well-being of healthcare workers, patients and visitors and the broader community at all levels of the healthcare system" (World Health Organization, 2017). With predictions that we may well face such threats, supply chain management and other disciplines must coordinate and collaborate in preparing for future pandemics (World Economic Forum, 2021). After all, predictions show that the next global pandemic, as severe as COVID-19, will occur with a 47–57% chance within the next 25 years (Smitham & Glassman, 2021). Supply chain risk identification, assessment, management, and monitoring are, indeed, necessary and ongoing sub-processes.

7 Conclusions

In this chapter, we provided an overview of the SCRM process. We presented methods to identify, assess, manage, and monitor supply chain risks. We also discussed factors, such as industry and supply chain complexity, which can constrain an enterprise in implementing specific supply chain risk mitigation approaches. Furthermore, we presented research that argues that SCRM is not merely a logical business process but can be influenced by behavioral and other non-objective factors. These factors may include attributes of the supply chain risk manager, the risk, or the disruption. Finally, we argued that SCRM should be everyone's job. Nonetheless, this process must be formalized in organizations so that it receives the attention it deserves. Figure 6 summarizes the discussion in this chapter.

Importantly, organizations must realize that they will be unable to identify and assess all possible risks that can disrupt their supply chain operations. There will be those "unknown unknowns," namely uncertainties of which management will be unaware and, thus, unprepared to manage (Ramasesh & Browning, 2014). As such, organizations need to go beyond SCRM and cultivate resilience in their supply chain systems. Building resilience goes beyond risk mitigation; it involves the identification of system capabilities that are not risk-specific but can address a variety of supply chain risks (Fiksel et al., 2015). Resilience, moreover, enables an organization to deal with supply chain risks and disruptions more effectively than its competition and thus gain a competitive advantage (Sheffi & Rice, 2005).

References

Akkermans, H. A., & Van Wassenhove, L. N. (2013). Searching for the grey swans: The next 50 years of production research. *International Journal of Production Research, 51*(23–24), 6746–6755.

Akkermans, H. A., & Van Wassenhove, L. N. (2018). A dynamic model of managerial response to grey swan events in supply networks. *International Journal of Production Research, 56*(1–2), 10–21.

Ambulkar, S., Blackhurst, J. V., & Grawe, S. (2015). Firm's resilience to supply chain disruptions: Scale development and empirical examination. *Journal of Operations Management, 33–34*(1), 111–122.

Azadegan, A., & Dooley, K. (2021). A typology of supply network resilience strategies: Complex collaborations in a complex world. *Journal of Supply Chain Management, 57*(1), 17–26.

Banker, S. (2016, May 31). General Motors embraces supply chain resiliency. *Forbes.* https://www.forbes.com/sites/stevebanker/2016/05/31/general-motors-embraces-supply-chain-resiliency/#4ca0e9f83684.

Blome, C., & Schoenherr, T. (2011). Supply chain risk management in financial crises—A multiple case-study approach. *International Journal of Production Economics, 134*(1), 43–57.

Bode, C., & Macdonald, J. R. (2017). Stages of supply chain disruption response: Direct, constraining, and mediating factors for impact mitigation. *Decision Sciences, 48*(5), 836–874.

Bode, C., & Wagner, S. M. (2015). Structural drivers of upstream supply chain complexity and the frequency of supply chain disruptions. *Journal of Operations Management, 36*(3), 215–228.

Bode, C., Wagner, S. M., Petersen, K. J., & Ellram, L. M. (2011). Understanding responses to supply chain disruptions: Insights from information processing and resource dependence perspectives. *Academy of Management Journal, 54*(4), 833–856.

Bozarth, C. C., Warsing, D. P., Flynn, B. B., & Flynn, E. J. (2009). The impact of supply chain complexity on manufacturing plant performance. *Journal of Operations Management, 27*(1), 78–93.

Bradsher, K. (2020, July 5). China dominates medical supplies, in this outbreak and the next. *The New York Times.* https://www.nytimes.com/2020/07/05/business/china-medical-supplies.html.

British Standards Institution (BS 31100). (2008). *Code of practice for risk management.* https://shop.bsigroup.com/ProductDetail/?pid=000000000030153956.

Business Continuity Institute. (2018). *Supply chain resilience report.* https://www.thebci.org/news/bci-supply-chain-resilience-report-2018.html.

Butt, A. S. (2021). Strategies to mitigate the impact of COVID-19 on supply chain disruptions: a multiple case analysis of buyers and distributors. *The International Journal of Logistics Management.* Ahead-of-print. https://doi.org/10.1108/IJLM-11-2020-0455

CAPS Research. (2015). *Managing supplier risk.* https://www.capsresearch.org/library.

Carter, C. R., Kaufmann, L., & Michel, A. (2007). Behavioral supply management: A taxonomy of judgment and decision-making biases. *International Journal of Physical Distribution and Logistics Management, 37*(8), 631–669.

Chapman, R. J. (2006). *Simple tools and techniques for enterprise risk management.* Wiley.

Chen, Y. S., Rungtusanatham, M. J., & Goldstein, S. M. (2019). Historical supplier performance and strategic relationship dissolution: Unintentional but serious supplier error as a moderator. *Decision Sciences, 50*(6), 1224–1258.

Choi, T. Y., & Krause, D. R. (2006). The supply base and its complexity: Implications for transaction costs, risks, responsiveness, and innovation. *Journal of Operations Management, 24*(5), 637–652.

Chopra, S., & Sodhi, M. (2014). Reducing the risk of supply chain disruptions. *MIT Sloan Management Review, 55*(3), 72–80.

Chopra, S., Sodhi, M., & Lücker, F. (2021). Achieving supply chain efficiency and resilience by using multi-level commons. *Decision Sciences, 52*(4), 817–832.

Chopra, S., & Sodhi, M. S. (2004). Supply-chain breakdown. *MIT. Sloan Management Review, 46*(1), 53–61.

Christopher, M., & Peck, H. (2004). Building the resilient supply chain. *International Journal of Logistics Management, 15*(2), 1–13.

Craighead, C. W., Blackhurst, J., Rungtusanatham, M. J., & Handfield, R. B. (2007). The severity of supply chain disruptions: Design characteristics and mitigation capabilities. *Decision Sciences, 38*(1), 131–156.

Cummings, S. (2020, November 1). A quick guide to insurance for supply-chain disruption. *Supply Chain Brain*. https://www.supplychainbrain.com/blogs/1-think-tank/post/32135-a-quick-guide-to-insurance-for-supply-chain-disruption.

DuHadway, S., Carnovale, S., & Kannan, V. R. (2018). Organizational communication and individual behavior: Implications for supply chain risk management. *Journal of Supply Chain Management, 54*, 3–19.

Eckerd, S., Boyer, K. K., Qi, Y., Eckerd, A., & Hill, J. A. (2016). Supply chain psychological contract breach: An experimental study across national cultures. *Journal of Supply Chain Management, 52*(3), 68–82.

Eckerd, S., Hill, J., Boyer, K. K., Donohue, K., & Ward, P. T. (2013). The relative impact of attribute, severity, and timing of psychological contract breach on behavioral and attitudinal outcomes. *Journal of Operations Management, 31*(7–8), 567–578.

Ellis, S. C., Henry, R. M., & Shockley, J. (2010). Buyer perceptions of supply disruption risk: A behavioral view and empirical assessment. *Journal of Operations Management, 28*(1), 34–46.

European Commission. (2020). Critical raw materials resilience: Charting a path toward greater security and sustainability. https://eur-lex.europa.eu/legal-content/EN/TXT/?uri=CELEX:52020DC0474.

Fan, Y., & Stevenson, M. (2018). A review of supply chain risk management: definition, theory, and research agenda. *International Journal of Physical Distribution and Logistics Management, 48*(3), 205–230.

Fiksel, J., Polyviou, M., Croxton, K. L., & Pettit, T. J. (2015). From risk to resilience: Learning to deal with disruption. *MIT Sloan Management Review, 56*(2), 79–86.

Finkenstadt, D. J., & Handfield, R. (2021). Blurry vision: Supply chain visibility for personal protective equipment during COVID-19. *Journal of Purchasing and Supply Management*. https://doi.org/10.1016/j.pursup.2021.100689

Hallikas, J., Karvonen, I., Urho, P., Virolainen, V.-M., & Tuominen, M. (2004). Risk management processes in supplier networks. *International Journal of Production Economics, 90*(1), 47–58.

Handfield, R., Finkenstadt, D. J., Schneller, E. S., Godfrey, A. B., & Guinto, P. (2020). A commons for a supply chain in the post-COVID-19 era: The case for a reformed strategic national stockpile. *The Milbank Quarterly, 98*(4), 1058–1090.

Harland, C., Brenchley, R., & Walker, H. (2003). Risk in supply networks. *Journal of Purchasing and Supply Management, 9*(2), 51–62.

Hendricks, K. B., & Singhal, V. R. (2003). The effect of supply chain glitches on shareholder wealth. *Journal of Operations Management, 21*(5), 501–522.

Hendricks, K. B., & Singhal, V. R. (2005). An empirical analysis of the effect of supply chain disruptions on long-run stock price performance and equity risk of the firm. *Production and Operations Management, 14*(1), 35–52.

Ho, W., Zheng, T., Yildiz, H., & Talluri, S. (2015). Supply chain risk management: A literature review. *International Journal of Production Research, 53*(16), 5031–5069.

ISO/ICE (ICE 31010:2019). Risk management—Risk assessment techniques. https://www.iso.org/standard/51073.html.

Kim, C. R. (2011, September 6). Toyota aims for quake-proof supply chain. *Reuters*. https://www.reuters.com/article/us-toyota/toyota-aims-for-quake-proof-supply-chain-idUSTRE7852RF20110906.

Kim, K., & Bui, L. (2019). Learning from Hurricane Maria: Island ports and supply chain resilience. *International Journal of Disaster Risk Reduction, 39*, 101244.

Kim, Y., Chen, Y.-S. S., & Linderman, K. (2015). Supply network disruption and resilience: A network structural perspective. *Journal of Operations Management, 33–34*, 43–59.

Knemeyer, A. M., Zinn, W., & Eroglu, C. (2009). Proactive planning for catastrophic events in supply chains. *Journal of Operations Management, 27*(2), 141–153.

Kraljic, P. (1983). Purchasing must become supply management. *Harvard Business Review, 61*(5), 109–117.

Lambert, D. M. (2008). *Supply chain management: processes, partnerships, performance* (3rd ed.). Supply Chain Management Institute.

Macdonald, J. R., & Corsi, T. M. (2013). Supply chain disruption management: Severe events, recovery, and performance. *Journal of Business Logistics, 34*(4), 270–288.

Manuj, I., & Mentzer, J. T. (2008). Global supply chain risk management. *Journal of Business Logistics, 29*(1), 133–155.

Mena, C., Melnyk, S. A., Baghersad, M., & Zobel, C. W. (2020). Sourcing decisions under conditions of risk and resilience: A behavioral study. *Decision Sciences, 51*(4), 985–1014.

Mir, S., Aloysius, J. A., & Eckerd, S. (2017). Understanding supplier switching behavior: The role of psychological contracts in a competitive setting. *Journal of Supply Chain Management, 53*(3), 3–18.

Mitroff, I. I., & Alpaslan, M. C. (2003). Preparing for evil. *Harvard Business Review, 81*(4), 109–115.

Murray, B., & Bradley, S. (2021). Semiconductor chip shortage hits medtech: Strategies to build resilient supply chains. *Deloitte Health Forward Blog*. https://www2.deloitte.com/us/en/blog/health-care-blog/2021/semiconductor-chip-shortage-hits-medtech-strategies-to-build-resilient-supply-chains.html.

Norrman, A., & Jansson, U. (2004). Ericsson's proactive supply chain risk management approach after a serious sub-supplier accident. *International Journal of Physical Distribution & Logistics Management, 34*(5), 434–456.

Olson, D. L., & Wu, D. (2010). *Enterprise risk management models*. Springer Texts in Business and Economics.

One Network. (2020). *What is a supply chain control tower?* https://www.onenetwork.com/supply-chain-management-solutions/supply-chain-control-towers/.

Paris, C., & Smith, J. (2021, September 26). Cargo piles up as California ports jostle over how to resolve delays. *The Wall Street Journal*. https://www.wsj.com/articles/cargo-delays-are-getting-worse-but-california-ports-still-rest-on-weekends-11632648602.

Peck, H. (2005). Drivers of supply chain vulnerability: An integrated framework. *International Journal of Physical Distribution & Logistics Management, 35*(4), 210–232.

Pettit, T. J., Croxton, K. L., & Fiksel, J. (2013). Ensuring supply chain resilience: Development and implementation of an assessment tool. *Journal of Business Logistics, 34*(1), 46–76.

Pfeffer, J., & Salancik, G. R. (1978). *The external control of organizations: A resource dependence perspective*. Harper & Row.

Polyviou, M. (2016). *Essays on supply chain disruptions: A schema, managerial reactions, and decision-making*. Doctoral dissertation, The Ohio State University.

Polyviou, M., Rungtusanatham, M. J., & Kull, T.-J. (2022). Supplier selection in the aftermath of a supply disruption and guilt: Once bitten, twice (not so) shy. *Decision Sciences, 53*(1), 28–50. https://doi.org/10.1111/deci.12528

Polyviou, M., Rungtusanatham, M. J., Reczek, R. W., & Knemeyer, A. M. (2018). Supplier non-retention post disruption: What role does anger play? *Journal of Operations Management, 61*, 1–14.

Ramasesh, R. V., & Browning, T. R. (2014). A conceptual framework for tackling knowable unknown unknowns in project management. *Journal of Operations Management, 32*(4), 190–204.

Rao, S., & Goldsby, T. J. (2009). Supply chain risks: A review and typology. *International Journal of Logistics Management, 20*(1), 97–123.

Ravindran, A. R., Ufuk Bilsel, R., Wadhwa, V., & Yang, T. (2010). Risk-adjusted multicriteria supplier selection models with applications. *International Journal of Production Research, 48*(2), 405–424.

Reimann, F., Kosmol, T., & Kaufmann, L. (2017). Responses to supplier-induced disruptions: A fuzzy-set analysis. *Journal of Supply Chain Management, 53*(4), 37–66.

Resilinc. (2021). Supply chain disruptions- Resioinc's mid-year report. https://www.resilinc.com/in-the-news/supply-chain-disruptions-resilincs-mid-year-report/.

Ritchie, B., & Brindley, C. (2007). An emergent framework for supply chain risk management and performance measurement. *Journal of the Operational Research Society, 58*(11), 1398–1411.

Sacks, C. A., Kesselheim, A. S., & Fralick, M. (2018). The shortage of normal saline in the wake of Hurricane Maria. *JAMA Internal Medicine, 178*(7), 885–886.

Sharma, A., Adhikary, A., & Borah, S. B. (2020). Covid-19's impact on supply chain decisions: Strategic insights from NASDAQ 100 firms using Twitter data. *Journal of Business Research, 117*, 443–449.

Sheffi, Y. (2005). *The resilient enterprise: overcoming vulnerability for competitive advantage*. MIT Press Books.

Sheffi, Y. (2020). *The new (ab)normal: Reshaping business and supply chain strategy beyond Covid-19*. MIT CTL Media.

Sheffi, Y., & Rice, J. B. (2005). A supply chain view of the resilient enterprise. *MIT Sloan Management Review, 47*(1), 41–48.

Simchi-Levi, D., Schmidt, W., & Wei, Y. (2014). From superstorms to factory fires: Managing unpredictable supply-chain disruptions. *Harvard Business Review, 92*(1/2), 96–101.

Simchi-Levi, D., Schmidt, W., Wei, Y., Zhang, P. Y., Combs, K., Ge, Y., Gusikhin, O., Sanders, M., & Zhang, D. (2015). Identifying risks and mitigating disruptions in the automotive supply chain. *Interfaces, 45*(5), 375–390.

Smith, J. (2020, April 15). Divided supply chains are challenging producers, retailers. *The Wall Street Journal.* https://www.wsj.com/articles/divided-supply-chains-are-challenging-producers-retailers-11586974088.

Smitham, E., & Glassman, A. (2021). The next pandemic could come soon and be deadlier. *Center for Global Development.* https://www.cgdev.org/blog/the-next-pandemic-could-come-soon-and-be-deadlier.

Sodhi, M. S., & Tang, C. S. (2012). *Managing supply chain risk* (Vol. 172). Springer Science & Business Media.

Suazo, M. (2011). The impact of affect and social exchange on outcomes of psychological contract breach. *Journal of Managerial Issues, 23*(2), 190–205.

Supply Chain Risk Leadership Council (SCRLC). (2011). *Supply chain risk management best practices.* http://www.scrlc.com/.

Taleb, N. N. (2007). *The black swan: The impact of the highly improbable* (Vol. 2). Random House.

Tang, C., & Tomlin, B. (2008). The power of flexibility for mitigating supply chain risks. *International journal of production economics, 116*(1), 12–27.

Tang, C. S. (2006). Robust strategies for mitigating supply chain disruptions. *International Journal of Logistics: Research and Applications, 9*(1), 33–45.

The White House. (2021a, February 24). Executive order on America's supply chains. https://www.whitehouse.gov/briefing-room/presidential-actions/2021/02/24/executive-order-on-americas-supply-chains/.

The White House. (2021b, October 13). Fact Sheet: Biden Administration efforts to address bottlenecks at ports of Los Angeles and Long Beach, moving goods from ship to shelf. https://www.whitehouse.gov/briefing-room/statements-releases/2021/10/13/fact-sheet-biden-administration-efforts-to-address-bottlenecks-at-ports-of-los-angeles-and-long-beach-moving-goods-from-ship-to-shelf/.

Timmermans, K., George M., & Lagunas, J. (2020). Building supply chain resilience: What to do now and next during COVID-19. *Accenture.* https://www.accenture.com/us-en/about/company/coronavirus-supply-chain-impact.

Tummala, R., & Schoenherr, T. (2011). Assessing and managing risks using the supply chain risk management process. *Supply Chain Management: An International Journal, 16*(6), 474–483.

US Department of Defense Standard Practice. (2012). *System safety*. https://www.dau.edu/cop/armyesoh/DAU%20Sponsored%20Documents/MIL-STD-882E.pdf.

US Department of Health and Human Services. (2020). *Strategic National Stockpile*. https://www.phe.gov/about/sns/Pages/default.aspx.

Van Weele, A. J. (2010). *Purchasing & supply chain management: Analysis, strategy, planning and practice*. Cengage Learning.

Wagner, S. M., & Bode, C. (2008). An empirical examination of supply chain performance along several dimensions of risk. *Journal of Business Logistics, 29*(1), 307–325.

Wiedmer, R., Rogers, Z. S., Polyviou, M., Mena, C., & Chae, S. (2021). The dark and bright sides of complexity: A dual perspective on supply network Resilience. *Journal of Business Logistics, 42*(3), 336–359.

Wieland, A., & Wallenburg, C. M. (2013). The influence of relational competencies on supply chain resilience: A relational view. *International Journal of Physical Distribution & Logistics Management, 43*(4), 300–320.

World Economic Forum. (2013). *Building resilience in supply chains*. http://www3.weforum.org/docs/WEF_RRN_MO_BuildingResilienceSupplyChains_Report_2013.pdf.

World Economic Forum. (2021). *Why disciplines must work together to prepare for future pandemics*. https://www.weforum.org/agenda/2021/07/disciplines-together-future-pandemics/.

World Health Organization. (2017). Asia Pacific Strategy for Emerging Diseases and Public Health Emergencies (APSED III): Advancing implementation of the International Health Regulations (2005). Manila, Philippines. World Health Organization Regional Office for the Western Pacific; 2017. License: CC BY-NC-SA 3.0 IG. https://iris.wpro.who.int/bitstream/handle/10665.1/13654/9789290618171-eng.pdf.

Wucker, M. (2016). *The gray rhino: How to recognize and act on the obvious dangers we ignore* (1st ed.). St. Martin's Press.

Yaffe-Bellany, D., & Corkery, M. (2020, April 11). Dumped milk, smashed eggs, plowed vegetables: Food waste of the pandemic. *The New York Times*. https://www.nytimes.com/2020/04/11/business/coronavirus-destroying-food.html.

Zinn, W., & Bowersox, D. J. (1988). Planning physical distribution with the principle of postponement. *Journal of Business Logistics, 9*(2), 117–136.

Zsidisin, G. A. (2007). Business and supply chain continuity. *CAPS Research Critical Issues Report*. https://www.capsresearch.org/library.

Zsidisin, G. A., Ellram, L. M., Carter, J. R., & Cavinato, J. L. (2004). An analysis of supply risk assessment techniques. *International Journal of Physical Distribution & Logistics Management, 34*(5), 397–413.

Zsidisin, G. A., & Henke, M. (2019). *Revisiting supply chain risk*. Springer Series in Supply Chain Management.

Zsidisin, G. A., Melnyk, S. A., & Ragatz, G. L. (2005). An institutional theory perspective of business continuity planning for purchasing and supply management. *International Journal of Production Research, 43*(16), 3401–3420.

Part II
Supply Chain Risk Strategies and Developments

Developing a Supply Chain Stress Test

Lan Luo and Charles L. Munson

Abstract Borrowing a concept from the financial sector, we propose an approach to developing a "stress test" to determine the ability of a given supply chain to deal with crises under extreme, but plausible, scenarios. Managers can use such a tool to assess the risk impacts of their respective supply chains. Using predictive global sensitivity analysis, we develop a single predictive structural equation that managers can use to estimate the percentage loss given a disruption scenario. Based on the estimation, managers can further categorize their current supply chains into five risk levels. The single structural equation allows managers to re-evaluate the chain promptly as conditions change. Managerial insights are observed from the proposed equation that can help mitigate the risk of a supply chain.

1 Motivation

Supply chain globalization can significantly benefit companies. Meanwhile, it exposes the supply chain to more risks, some of which could cause supply chain disruptions. The COVID-19 pandemic affected all corners of the globe and impacted supply chains in many ways. Manufactures in China, where the novel coronavirus first spread, shut down their plants for months. This cut supplies to many industries globally. When trying to reopen their businesses, new challenges arose as rolling shutdowns around the world caused those manufacturers to face higher default risks and much higher costs.

In 2019, deadly Typhoon Mangkhut landed in south China and caused the evacuation of millions of people along with the closure of all southern Chinese

L. Luo
Barney School of Business, University of Hartford, West Hartford, CT, USA
e-mail: lluo@hartford.edu

C. L. Munson (✉)
Carson College of Business, Washington State University, Pullman, WA, USA
e-mail: munson@wsu.edu

airports. Most factories in Guangdong, which are the major suppliers or manufacturers for many overseas companies, were closed. In 2010, 297,000 people died in natural disasters such as earthquakes, floods, and typhoons. The global direct economic losses exceeded $123.9 billion (Guha-Sapir et al., 2011). During that year, Indonesia suffered from a series of volcanic eruptions, and Southeast Asia was ravaged by Typhoon Megi. In March 2011, an 8.9-magnitude earthquake coupled with an ensuing tsunami completely cut off the supplies of many critical components and materials from Japan, greatly affecting a wide spectrum of manufacturers, ranging from Korean shipbuilder Hyundai Heavy Industries to US solar panel maker SunPower Corp (Kim & Jim, 2011). In addition to pandemics and natural disasters, other uncontrollable risks, for example, strikes, wars, and terrorism, can trigger supply chain breakdowns. In 2007, railway workers in Canada went on a week-long strike, disrupting the supply of many important materials and affecting several markets in North America (Conkey et al., 2007). In 2011, an armed uprising in Libya caused an unexpected major oil supply disruption. These risks are all extreme but plausible.

Some researchers focus on how to manage supply chains under such risks, for instance, sourcing from both reliable and unreliable suppliers (Hu & Kostamis, 2015). Meanwhile, others are interested in how to assess the capability of bearing the "worst scenario" of the supply chain (Jain & Leong, 2005; Schmitt & Singh, 2009). In finance and banking, a concept known as a *stress test* is popular. Basically, a stress test examines the "breaking point" of a financial institution. With the same spirit, we apply a stress test to a supply chain to see under what extreme but plausible conditions the current supply chain would break down. Stress tests represent one of the emerging needs of supply chain management in this era of pandemics and disasters (Simchi-Levi & Simchi-Levi, 2020). Existing literature favors a simulation approach, such as Jain and Leong (2005) and Schmitt and Singh (2009). Investigators first estimate the probability of extreme but plausible risks. Then, based on those estimations, a simulation approach is utilized to determine the impact of the risks. Alternatively, instead of estimating the probability of those high-impact, low-probability disruption risks, Simchi-Levi et al. (2014) invoke a linear programming technique to estimate a firm's vulnerability given that a disruption does occur. Both simulation and linear programming can be very time-consuming when real-time information is required for decision making. In response, our main research goal in this chapter is to develop a managerially friendly mechanism that can stress test a supply chain to see how the supply chain would behave under extreme conditions. Using predictive global sensitivity analysis (PGSA), we are able to generate a single equation that could be implemented in Excel to allow managers to estimate the risk of the supply chain based on the values of just a few key independent variables. Managers could further use the equation to quickly perform "what-if" analyses to gauge the potential risk reduction or enhancement impacts of various decisions or changes in circumstances.

The rest of this chapter is organized as follows. Section 2 reviews the related literature. In Sect. 3, we describe an adapted linear programming model as our baseline supply chain risk model. In Sect. 4, we develop our structural equation model. We provide model validation and analysis in Sect. 5. Finally, we provide some managerial insights for mitigating supply chain risk and conclude the chapter in Sect. 6.

2 Literature Review

2.1 Finance and Banking Stress Testing

Because the term *stress test* is inspired by stress testing in banks, we examine the finance and banking literature to see how the stress test framework developed there. Berkowitz (1999) is one of the earliest papers that recommends folding stress tests into risk models, thereby requiring all scenarios to be assigned probabilities. Alexander and Sheedy (2008) backtest eight risk models, including both conditional and unconditional models, and four possible return distributions in order to identify the most suitable risk models for which to conduct a stress test. They further develop a methodology for conducting stress tests in the context of a risk model. Other than using an existing risk model, Burrows et al. (2012) develop a Risk Assessment Model of Systemic Institutions to conduct a top-down stress test for the Bank of England, which provides another angle for developing a stress testing framework for supply chains.

2.2 Supply Chain Risk Management

A stress test usually incorporates risk evaluation and risk management as well as the evaluation of supply chain performance. Chopra and Sodhi (2004) provide insights on managing risk by understanding the variety and interconnectedness of supply chain risk. Gunasekaran et al. (2001) attempt to develop a framework for measuring the strategic, tactical, and operational-level performance in a supply chain, and they provide a list of key performance metrics. Hu and Kostamis (2015) create a model to study the management of supply chain disruptions when sourcing from both reliable and unreliable suppliers. Namdar et al. (2018) also look into the usage of sourcing strategies to achieve supply chain resilience.

Other papers use a simulation approach to study supply chains. For example, Ouzrout et al. (2009) analyze the Value Chain Operations Reference model. Another relevant paper by Jain and Leong (2005) applies a stress test to a supply chain. That paper focuses on a small company's point of view and the stress derived from a defense contractor. The supply chain is tested at surge and mobilization volume levels to meet requirements. Schmitt and Singh (2009) also use a simulation approach to provide a way to quantify supply chain disruption risk. Oliveira et al. (2019) review literature that uses simulation and optimization (S&O) methods in supply chain risk management and reveals the gap between risk management phases and S&O methods. Based on a 3-year research engagement with Ford Motor Company, Simchi-Levi et al. (2015) develop a novel risk-exposure model that assesses the impact of a disruption originating anywhere in a firm's supply chain. The advantage of their method is that it avoids the limitations of the legacy risk-analysis approach, i.e. information availability and the difficulty in estimating the probability of an extreme event, while taking risks at all phases into consideration.

2.3 Predictive Global Sensitivity Analysis

Wagner (1995) introduced the concept of global sensitivity analysis. As applied to a deterministic mathematical model, it requires the identification of dependent variables of interest along with a list of potential summary independent variables that the modeler believes may be the primary drivers of the math model. Next, after solving the model with a wide range of parameter values, the modeler uses stepwise regression to obtain equations that can predict model outcomes based on the values of the select few key independent variables. PGSA works well for real-time applications, especially when what-if analysis and quick decision making are desired. The technique has been applied in several areas. Kouvelis et al. (2013) use PGSA to design and monitor strategic international facility networks based on eight industry groups. Tian et al. (2015) develop equations to estimate the capacity cost and material waste of a continuous process line in real time. Lee and Munson (2015) apply PGSA to monitor and modify operational hedging strategies based on four key summary independent variables: exchange rate change, average exchange rate change, regional demand change, and worldwide demand change.

In response to the need for real-world/real-time applications proposed in Oliveira et al. (2019), we adopt PGSA in this chapter to develop a structural equation that provides a tool at the manager's fingertips that he or she can quickly apply to evaluate supply chain risk as conditions change.

3 Model Development

3.1 Model Description

3.1.1 Baseline Supply Chain Model

We construct our baseline model by adapting the Time to Recover (TTR) model found in Simchi-Levi et al. (2015). As with the TTR model, we estimate a firm's vulnerability given that a disruption has occurred. However, instead of integrating all parties in the supply chain after disruption, we only focus on the manufacturer to see what its performance would be if it behaved optimally after disruption.

Developing a Supply Chain Stress Test

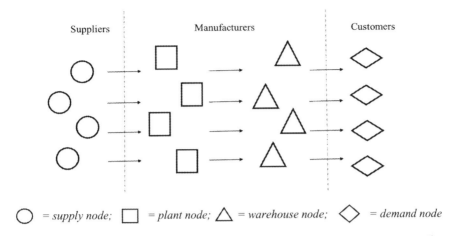

Fig. 1 Configuration of supply chains

Our baseline model considers disruption in multi-level supply chains with multiple products. Moreover, the model accommodates simultaneous disruptions in multiple supply chain nodes. The mechanism of the linear optimization model is that, given the structure of a supply chain (e.g., products, suppliers, plants and distribution centers, etc.) and a disruption scenario (interrupted suppliers and/or plants and/or distribution centers), the model determines the optimal reallocations of the firm's current resources to satisfy exogenous demand. More importantly, the model also determines the performance impact (*lost profit* in our model) assuming that the firm responds optimally, which is our dependent variable in the predictive equation model.

Figure 1 shows the configuration of a typical supply chain with four layers: supply nodes, plant nodes, warehouse nodes, and demand nodes. Manufacturers are responsible for plant nodes and warehouse nodes. They need to make capacity, inventory, and production decisions based on downstream demand. All upstream nodes, e.g., suppliers of the suppliers, can be embedded in the supply layer, and all downstream nodes, e.g., customers of customers, can be embedded in the customer layer. Disruption can occur at any node along the supply chain. Following the TTR model, once a disruption has been identified, manufacturers would respond optimally, i.e., forecast demand and reallocate materials across all surviving plants to generate the optimal production plan.

3.1.2 Baseline Model Assumptions

1. Firms respond optimally to disruptions. In other words, for the given conditions following a disruption, firms will allocate resources, plan production, etc., in order to maximize their profit.
2. Capacity cannot be shared across products at each plant.

3. The TTR of all disrupted nodes is identical in a given scenario.
4. Once a node is interrupted, its productivity becomes zero, and the node loses all of its inventory. (This mimics the worst-case scenario that "knocks out" a node until the TTR has passed. In cases where not all inventory is lost, the leftover inventory could be transferred to other nodes and incorporated into the *Weighted Average Inventory* independent variable in our structural equation model.)
5. Demand for each period is stochastic and follows a normal distribution.
6. Parts will be delivered to the manufacturer at the beginning of each period, and the amounts are the same for each period under contracts based on forecasts.
7. No backorders are allowed.
8. Reallocation cost has been estimated and reflected in marginal profit.

3.2 Notation

Indices

i: Suppliers
j: Products
m: Plants
k: Parts
w: Warehouses
n: Scenarios
t: Time periods

Sets

I: Set of all suppliers
I_k: Set of all suppliers that supply part k
I_n: Set of all disrupted suppliers under scenario n
J: Set of all final products
M: Set of all plants
M_n: Set of all disrupted plants under scenario n
K: Set of all parts needed for the company
W: Set of all warehouses of finished goods
W_n: Set of all disrupted warehouses under scenario n

Parameters

f_j: Marginal profit of product j, $f_j > 0$
r_{jk}: Number of part k needed for product j, $r_{jk} > 0$
y_{ik}: Number of part k supplied by supplier i, $y_{ik} \geq 0$
μ_{mkt}: Current inventory at plant m of part k in period t, $\mu_{mkt} \geq 0$
δ_{jwt}: Current inventory at warehouse w of finished product j in period t, $\delta_{jwt} \geq 0$
a_j: Mean demand of product j per period, $a_j > 0$
b_j: Standard deviation of demand of product j per period, $b_j > 0$

d_{jt}: Realized demand of product j in period t, $d_{jt} > 0$
c_{jm}: Production capacity of product j at plant m per period, $c_{jm} \geq 0$
t_n: Time for the whole supply chain to recover to full functionally after disruption, $t_n > 0$
D: Time period of disruption

Decision variables

u_{jm}: Production level of product j produced at plant m
l_{jt}: Lost demand of product j in period t

3.3 Formulation of the Baseline Model

$$\text{Minimize} \sum_{t=1}^{t_n} \sum_{j \in J} f_j l_{jt}$$

Subject to

$$c_{jm} = 0, \quad \forall m \in M_n \tag{1}$$

$$\sum_{k \in K} y_{ik} = 0, \quad \forall i \in I_n \tag{2}$$

$$\sum_{k \in K} \mu_{mkt} = 0, \quad \forall m \in M_n \tag{3}$$

$$\sum_{j \in J} \delta_{jwt} = 0, \quad \forall w \in W_n \tag{4}$$

$$\sum_{m \in M} \mu_{mkt} + \sum_{i \in I_k} y_{ik} \geq \sum_{j \in J} \sum_{m \in M} \mu_{jmt} \times r_{jk}, \quad \forall k \in K, t \in \{1, 2, \ldots, t_n\} \tag{5}$$

$$u_{jmt} \leq c_{jm}, \quad \forall m \in M, j \in J, t \in \{1, 2, \ldots, t_n\} \tag{6}$$

$$\sum_{w \in W} \delta_{jwt} + \sum_{m \in M} \mu_{jmt} + l_{jt} = d_{jt} + \sum_{w \in W} \delta_{jw(t+1)}, \quad \forall j \in J, t \in \{1, 2, \ldots, t_n\} \tag{7}$$

$$l_{jt} \leq d_{jt}, \quad \forall j \in J, t \in \{1, 2, \ldots, t_n\} \tag{8}$$

$$\delta_{jwt} \geq 0, \quad \forall w \in W, j \in J, t \in \{1, 2, \ldots, t_n\} \tag{9}$$

$$u_{jmt}, l_{jt} \geq 0, \quad \forall m \in M, j \in J, t \in \{1, 2, \ldots, t_n\} \tag{10}$$

Constraints (1)–(4) ensure that the capacity, inventory, and/or supply at disruption nodes equal zero. Constraint (5) is the material consumption constraint, i.e., the usage of materials for each period will not exceed the inventory at the beginning of the period plus the shipment from suppliers for that period. Constraint (6) ensures that the production level should be less than or equal to the capacity at each plant. Constraint (7) is a balance equation that determines lost demand in each period, if any. As formulated in constraint (7), the demand in period t together with the leftover inventory, i.e., the beginning inventory of period $t + 1$, equals the beginning inventory in period t plus the production level plus the lost demand in period t. Because the objective function strives for low demand, either the lost demand or the beginning inventory in period $t + 1$ has to be zero. Constraint (8) states that the lost demand should not exceed the actual demand in period t. With all beginning inventory non-negative, constraint (9) ensures that there are no backorders. Both the production level and the lost demand should be non-negative numbers as shown in the last constraint.

4 Structural Equation Model

We develop a set of summary independent variables that we believe can accurately predict the overall lost profit in the baseline model. Then we run our baseline model many times using multiple sets of data representing various supply chain structures and disruption scenarios. After each run, the corresponding performance impact resulting from the baseline model is recorded, along with the values of the independent variables from that run. This provides us with a rich set of data points with which we can run regressions. Via stepwise regression, we obtain a single equation that allows us to estimate supply chain stress for any supply chain that fits the configuration of Fig. 1, without the need to populate and run the baseline model each time. Furthermore, the equation can tell us how and in what respect the independent variables affect the performance of the supply chain in extreme scenarios.

4.1 Dependent Variable

We measure the stress of the supply chain by looking at the *percentage loss* (PL) of the given scenario, i.e., we divide the lost profit obtained from the baseline LP model by the profit that the company would obtain if no disruption occurs. If the ratios fall within the range 0–20%, we categorize the stress as *low risk*, between 20 and 40% as *mid-low risk*, between 40 and 60% as *medium risk*, and between 60 and 80% as *mid-high risk*. Any ratio above 80% will be a disaster and marked as *high risk*. Our dependent variable of interest for the predictive equation is the PL. Based on the PL, managers can easily categorize the supply chain into the corresponding risk level.

4.2 Independent Variables

In order to implement the predictive structural equation, we initially constructed some potential summary independent variables that we surmised might have the strongest impact on the dependent variable. We looked at each potential variable X separately by regressing the dependent variable PL on the terms X^Δ, $\Delta \in \{-3, -2, -1, 1, 2, 3\}$. The potential variables producing high enough R^2 values were kept in our final set. After testing these variables using these "one-at-a-time regressions," we identified the following as independent variables to include in the subsequent stage:

1. Total capacity of the most profitable product (TCMPP) after disruption:

$$\text{TCMPP} = \sum_{m \in M} c_{j^*m}, \quad \text{where } j^* = \arg\max_j f_j$$

2. Number of missing part categories (NMP) after disruption:

$$\text{NMP} = \left\{ \sum_{k \in K} \text{Count}_k \,\middle|\, \text{Count}_k = 1 \text{ if } \sum_{i \in I_k} y_{ik} = 0 \text{ and } 0 \text{ otherwise}, \forall k \in K \right\}$$

3. Weighted average inventory (WAI) after disruption:

$$\text{WAI} = \sum_{j \in J} \frac{a_j \sum_{w \in W} \delta_{jwD}}{\sum_{j \in J} a_j}$$

4. Minimum number of suppliers per part (MNSPP) before disruption:

$$\text{MNSPP} = \min_{k \in K} |I_k|$$

5. Weighted average capacity/demand slack (WCDS) after disruption:

$$\text{WCDS} = \sum_{j \in J} \frac{f_j}{\sum_{j \in J} f_j} \times \frac{\sum_{m \in M} c_{jm}}{a_j}$$

6. Weighted average standard deviation of demand (WSTD):

$$\text{WSTD} = \sum_{j \in J} \frac{a_j b_j}{\sum_{j \in J} a_j}$$

7. Number of suppliers: $|I|$
8. Number of plants: $|M|$
9. Number of parts: $|K|$
10. Number of warehouses: $|W|$
11. Number of products: $|J|$
12. TTR: t_n

By taking different combinations of the above summary independent variables, we have introduced interaction effects and nonlinearities to the regression model. The combination terms were constructed as follows: (1) for each independent variable A, we have terms as A^{Δ_1}, where $\Delta_1 \in \{-3, -2, -1, 1, 2, 3\}$; (2) for any two independent variables A and B we have $A^{\Delta_2} B^{\Delta_3}$, where $\Delta_2, \Delta_3 \in \{-2, -1, 1, 2\}$; and (3) for any three independent variables A, B, and C we use $A^{\Delta_4} B^{\Delta_5} C^{\Delta_6}$, where $\Delta_4, \Delta_5, \Delta_6 \in \{-1, 1\}$. We incorporated all of these terms into a single model and conducted a stepwise regression in SAS keeping the default alpha values of 0.05 to enter and 0.05 to remove, where alpha is the significance level for a potential independent variable to be incorporated and stay in the model during selection process (SAS Help Center, 2021).

4.3 Creating the Dataset for Regression

To create the dataset to run regressions, we varied parameters as shown in Table 1. We primarily used uniform distributions to generate the data. Importantly, that distribution choice is in no way intended to model the actual probabilities of a supply chain possessing those specific characteristics. Rather, we seek to create a very wide range of possible inputs (and associated outputs) to generate a robust regression model.

We randomly allocated the capacity of each product and the total inventory of each part to different plants. Similarly, the inventory of finished goods was randomly allocated to each warehouse. The supply of each part from different suppliers was also determined randomly. We ran 6000 base cases (supply chains) with 10 variations (disruption scenarios) for each case. We ran our stepwise regression in SAS using these 60,000 observations.

Table 1 Parameters used to create the dataset

Parameter	Value
Number of plants $|M|$	DU^a (1,10)
Number of products $|J|$	DU(1,100)
Number of suppliers for part k $|I_k|$	1 (15%), DU(2,5) (38%), DU(6,10) (47%)[b]
Number of parts $|K|$	DU(1,200)
Number of the same parts used	DU(0,5)
Number of warehouses $|W|$	DU(1,10)
Marginal Profit f_j	DU(1,6000) × 5
Mean demand a_j	DU(1,100) ×50 × (1+Uniform(−20%,20%))
Standard deviation of demand b_j	Uniform(0,$\frac{1}{3}a_j$)
Total capacity	$a_j + 4 \times b_j$
Finished goods inventory	Uniform(0,2)×a_j
Parts inventory	Uniform(0, 1) × $\sum_{j \in U} r_{jk}a_j$
Parts supply	$\sum_{j \in U} r_{jk}(a_j + 3 \times b_j)$
TTR (in semi-months)	DU(1,26)
Disruption probability at each node	50% chance of complete disruption

[a]DU, discrete uniform distribution
[b]Based on the survey in Munson and Jackson (2014)

4.4 Structural Equation

The final model selected by stepwise regression contains more than 100 terms with an adjusted R^2 of 97.9%. However, we observed that we could include just the 15 most significant terms and still have an adjusted R^2 of 97.53% (see Table 2). We then ran a regular regression of these 15 terms on the dependent variable PL to obtain the estimates shown in Table 3.

From Table 3, the single structural equation for estimating PL can be written as:

$$PL = 1.21791 - 1.59164 \text{WCDS} + 0.67847 \text{WCDS}^2 - 0.00014015 \frac{\text{WAI}}{t_n}$$

$$+ 0.33965 \frac{\text{WCDS}}{t_n} - 0.01669 \frac{|M|}{t_n} - 0.00000000643081 \frac{\text{WAI} \times \text{WSTD}}{\text{WCDS}}$$

$$- 0.01037 \frac{|I|}{t_n} - 0.0086 \frac{|W|}{t_n} - 0.08754 \text{WCDS}^3 + 0.00104 t_n$$

$$+ 0.000000004474709 \frac{\text{TCMPP} \times \text{NMP}}{\text{WCDS}} - 0.04008 \frac{\text{MNSPP}}{|J|} - 0.02557 |M|$$

$$+ 0.00148 |M|^2 - 0.0000216 \frac{\text{WAI}}{|M|}$$

The coefficient estimations are based on minimizing the sum of squared errors using a regression model. As the structural equation is ultimately the estimation of a line designed to produce values between 0 and 100%, some of the predicted values

Table 2 Stepwise selection summary

Step	Effect entered	Effect removed	Number of effects in	Adjusted R^2	F Value	$P(>F)$
0	Intercept		1	0	0	1
1	WCDS		2	0.7966	234,947.00	<.0001
2	WCDS^2		3	0.9305	115,536.00	<.0001
3	$\frac{WAI}{t_n}$		4	0.9493	22,275.20	<.0001
4	$\frac{WCDS}{t_n}$		5	0.9603	16,689.20	<.0001
5	$\frac{|M|}{t_n}$		6	0.9670	12,154.50	<.0001
6	$|M| \times$ WCDS		7	0.9695	4979.57	<.0001
7	$\frac{WAI \times WSTD}{WCDS}$		8	0.9709	2702.87	<.0001
8	$\frac{|J|}{t_n}$		9	0.9719	2205.66	<.0001
9	$\frac{|W|}{t_n}$		10	0.9724	1106.53	<.0001
10	WCDS^3		11	0.9729	1019.23	<.0001
11	t_n		12	0.9732	845.33	<.0001
12	$\frac{TCMPP \times NMP}{WCDS}$		13	0.9736	836.82	<.0001
13	$\frac{MNSPP}{|J|}$		14	0.9739	700.03	<.0001
14	$|M|$		15	0.9742	771.38	<.0001
15	$|M|^2$		16	0.9748	1282.90	<.0001
16		$|M| \times$ WCDS	15	0.9748	0.64	0.4236
17	$\frac{WAI}{|M|}$		16	0.9753	1209.49	<.0001

Table 3 Parameter estimates

Variable	Parameter estimate	Standard error	t Value	$P(>|t|)$
Intercept	1.21791	0.00271	448.71	<.0001
WCDS	−1.59164	0.00767	−207.64	<.0001
WCDS^2	0.67847	0.00796	85.21	<.0001
$\frac{WAI}{t_n}$	−0.00014015	1.00E−06	−140.17	<.0001
$\frac{WCDS}{t_n}$	0.33965	0.00178	191.17	<.0001
$\frac{|M|}{t_n}$	−0.01669	0.0002786	−59.89	<.0001
$\frac{WAI \times WSTD}{WCDS}$	−6.43E−09	2.57E−10	−25.03	<.0001
$\frac{|J|}{t_n}$	−0.01037	0.000242	−42.86	<.0001
$\frac{|W|}{t_n}$	−0.0086	0.0002321	−37.07	<.0001
WCDS^3	−0.08754	0.00256	−34.15	<.0001
t_n	0.00104	0.0000318	32.61	<.0001
$\frac{TCMPP \times NMP}{WCDS}$	4.47E−09	2.45E−10	18.25	<.0001
$\frac{MNSPP}{|J|}$	−0.04008	0.00138	−29.15	<.0001
$|M|$	−0.02557	0.0003932	−65.04	<.0001
$|M|^2$	0.00148	3.106E−05	47.6	<.0001
$\frac{WAI}{|M|}$	−0.0000216	6.21E−07	−34.78	<.0001

of PL could possibly exceed 100% or be less than zero. In those cases, we assume that the estimate represents the extreme cases, so we simply truncate the estimated value from the structural equation between 0 and 1 using:

$$\text{Estimated percentage loss} = \max\{\min\{PL, 1\}, 0\}$$

5 Analysis

5.1 Model Validation

In addition to the extremely high adjusted R^2 value for our structural equation, we sought further validation of the model by examining actual errors of individual observations. Specifically, we generated another 2000 random cases using the same experimental design (Table 1) and computed the absolute deviation (AD) of both the main sample (60,000 observations) and the hold-out sample (2000 observations). Table 4 shows the percentage of observations having an absolute deviation within the given range. For example, 86.64% of observations fell within ±5% of the actual lost profit, and 96.72% of predictions had less than a 10% error. The numbers are very similar for the hold-out sample.

Tables 5, 6, and 7 examine the ability of the model to predict general risk levels as defined by our five risk categories. After all, the main purpose of a stress test is to gauge robustness to disruptions, rather than to predict a *precise* profit loss. Consequently, we further calculate the mean absolute deviation (MAD) and percentage of miscategorized observations for each risk level in Table 5. The overall MAD with our predictive equation is 2.66%. It is noteworthy that the MAD within the *high*-risk level is only 1.97%, which is less than the overall MAD. Furthermore, the percentage of miscategorized observations is only 1.49% in the *high-risk* level range, indicating that the structural model has especially good predictability for devastating disasters.

Table 6 shows that 90.87% of observations correctly classified the firm into the proper risk category. And of those that missed, nearly all just missed by one level (e.g., predicted *mid-low* instead of *low*). Only 0.16% of the observations were placed

Table 4 Cumulative distribution of observations based on absolute deviation (AD)

	<0.05	<0.1	<0.15	<0.2	<0.25	<0.3	<0.5
Main sample							
Cumulative observations	86.64%	96.72%	98.69%	99.39%	99.71%	99.88%	99.99%
Hold-out sample							
Cumulative observations	86.35%	96.80%	98.85%	99.50%	99.75%	99.75%	100.00%

Table 5 Mean absolute deviation (MAD) for different risk categories

	Overall	Low	Mid-low	Medium	Mid-high	High
Main sample						
MAD	2.66%	2.35%	2.97%	3.47%	3.53%	1.97%
# observations	60,000	29,535	14,191	7528	2985	5761
% miscategorized	9.13%	6.24%	10.65%	17.91%	23.18%	1.49%
Hold-out sample						
MAD	2.64%	2.42%	2.75%	3.43%	3.55%	2.11%
# observations	2000	980	506	224	90	200
% miscategorized	9.50%	6.22%	12.45%	20.54%	20.00%	1.00%

Table 6 Number of miscategorized observations

	Correctly classified	Off by 1 category	Off by ≥ 2 categories
Main sample			
Overall number of observations	54,520	5387	93
(%)	90.87%	8.98%	0.16%
Under-estimated	0	2472	55
(%)	0.00%	4.12%	0.09%
Over-estimated	0	2915	38
(%)	0.00%	4.86%	0.06%
Hold-out sample			
Overall number of observations	1809	186	5
(%)	90.45%	9.30%	0.25%
Under-estimated	0	88	4
(%)	0.00%	4.40%	0.20%
Over-estimated	0	98	1
(%)	0.00%	4.90%	0.05%

Table 7 Percentage of miscategorized observations with high absolute deviation (AD)

	Total	>0.05 AD	>0.1 AD
Main sample			
Overall	9.13%	4.56%	2.10%
Under-estimated	4.21%	2.23%	1.04%
Over-estimated	4.92%	2.33%	1.06%
Hold-out sample			
Overall	9.50%	4.80%	1.85%
Under-estimated	4.55%	2.25%	0.75%
Over-estimated	4.95%	2.55%	1.10%

into a risk level that deviated two or more levels from where they should have been classified. The numbers are almost the same in the hold-out sample. Among those 9.13% miscategorized observations, about 50% (4.21% in the whole sample) were underestimated (Table 7). (Underestimating risk is more dangerous to a firm than overestimating risk, which would lead to conservative decision making.)

Table 7 also shows that among the miscategorized observations, about half had an AD of less than 5%, suggesting that the actual profit loss was near the cut-off point of two categories anyway. We also examined the miscategorized observations and found that the largest deviations were mainly caused by zero capacity and demand slacks. When the capacity is zero, managers can easily capture the actual PL by using only inventory to fulfill the upcoming demand. In such cases, there would be no need to use linear programming to find the optimal resource allocation and production plan.

5.2 A Numerical Example

In this section, we provide a numerical example illustrating how to apply the structural equation to estimate risk. Suppose that a company currently has two plants and produces three different products. The parts are supplied by five different suppliers. All finished goods are stored at three warehouses. The manager of the company has identified a potential risk scenario, the TTR from which is six periods (three months), and she wants to assess the performance of the current supply chain.

Based on the identified scenario, the manager calculates the independent variables as follows: TCMPP = 1346, NMP = 0, WAI = 3682.669, MNSPP = 1, WCDS = 0.6063, and WSTD = 212.5899. Then we have:

$$PL = 1.21791 - 1.59164 \times 0.6063 + 0.67847(0.6063)^2$$

$$- 0.00014015 \frac{3682.669}{6} + 0.33965 \frac{0.6063}{6} - 0.01669 \frac{2}{6}$$

$$- 0.00000000643081 \frac{3682.669 \times 212.5899}{0.6063} - 0.01037 \frac{5}{6} - 0.0086 \frac{3}{6}$$

$$- 0.08754(0.6063)^3 + 0.00104 \times 6 + 0.000000004474709 \frac{1346 \times 0}{0.6063}$$

$$- 0.04008 \frac{1}{3} - 0.02557 \times 2 + 0.00148(2)^2 - 0.0000216 \frac{3682.669}{2} = 31.22\%$$

Thus, Estimated Percentage Loss = max {min{0.3122,1},0} = 31.22%. According to our preset risk-level scheme, the risk level of the current supply chain under the identified disruption is *mid-low risk*. We find that the actual value of the dependent variable PL from the baseline LP model is 31.72% for the current supply chain, and the risk level is also *mid-low risk*. The absolute deviation in this example is 0.50%.

5.3 Discussion

From the structural equation, we can see that WCDS plays a heavy role in determining the PL for a given a disruption scenario. Accordingly, if the plants have been identified as vulnerable, the manager may consider implementing backup capacity at some location ("safe house") that would not be affected by the disruption, or perhaps increasing the capacity at the current invulnerable plants. In fact, from the simulations we find that the cases with zero capacity but low risks have some features in common: high finished goods inventory and short TTR. By increasing the inventory of finished goods and investing in shortening the recovery time, companies with a lower risk can benefit as well.

Developing a Supply Chain Stress Test

Inventory is also a crucial driver for the PL of a company. If warehouses easily fail under a certain scenario, then adding capacity to the current plants should be taken into consideration. Or the manager can spread the risk by building additional warehouses.

A shorter TTR would help reduce lost sales if it is difficult for the company to fulfill demand in each period. Managers can choose one or mix the above strategies in order to lower the risk level of the current supply chain under certain disruption scenarios. The choice should be based on available budgets.

5.4 Numerical Example Revisited

Let us revisit the numerical example from Sect. 5.2, which produced an estimated profit loss of 31.22% (*mid-low risk*). Assume that the company has three possible options to lower its risk: (1) have a backup plant with 50% total capacity for each product, resulting in three plants and a new WCDS value of 1.3207; (2) invest in reducing the TTR to 1 period; and (3) invest in reducing the TTR to 4 periods and increasing initial inventory to 1.5 times the original case, resulting in a WAI value of 5524.0035. By using the proposed structural equation, managers can quickly estimate the change in PL. From the equation, we have the estimated percentage loss for the above 3 options as follows:

Option 1:

$$
\begin{aligned}
PL = {} & 1.21791 - 1.59164 \times 1.3207 + 0.67847(1.3207)^2 \\
& - 0.00014015 \frac{3682.669}{6} + 0.33965 \frac{1.3207}{6} - 0.01669 \frac{3}{6} \\
& - 0.00000000643081 \frac{3682.669 \times 212.5899}{1.3207} - 0.01037 \frac{5}{6} - 0.0086 \frac{3}{6} \\
& - 0.08754(1.320693)^3 + 0.00104 \times 6 + 0.000000004474709 \frac{1346 \times 0}{1.3207} \\
& - 0.04008 \frac{1}{3} - 0.02557 \times 3 + 0.00148(3)^2 - 0.0000216 \frac{3682.669}{3} \\
= {} & -3.58\%
\end{aligned}
$$

Estimated Percentage Loss = max $\{\min\{-3.58\%, 1\}, 0\} = 0\%$, and the identified risk level is *low*.

Option 2:

$$PL = 1.21791 - 1.59164 \times 0.6063 + 0.67847(0.6063)^2$$
$$- 0.00014015 \frac{3682.669}{1} + 0.33965 \frac{0.6063}{1} - 0.01669 \frac{2}{1}$$
$$- 0.00000000643081 \frac{3682.669 \times 212.5899}{0.6063} - 0.01037 \frac{5}{1} - 0.0086 \frac{3}{1}$$
$$- 0.08754(0.6063)^3 + 0.00104 \times 1 + 0.000000004474709 \frac{1346 \times 0}{0.6063}$$
$$- 0.04008 \frac{1}{3} - 0.02557 \times 2 + 0.00148(2)^2 - 0.0000216 \frac{3682.669}{2}$$
$$= -4.41\%$$

Estimated Percentage Loss = max {min{−4.41%, 1}, 0} = 0%, and the identified risk level is *low*.

Option 3:

$$PL = 1.21791 - 1.59164 \times 0.6063 + 0.67847(0.6063)^2$$
$$- 0.00014015 \frac{5524.0035}{4} + 0.33965 \frac{0.6063}{4} - 0.01669 \frac{2}{4}$$
$$- 0.00000000643081 \frac{5524.0035 \times 212.5899}{0.6063} - 0.01037 \frac{5}{4} - 0.0086 \frac{3}{4}$$
$$- 0.08754(0.6063)^3 + 0.00104 \times 4 + 0.000000004474709 \frac{1346 \times 0}{0.6063}$$
$$- 0.04008 \frac{1}{3} - 0.02557 \times 2 + 0.00148(2)^2 - 0.0000216 \frac{5524.0035}{2}$$
$$= 18.64\%$$

Estimated Percentage Loss = max {min{18.64%, 1}, 0} = 18.64%, and the identified risk level is *low*.

When running the baseline LP model itself, the actual value of PL and corresponding risk categories for the above three options are 0% with *low risk*, 0% with *low risk* and 11.67% with *low risk* respectively. Therefore, our predictions are very close to the actual values. By using our predictive equation in this way, the manager can have a quick evaluation of options such as these by simply changing the inputs in the equation.

The results indicate that option 1, adding extra capacity, and option 2, reducing TTR to 1 period, can completely eliminate expected lost profits, while option 3, a combination of adding 50% more inventory and investing to reduce the TTR to four periods, is able to achieve a lower PL and also move the firm into a low risk level. If the budget allows, option 1 or 2 should be pursued. If not, option 3 or a similar strategy should be taken into consideration.

6 Conclusions

This chapter provides a tool that can help managers evaluate the risk of their current supply chain with only a few simple calculations. The single prediction equation has an adjusted R^2 of 97.53% and a MAD of 2.66%, indicating acceptable accuracy. The estimated percentage loss produced from the model neatly maps a firm's risk into one of five risk categories. Importantly, in addition to providing a snapshot of current risk exposure, the model allows managers to quickly test the impact on risk exposure of various corporate actions. A straightforward recalculation of independent variables produces the predicted new risk exposure immediately. In this way, managers can discuss options, costs, and trade-offs during meetings in real time.

With the equation, we have also discovered some managerial insights about mitigating risk: (1) extra accessible capacity will help reduce the PL when warehouses are more vulnerable; (2) more finished goods inventory can improve the performance of the supply chain under a certain disruption if plants fail; and (3) in either case, by investing in shortening the TTR, managers can expect a lower PL.

Though the MAD of our predictive model is small, the study is limited to our experimental design. The equation could become even more robust with a more comprehensive design, wider parameter values, and a larger simulated population. Alternatively, by describing the background on how the model was developed, we have provided a roadmap for firms to create their own equation based on input values from their specific company or industry.

For future studies, different independent variables could be formulated to test supply chain resiliency from different angles. Some assumptions could be relaxed. For example, the demand could follow a more general distribution. Or instead of the assumption that all capacity and inventory is lost under disruption, a disrupted node might face different levels of disruption.

References

Alexander, C., & Sheedy, E. (2008). Developing a stress testing framework based on market risk models. *Journal of Banking and Finance, 32*(10), 2220–2236.
Berkowitz, J. (1999). A coherent framework for stress-testing. *FEDS working paper*.
Burrows, O., Learmonth, D., McKeown, J., & Williams, R. (2012). RAMSI: A top-down stress-testing model developed at the Bank of England. *Bank of England Quarterly Bulletin, Sep. 13*, Q3.
Chopra, S., & Sodhi, M. S. (2004). Managing risk to avoid supply-chain breakdown. *MIT Sloan Management Review, 46*(1), 53–61.
Conkey, C., Machalaba, D., & Belkin, D. (2007). Bridge collapse could spur infrastructure fixes. *The Wall Street Journal, Aug. 3*, B1.
Guha-Sapir, D., Vos, F., Below, R., & Ponserre, S. (2011). *Annual disaster statistical review 2010: The numbers and trends*. Centre for Research on the Epidemiology of Disasters.
Gunasekaran, A., Patel, C., & Tirtiroglu, E. (2001). Performance measure and metrics in a supply chain environment. *International Journal of Production & Operations Management, 21*(1/2), 71–87.

Hu, B., & Kostamis, D. (2015). Managing supply disruptions when sourcing from reliable and unreliable suppliers. *Production and Operations Management, 24(5)*, 808–820.

Jain, S., & Leong, S. (2005). Stress testing a supply chain using simulation. In *Proceedings of the Winter Simulation Conference,* IEEE, 8.

Kim, M., & Jim, C. (2011). Global supply chain rattled by Japan quake, tsunami. *Reuters, Mar. 12.*

Kouvelis, P., Munson, C. L., & Yang, S. (2013). Robust structural equations for designing and monitoring strategic international facility networks. *Production and Operations Management, 22(3)*, 535–554.

Lee, C., & Munson, C. L. (2015). A predictive global sensitivity analysis approach to monitoring and modifying operational hedging positions. *International Journal of Integrated Supply Management, 9(3)*, 178–201.

Munson, C. L., & Jackson, J. (2014). Quantity discounts: An overview and practical guide for buyers and sellers. *Foundations and Trends in Technology, Information and Operations Management, 8(1–2)*, 1–132.

Namdar, J., Li, X., Sawhney, R., & Pradhan, N. (2018). Supply chain resilience for single and multiple sourcing in the presence of disruption risks. *International Journal of Production Research, 56(6)*, 2339–2360.

Oliveira, J. B., Jin, M., Lima, R. S., Kobza, J. E., & Montevechi, J. A. B. (2019). The role of simulation and optimization methods in supply chain risk management: Performance and review standpoints. *Simulation Modelling Practice and Theory, 92*, 17–44.

Ouzrout, Y., Savino, M. M., Bouras, A., & Domenico, C. D. (2009). Supply chain management analysis: A simulation approach to the value chain operations reference (VCOR) model. *International Journal of Value Chain Management, 3(3)*, 263–287.

SAS Help Center. (2021). https://documentation.sas.com/doc/en/pgmsascdc/9.4_3.5/stathpug/stathpug_introcom_stat_sect032.htm. Retrieved 25 June 2021.

Schmitt, A. J., & Singh, M. (2009). Quantifying supply chain disruption risk using Monte Carlo and discrete-event simulation. In *Proceedings of the 2009 Winter Simulation Conference (WSC),* IEEE, 1237–1248.

Simchi-Levi, D., Schmidt, W., & Wei, Y. (2014). From superstorms to factory fires—Managing unpredictable supply-chain disruptions. *Harvard Business Review, Jan.–Feb.,* 3–7.

Simchi-Levi, D., Schmidt, W., Wei, Y., Zhang, P., Combs, K., Ge, Y., Gusikhin, O., Sanders, M., & Zhang, D. (2015). Identifying risks and mitigating disruptions in the automotive supply chain. *Interfaces, 45(5)*, 375–390.

Simchi-Levi, D., & Simchi-Levi, E. (2020). We need a stress test for critical supply chains. *Harvard Business Review, Apr. 28.*

Tian, Z., Kouvelis, P., & Munson, C. L. (2015). Understanding and managing product line complexity: Applying sensitivity analysis to a large-scale MILP model to price and schedule new customer orders. *IIE Transactions, 47(4)*, 307–328.

Wagner, H. M. (1995). Global sensitivity analysis. *Operations Research, 43(6)*, 948–969.

Retail Supply Chain Risk and Disruption: A Behavioral Agency Approach

Raul Beal Partyka, Fernando Gonçalves Picasso, and Ely Laureano Paiva

Abstract Relationships between consumers and retailers are frequently the subject of studies in the context of operations management. The traditional agency theory is widespread, especially in the principal–agent relationship. This chapter describes retail consumer behavior during the COVID-19 pandemic and the unexpected consumer demands experienced by retailers from the perspective of behavioral agency theory. This chapter discusses that this shift in the retail supply chain can be explained by the behavioral agency theory, above all because of its mitigation mechanisms. Supply chain disruption is also discussed by pointing out the main activities of retailers and consumers and the respective mitigation mechanisms they adopt. This chapter contributes to the evolution of the supply chain studies' field by providing analyses from the retail supply chain debate about risk and disruption implications in the context of the COVID-19 pandemic. It also helps broaden the applicability of the behavioral agency theory in operations and supply chain management.

1 Introduction

Increased attention is being paid to supply chain risk management, as many companies have faced more supply chain management challenges (Kildow, 2011). For instance, several companies have faced harsh times by relying on single sources of critical supplies. (Linton & Vakil, 2020). The main objective of supply chain risk management is guaranteeing the continual flow of materials or services from the source to the customer when a disruption occurs in the supply chain. Risk mitigation and contingency planning capabilities are crucial in this context (Matsuo, 2015; Tomlin, 2006), because they enable companies to plan and properly respond in the

---

R. B. Partyka (✉) · F. G. Picasso · E. L. Paiva
Department of Operations Management, Fundação Getulio Vargas's Sao Paulo School of Business Administration (FGV EAESP), Sao Paulo, Brazil
e-mail: raul.partyka@fgv.edu.br

© The Author(s), under exclusive license to Springer Nature Switzerland AG 2022
Y. Khojasteh et al. (eds.), *Supply Chain Risk Mitigation*, International Series in Operations Research & Management Science 332,
https://doi.org/10.1007/978-3-031-09183-4_4

case of a supply chain disruption (Ambulkar et al., 2015; Blackhurst et al., 2005; Bode et al., 2011; Brandon-Jones et al., 2014; Craighead et al., 2007). Mitigation capabilities are the actions that are taken before a disruption occurs, while contingency capabilities are measures taken after and in response to a disruption (Matsuo, 2015).

The recent spread of the COVID-19 virus has exposed the vulnerability of supply chains as never before (Choi et al., 2020; Esper, 2020). Although information about the virus was available months before its worldwide spread, when lockdown rules resulted in businesses being shut down the effects were unexpected. Consumers, for instance, went into a panic-buying mode for essential goods, leading to stockouts of items such as toilet paper, hand sanitizer, and cleaning products. They also started hoarding all kinds of grocery and non-perishable products that they thought they might need, leading to empty supermarket shelves and disrupting the retail supply chain. A recent survey showed that the main problem members of the retail supply chain are facing is a lack of clarity with regard to consumer demand (Leonard, 2020), and because these particular supply chains were unable to anticipate consumer behaviors, they implemented contingency measures, such as limiting the number of items consumers could buy in an effort to help the retail supply chain recover. The way in which retailers responded to this supply chain disruption can be analyzed through the lens of agency theory.

Behavioral agency theory applies to "analyses of executive decision-making when the personal reputation, morality, wealth, and well-being of various company stakeholders are at stake" (Gomez-Mejia et al., 2021). So far, only two works have tried to apply the behavioral agency theory to these relationships (Cole & Aitken, 2019; Villena et al., 2009). "Agents are not rational 100% of the time" (Massa et al., 2020, p. 222). Therefore, because of their different desires and goals, consumers act as agents in co-creative results and may involve selfish behaviors to satisfy their self-interests (Leo et al., 2019). Based on the behavior of agents, Pepper and Gore (2015) developed the behavioral agency theory, which they tested empirically and in which they identified other implications. The client, as principal, delegates work to an agent that does the work (Ngamvichaikit & Beise-Zee, 2014). The agent can then choose (or may not have the option) and own their service process that has a lot of customer interaction. Even if the client as beneficiary has never formally agreed to this relationship, past studies have focused primarily on event management and the consequent impact on performance, such as a supplier that does not carry out a specific activity that follows a defined pattern and within a determined period. There are no studies that recognize the interaction between consumers from the viewpoint of behavioral agency theory for mitigating conflict (principal vs. agent).

There is no denying the applicability of the agency theory as a lens for understanding the agency's problem and the risk-sharing problem that arises when the retailer delegates responsibility for and the maintenance of retail stock to the supplier (Rungtusanatham et al., 2007). The relationship, however, is not one-sided in relation to delegating a task to an agent. In fact, the contract is bilateral. At the very least the agent expects some monetary compensation for its efforts. Behavioral agency theory predicts that the different goals and intrinsic preferences of agents will lead to different mechanisms for mitigating agency problems (Pepper & Gore, 2015). "Retailers can limit such opportunistic behavior by investing in their supplier monitoring capabilities." (Morgan et al., 2007, p. 525). Behavioral agency theory makes it possible to understand the conflicts between principal (retailer) and agent (consumer), and to capture the emotions of this relationship. Application of this construct may have implications, especially in smaller companies where there is little corporate policy and few guidelines, and where negotiations are at the mercy of centralized and, consequently, emotional decision-making.

The purpose of this chapter is to describe how the disruption in the retail supply chain during COVID-19 can be explained by the mechanisms of behavioral agency theory. The consumer–retailer relationship is used to better understand the contracts and roles of both. Retail as a service comprises, in its essence, the relationship between retailers and their customers, in which retailers constantly attempt to add value for these customers.

Although, in the behavioral operations literature, research integrated with behavioral agency theory is not consolidated, there is a growing concern about sustainability that suggests that operations and supply chain management (OSCM) professionals have responsibilities to stakeholders, that is, to consumers and clients (Gomez-Mejia et al., 2021). This principal–agent relationship is important in the retail context, as it leads to situations in which an actor behaves like an agent, even in the absence of a binding contract (Kim & Mahoney, 2005). These agents, as individuals, are also loss-averse, and their risk behavior and preferences depend on the context (Martin et al., 2016). The behavioral agency theory is relevant in such circumstances and formulated in the context of the retailer–customer relationship, even though there are no formal contracts in this relationship between retail managers and consumers.

Therefore, as OSCM professionals may have to manage between meeting goals, investing in suppliers, ensuring employee safety, and also satisfying the desire of consumers, as stakeholders, the behavioral agency theory seeks to examine the actions, expecting payoffs, harm supply chain partners, and what motivates them to act with those choices, and how to create incentives for possible ways to reduce risks (Gomez-Mejia et al., 2021). Therefore, this study considers that the customer as a consumer in retail is a motivator of possible conflicts in the process, thereby contributing primarily to the supply chain risk and disruption literature by aggregating behavioral agency theory into the construct. We also contribute to the behavioral agency theory by exploring its mechanisms in different contexts.

2 Supply Chain Risk and Disruption

Supply chain disruption occurs when the flow of materials or services is halted either by a problem at the node level (facilities), arc level (transportation), or network level (Kim et al., 2015). Depending on the type of disruption, it can be anticipated and a proactive supply chain response can follow (Craighead et al., 2007; Knemeyer et al.,

2009). But supply chains are usually forced to respond reactively because an interruption in the supply chain is completely unexpected (Craighead et al., 2007). For example, the 2011 floods in Thailand caused an unexpected disruption in the auto supply chain (Haraguchi & Lall, 2015). Despite the nature of a supply chain disruption, the chance of one occurring has led supply chains to develop two different types of recovery capability: mitigation and contingency (Brandon-Jones et al., 2014; Durach et al., 2020; Matsuo, 2015; Tomlin, 2006)

Mitigation capabilities focus on initiatives that are developed well in advance of a supply chain disruption (Matsuo, 2015). For example, Nissan regularly carries out training drills to deal with supply chain disruptions that might eventually be caused by a natural disaster (Aggarwal & Srivastava, 2016). These initiatives involve different inventory management, flexible production processes, alternative product designs, and supply base redundancy (Bode et al., 2011). Contingency capabilities, on the other hand, focus on initiatives that are implemented after a supply chain disruption occurs (Matsuo, 2015). For instance, Nokia responded contingently to the 2001 fire at the Philips plant by understanding the disruption and subsequently developing appropriate plans (Walker & Wilson, 2012). The wide spread of COVID-19 also caused an unexpected lockdown that forced supply chains to quickly change their structures to respond to specific needs; the food service supply chain, for instance, had to change completely to online orders only (Wollenhaupt, 2020). A contingent response involves the ability to sense the disruption's potential impact and mobilize critical disruption-response resources (Ambulkar et al., 2015; Sheffi & Rice, 2005).

3 Behavioral Agency Theory

First, the principal's and agent's goals and desires may differ, and the principal may not fulfill the agent's requirements. Consumers may have different goals and may not always perform as expected; some may choose to spend proportionally more of their income than others. Second, there may be information asymmetry, in which there is an imbalance in the availability of initial information, which leads to risk asymmetry, whereby the principal and the agent share risks, but may have different attitudes toward them. Finally, consumers may be motivated by self-interest and are likely to stop contributing to the process if they perceive the other party's opportunism.

Customer inputs are the root cause of the unique issues and challenges of service management (Sampson & Froehle, 2006). In many services, when the customer is an end consumer (for example, a consumer of healthcare, education, personal care, or legal services), customers themselves have vital roles to play in creating service outcomes, that is, with their various behaviors they participate at some level in the creation of the service and can guarantee their own satisfaction (Bitner et al., 1997). The agents will then perform the service if they have the capacity (necessary knowledge, skills, and aptitudes), the motivation and also the right opportunities (structures and work environment) (Pepper & Gore, 2014); if not, they may become embroiled in agency issues with the principal. This is particularly true for Brazil as a country in Latin America, where in contrast to developed countries, culture and traditions are incorporated into corporate behavior in order to protect the company's resources for future generations (Gomez-Mejia & Wiseman, 2007).

In behavioral agency theory, incentives are no longer the best way to motivate agents (Massa et al., 2020). Three mechanisms are usually used to develop behavioral agency management: monitoring, sharing, and trust (Reim et al., 2018). Using management from the perspective of behavioral agency theory, we have excluded monitoring from this list. Behavioral agency theory suggests that the inclusion of trust and reciprocity can modify experimental results, that is, individuals seem to respond negatively to controls in the workplace that restrict their feelings of autonomy (Beccerra & Gupta, 1999; Falk & Kosfeld, 2006; Kuang & Moser, 2009). When taken together, studies based on behavioral agency theory suggest that agents behave positively to trust and negatively to monitoring and control. This is the reason why "auto-activated" monitoring and control need to be continuous in order to change the agent's behavior (Laird & Bailey, 2016).

The agency's problems with the client's adverse behavior are a fundamental risk for the case companies. When companies in the case become responsible for repairs and maintenance, thereby making products available, they are exposed to the consequences of customer use. For example, overloading a truck, driving too fast, or changing gear inappropriately will lead to critical components wearing out, so more repairs will be needed to ensure product availability (Reim et al., 2018).

Whenever two parties interact, as in the case of a vendor and a retailer, each party may potentially assume both the principal and agent roles, but for different performance outcomes (Rungtusanatham et al., 2007). On the other hand, for example, the focus in a vendor-managed inventory (VMI arrangement), is on the vendor being responsible for making decisions as to what quantities to order, where the ordered inventory should be stored, and when the order's quantities should arrive at customer locations (Çetinkaya & Lee, 2000). In this setting, the retailer is the principal with sales as the performance outcome, but this retailer is also the agent as far as the availability of retail items for sale is concerned (Rungtusanatham et al., 2007).

4 Retail Supply Chain Disruption Through the Behavioral Agency Lens

The logic of service system offerings increases the likelihood of unintended and unpredictable customer behavior that affects the supplier's operations, and the risk of adverse effects or opportunistic customer behavior increases (Ng et al., 2013; Sakao et al., 2013). Conflict is the culmination of a project or its execution when there are diverging interests. Service facilities that are affected by frequent contact with customers are perceived as inherently limited in their production efficiency due to the uncertainty that customers, as human beings, introduce into the service creation process. This uncertainty stems from individual differences in customers' attitudes and behaviors (Chase, 1981). It is true that human beings generally act as agents, that is, not only when they are stimulated to act, but also when they want to look for new experiences and explore their environment (Dobscha & Foxman, 2012). Customers are the main inputs to the production process, particularly in service companies, but they are, by nature, emotional (Frei, 2006).

Retail managers' fears of the risks associated with allowing a supplier to significantly influence their category's management efforts are well known. Contrary to the expectations of the retailer and the supplier's manager, this practice does not seem to increase the retailer's or supplier's dependence on each other, the buyer–supplier interdependence. It is clearly in the retailer's interest to protect itself from the opportunism of the focal supplier in its strategy of segmenting expenses in areas that contain similar products, allowing for greater focus, consolidation, and efficiency (Morgan et al., 2007). The intensified attention of the client and the desire for autonomy in decision-making are contrary to the model of the passive and compatible client in a common service encounter with an authoritarian service provider, which can further increase the client's suspicion about the service provider's motives, whether this has to do with the authoritarian behavior or decision-making of the service provider. Both situations will not only create customer dissatisfaction for preferring decision-making autonomy but will also provide samples of the attempt to exploit the quality and credibility of the service to gain unjustified benefits.

Offering customers more decision-making authority, however, may not lead to greater customer satisfaction, since the benefits of service credibility are the ability to solve problems and the specialized or tacit knowledge of the service provider, as opposed to merely well-structured data (Debo et al., 2008; Wu, 2011). The demand for professional services typically derives from customers who do not have sufficient experience and/or competence to make decisions in this service area (Ngamvichaikit & Beise-Zee, 2014).

When assuming maintenance of and/or responsibility for the operation, the supplier's performance becomes more deeply involved with the production process, potentially resulting in better continuity in the agency relationship (Bullinger et al., 2004). For the service provider, any potential cost savings for the customer are—from the agency theory viewpoint—based on aligning preferences between the parties, because the provider is aiming to maximize their efficiency. The customer, in its turn, shares this interest, because it forms the basis of their "profit" from their perspective (Hypko et al., 2010).

In terms of trust, evidence has already been found of increased customer perceptions of empathy and agency that lead to greater customer satisfaction, by emphasizing how "we" (the company) serve "you" (the customer), and the "me" (the agent) in these customer-company interactions (Packard et al., 2018). In the absence of trust, however, the retailer, on the one hand, may not be as willing to give the vendor ready access to relevant information, or to relinquish control of vital decisions pertaining to inventory replenishment and maintenance, while the vendor, on the other, may not be as willing to accept partial (or full) responsibility for shrinkage as required in a Pay-On-Scan (or a Scan-Based Trading) arrangement. (Rungtusanatham et al., 2007).

COVID-19 caused a variety of unexpected supply chain disruptions, forcing supply chains to respond contingently. For example, when consumers started panic buying in supermarkets and hoarding goods, they disrupted the retail supply chain, causing a shortage of some essential items. The unexpected behaviors of these consumers produced several effects in the relationship between retail supply chains and customers. Such behaviors reduced the confidence of supply chain members in consumers because of volatile customer demands (Leonard, 2020). The result was low levels of information-sharing because retailers were unable to forecast demand with any degree of accuracy, which forced them to monitor customer behaviors closely in an attempt to control their impulses. For instance, retailers had to limit the quantity of flu-related products per customer due to a shortage of products (D'Innocenzio, 2020).

The lack of trust, an inability to share accurate information, and monitoring customers' behavior can be seen through the lens of behavioral agency theory. Table 1 presents some examples of real cases, showing actions and behavioral agency mechanisms for mitigating interruptions in the supply chain, such as tsunamis, hurricanes, shuttered factories, port congestion, trucker strikes, and the COVID-19 pandemic. Above all, however, the table shows potential mechanisms for mitigating adverse customer behavior. The perceptions (of retailers and consumers) were used to develop a framework that combines agency problems with agency mechanisms.

5 Final Considerations

Principal–agent problems, such as opportunism, differences in objectives, and information asymmetry can compromise the effectiveness of risk mitigation practices (Li et al., 2015). The analysis of this study indicates how important it is for behavioral theorists to incorporate the intrinsic motivations and goals of individuals in principal–agent relationships, in order not only to understand the agent's behavior but also how it can be managed. Behavioral agency theory, therefore, provides the conceptual basis for understanding why the divergent intrinsic motivations and goals of individuals and firms and their backgrounds are likely to lead to agency-specific problems and result in different mechanisms being required for aligning goals.

Different actions are needed for overcoming two behavioral agency problems in retail supply chains. First, when there are conflicting goals supermarket prices increase, information sharing decreases (between retailers and consumers) and more specifically, the number of items each consumer can purchase is limited.

Second, to solve problems related to monitoring, managers sought to hold frequent meetings with suppliers. They also tried to play fair, showing that trust had been broken (because of agent-consumer actions), distributed online demands

Table 1 Application of the behavioral agency theory in the COVID-19 pandemic

Problem categories	Problem dimensions	Agent actions (consumer)	Principal actions (retailer)	Behavioral agency mechanisms
Conflicting goals	Customers overly cost-focused	Short-term focus, price was not a primary concern	Prices were increased mostly for production causes (i.e., reduction in productivity)	Sharing (results-based pricing)
	Adverse selection	Consumers could not find the products they needed online	Information about stockouts was shared, substitutable products offered	Sharing (incentives for good performance)
	Myopic behavior	Consumers went into a panic-buying mode, hoarding items such as toilet paper, canned goods, and cleaning products	Grocery stores started limiting the quantity customers could buy per purchase	Monitoring (conditional contracting)
	Careless behavior			
Difficulties in monitoring	Monitoring impracticality	Unexpected surge in demand	Frequent meetings with suppliers to replenish bare shelves	Monitoring (information management)
	Monitoring and contract aversion	Consumers wanted to buy as many items as they were available	Trust was broken since retailers could not rely on consumers' behaviors	Trust (reliance on customer relationship)
	Data overload	The surge in online demand caused systems to collapse	Retailers had to pulverize consumers' online demand to re-establish their system	Sharing (profit and risk calculations)
	Data reliability	Uncommon behavior brought unreliable historical data	Attempted to use historical data to forecast future consumer behavior	Monitoring (information management)

through their distribution channels until normality was re-established, and used historical data to try to predict consumer behavior.

If corporations can mitigate risks by selecting their agents, manufacturing company customers may not even be a part of the process, while the customers of service companies, such as retail stores, become very closely involved with the processes, and this is a challenge for managers in service companies when compared to manufacturing companies (Sampson & Froehle, 2006). Customers are mainly interested in the result of the service encounter, and possibly in the process that produces this result, which mainly occurs, in fact, when the exchange of services is an important element (Van der Valk & Van Iwaarden, 2011).

Because of the way people live today, it is almost inevitable that new events like COVID-19 will occur in the future. This kind of event causes great supply chain disruption (Craighead et al., 2020), which can be mitigated by employing the mechanisms of behavioral agency theory. First, unlike other supply chain disruptions that affect only a specific region, such as a localized earthquake, pandemics occur across the globe at the same time, and affect different regions of the globe simultaneously. Second, a common supply chain disruption may affect the nodes of a supply chain differently, while a pandemic, such as COVID-19, affects all the members of a supply chain in a similar way. This is not the same, for example, as the earthquake that hit Japan in 2011 and that affected particularly auto suppliers (Aggarwal & Srivastava, 2016). Third, an ordinary supply chain disruption affects only one side of supply and demand, like the Japanese earthquake that affected only auto-suppliers but had no effect at all on the behavior of consumers about buying automobiles. On the other hand, pandemic events tend to affect both the supply and demand sides. For instance, lockdown measures impacted how the food service supply chain should offer its products and how consumers should access these products.

A very clear context is that of the recent global financial crisis in 2008, for example, when many financial services customers lost money after the turmoil, and many of them blamed their losses on the sellers because of what they had recommended and for their poor judgment. In short, the agency guarantees a comprehensive investigation as a design, behavioral and social dimension (Kucirkova, 2018).

As a result of the initial gap—how the disruption in the retail supply chain during COVID-19 can be explained by the mechanisms of behavioral agency theory—there are other consumer agency activities to be investigated. Consumers tend to complain and spread negative word-of-mouth communication within their social networks instead of breaking off the relationship altogether, or broadcasting their feeling of frustration to third parties (e.g., media and news agency consumers) (Legocki et al., 2020). Based on the findings of the mismatch between consumer expectations and actual demand, there is a misalignment between the supplier's goal and the retailer's goal, which is generally assumed to exist in supply chain management paradigms (see bullwhip effect). Therefore, it is not only in the dyadic relationship between a buyer and a supplier that opportunism has negative consequences in supply chains (Morgan et al., 2007).

The application of organization theories is particularly valuable for developing this field. Research that is guided by such theories can help researchers develop an understanding of the problems that behavioral agency theory tends to solve. This study may also be relevant for service triads; for example, a manufacturing company with services that employs a maintenance partner (a third party), which is in frequent contact with the end customer, and which, therefore, plays a significant part in influencing the main provider's service output (Heaslip & Kovács, 2019). Collaboration can be seen as an integration between key players and agents for: exchanging information, matching goals and aligning incentives that have the potential to build trust and reduce uncertainty for achieving effective supply chain management (Fayezi et al., 2012).

Future research investigations could start by understanding effective management, especially reacting to risk impacts, and particularly those that affect the principal, and risk recovery options (Cheng & Kam, 2008). Although exploratory studies can identify whether this reaction is observed and how much is understood, as in the action to improve financial performance, and the speed of sharing is strengthened by the duration of the relationship and the trust of the supplier (Li et al., 2015), it is important to measure the speed of information sharing, the duration of the relationship and the trust of the supplier. These are the financial or organizational performance mechanisms of companies and are effective approaches for improving a company's sustainability performance (Shafiq et al., 2017)

Risk assessment techniques make it easier for purchasing organizations to obtain information for verifying supplier behavior, thus promoting congruent goals between buying and selling companies and reducing uncertainty (Zsidisin et al., 2004). A fruitful line of research would be to investigate the risk assessment techniques employed by the supplier for promoting congruence between the behavior of the retail consumer and the retailer's service, thereby mitigating risk and uncertainty. The investigation could explore whether the performance objective is more important than the ability of a supplier (such as the retailer) to transfer the risk to agents (or those without ties), or whether the balance between risk and reward is a necessary condition for suppliers to accept the risk (Selviaridis & Norrman, 2014). The principal can use the flexibility of the network (either pre-existing or shaped by it) for risk recovery. Assessing and balancing these commitments, and meeting supply delivery needs are key facets of a resilient and successful network (Cheng & Kam, 2008).

Although various locations have for decades suffered disasters (e.g., tsunamis, hurricanes, shuttered factories, port congestion, and trucker shortages), consumer behavior in retail supply chains is still an open challenge for current supply chain management. A future agenda for thematic risk mitigation in retail supply chains could include: the mapping out of the supply network, the adoption of blockchain technologies, machine learning and artificial intelligence, data analysis and production reordering. All topics must be covered, without forgetting to address ethics, communication, and sustainability jointly. Many of the above-mentioned measures may require a lot of resources and be difficult to implement, but since they are risk mitigation measures, they should be assessed, above all for their ability to save costs, rather than solely for their capacity to guarantee revenue.

References

Aggarwal, S., & Srivastava, M. K. (2016). *Nissan: Recovering supply chain operations* (pp. 1–6). Ivey Publishing, W16331.

Ambulkar, S., Blackhurst, J., & Grawe, S. (2015). Firm's resilience to supply chain disruptions: Scale development and empirical examination. *Journal of Operations Management, 33–34*, 111–122. https://doi.org/10.1016/j.jom.2014.11.002

Beccerra, M., & Gupta, A. K. (1999). Trust within the organization: Integrating the trust literature with agency theory and transaction costs economics. *Public Administration Quarterly, 23*(2), 177–203.

Bitner, M. J., Faranda, W. T., Hubbert, A. R., & Zeithaml, V. A. (1997). Customer contributions and roles in service delivery. *International Journal of Service Industry Management, 8*(3), 193–205. https://doi.org/10.1108/09564239710185398

Blackhurst, J., Craighead, C. W., Elkins, D., & Handfield, R. B. (2005). An empirically derived agenda of critical research issues for managing supply-chain disruptions. *International Journal of Production Research, 43*(19), 4067–4081. https://doi.org/10.1080/00207540500151549

Bode, C., Wagner, S. M., Petersen, K. J., & Ellram, L. M. (2011). Understanding responses to supply chain disruptions: Insights from information processing and resource dependence perspectives. *Academy of Management Journal, 54*(4), 833–856. https://doi.org/10.1007/BF00859739

Brandon-Jones, E., Squire, B., Autry, C. W., & Petersen, K. J. (2014). A contingent resource-based perspective of supply chain resilience and robustness. *Journal of Supply Chain Management, 50*(3), 55–73. https://doi.org/10.1111/jscm.12050

Bullinger, H.-J., Spath, D., Schuster, E., & Meiren, T. (2004). Operator models: a more advanced form of service model. *Economic Bulletin, 41*(3), 103–106. https://doi.org/10.1007/s10160-004-0261-2

Çetinkaya, S., & Lee, C. Y. (2000). Stock replenishment and shipment scheduling for vendor-managed inventory systems. *Management Science, 46*(2), 217–232. https://doi.org/10.1287/mnsc.46.2.217.11923

Chase, R. B. (1981). The customer contact approach to services: theoretical bases and practical extensions. *Operations Research, 29*(4), 698–706. https://doi.org/10.1287/opre.29.4.698

Cheng, S. K., & Kam, B. H. (2008). A conceptual framework for analysing risk in supply networks. *Journal of Enterprise Information Management, 21*(4), 345–360. https://doi.org/10.1108/17410390810888642

Choi, T. Y., Rogers, D., & Vakil, B. (2020). Coronavirus is a wake-up call for supply chain management. *Harvard Business Review*, (February 2019), 1–13.

Cole, R., & Aitken, J. (2019). Selecting suppliers for socially sustainable supply chain management: post-exchange supplier development activities as pre-selection requirements. *Production Planning and Control, 30*(14), 1184–1202. https://doi.org/10.1080/09537287.2019.1595208

Craighead, C. W., Blackhurst, J., Rungtusanatham, M. J., & Handfield, R. B. (2007). The severity of supply chain disruptions: design characteristics and mitigation capabilities. *Decision Sciences, 38*(1), 131–156. https://doi.org/10.1080/00207540500151549

Craighead, C. W., Ketchen, D. J., & Darby, J. L. (2020). Pandemics and supply chain management research: toward a theoretical toolbox. *Decision Sciences, 51*(4), 838–866. https://doi.org/10.1111/deci.12468

Debo, L. G., Toktay, L. B., & Wassenhove, L. N. V. (2008). Queuing for expert services. *Management Science, 54*(8), 1497–1512. https://doi.org/10.1287/mnsc.1080.0867

D'Innocenzio, A. (2020). Fear of coronavirus sends consumers into a grocery-hoarding frenzy. *Fortune*.

Dobscha, S., & Foxman, E. (2012). Mythic agency and retail conquest. *Journal of Retailing, 88*(2), 291–307. https://doi.org/10.1016/j.jretai.2011.11.002

Durach, C. F., Wiengarten, F., & Choi, T. Y. (2020). Supplier–supplier coopetition and supply chain disruption: first-tier supplier resilience in the tetradic context. *International Journal of Operations and Production Management*. https://doi.org/10.1108/IJOPM-03-2019-0224

Esper, T. L. (2020). Supply chain management amid the coronavirus pandemic. *Journal of Public Policy and Marketing*, 1–2. https://doi.org/10.1177/0743915620932150

Falk, A., & Kosfeld, M. (2006). The hidden costs of control. *The American Economic Review, 96*(5), 1611–1630. https://doi.org/10.1257/aer.96.5.1611

Fayezi, S., O'Loughlin, A., & Zutshi, A. (2012). Agency theory and supply chain management: A structured literature review. *Supply Chain Management, 17*(5), 556–570. https://doi.org/10.1108/13598541211258618

Frei, F. X. (2006). Breaking the trade-off between efficiency and service. *Harvard Business Review, 84*(11), 1–12.

Gomez-Mejia, L. R., Martin, G., Villena, V. H., & Wiseman, R. M. (2021). The behavioral agency model: Revised concepts and implications for operations and supply chain research. *Decision Sciences, 52*(5), 1026–1038. https://doi.org/10.1111/deci.12547

Gomez-Mejia, L. R., & Wiseman, R. M. (2007). Does agency theory have universal relevance ? A reply to Lubatkin, Lane, Collin, and Very. *Journal of Organizational Behavior, 28*(1), 81–88. https://doi.org/10.1002/job.407

Haraguchi, M., & Lall, U. (2015). Flood risks and impacts: A case study of Thailand's floods in 2011 and research questions for supply chain decision making'. *International Journal of Disaster Risk Reduction, 14*(January 2012), 256–272. https://doi.org/10.1016/j.ijdrr.2014.09.005

Heaslip, G., & Kovács, G. (2019). Examination of service triads in humanitarian logistics. *International Journal of Logistics Management, 30*(2), 595–619. https://doi.org/10.1108/IJLM-09-2017-0221

Hypko, P., Tilebein, M., & Gleich, R. (2010). Benefits and uncertainties of performance-based contracting in manufacturing industries: An agency theory perspective. *Journal of Service Management, 21*(4), 460–489. https://doi.org/10.1108/09564231011066114

Kildow, B. (2011). *A supply chain management guide to business continuity*. AMACOM.

Kim, J., & Mahoney, J. T. (2005). Property rights theory, transaction costs theory, and agency theory: an organizational economics approach to strategic management. *Managerial and Decision Economics, 26*(4), 223–242. https://doi.org/10.1002/mde.1218

Kim, Y., Chen, Y.-S., & Linderman, K. (2015). Supply network disruption and resilience: A network structural perspective. *Journal of Operations Management, 33–34*, 43–59. https://doi.org/10.1016/j.jom.2014.10.006

Knemeyer, A. M., Zinn, W., & Eroglu, C. (2009). Proactive planning for catastrophic events in supply chains. *Journal of Operations Management, 27*(2), 141–153. https://doi.org/10.1016/j.jom.2008.06.002

Kuang, X., & Moser, D. V. (2009). Reciprocity and the effectiveness of optimal agency contracts. *The Accounting Review, 84*(5), 1671–1694. https://doi.org/10.2308/accr.2009.84.5.1671

Kucirkova, N. (2018). Children's agency and reading with story-apps: considerations of design, behavioural and social dimensions. *Qualitative Research in Psychology*, 1–25. https://doi.org/10.1080/14780887.2018.1545065

Laird, B. K., & Bailey, C. D. (2016). Does monitoring reduce the agent's preference for honesty? *Research on Professional Responsibility and Ethics in Accounting, 20*, 67–94. https://doi.org/10.1108/S1574-076520160000020003

Legocki, K. V., Walker, K. L., & Kiesler, T. (2020). Sound and fury: digital vigilantism as a form of consumer voice. *Journal of Public Policy and Marketing, 39*(2), 169–187. https://doi.org/10.1177/0743915620902403

Leo, W. W. C., Chou, C. Y., & Chen, T. (2019). Working consumers' psychological states in firm-hosted virtual communities. *Journal of Service Management, 30*(3), 302–325. https://doi.org/10.1108/JOSM-03-2018-0077

Leonard, M. (2020). *Consumer demand: The big supply chain question mark*. Supply Chain Dive.

Li, G., Fan, H., Lee, P. K., & Cheng, T. C. (2015). Joint supply chain risk management: An agency and collaboration perspective. *International Journal of Production Economics, 164*, 83–94. https://doi.org/10.1016/j.ijpe.2015.02.021

Linton, T., & Vakil, B. (2020). Coronavirus is proving we need more resilient supply chains. *Harvard Business Review*. Available at: https://hbr.org/2020/03/coronavirus-is-proving-that-we-need-more-resilient-supply-chains.

Martin, G. P., Wiseman, R. M., & Gomez-Mejia, L. R. (2016). Going short-term or long-term? CEO stock options and temporal orientation in the presence of slack. *Strategic Management Journal, 37*(12), 2463–2480. https://doi.org/10.1002/smj.2445

Massa, R. M., Partyka, R. B., & Lana, J. (2020). Behavioral agency research and theory: a review and research agenda. *Cadernos EBAPE.BR, 18*(2), 220–236. https://doi.org/10.1590/1679-395177017

Matsuo, H. (2015). Implications of the Tohoku earthquake for Toyota's coordination mechanism: Supply chain disruption of automotive semiconductors. *International Journal of Production Economics, 161*, 217–227. https://doi.org/10.1016/j.ijpe.2014.07.010

Morgan, N. A., Kaleka, A., & Gooner, R. A. (2007). Focal supplier opportunism in supermarket retailer category management. *Journal of Operations Management, 25*(2), 512–527. https://doi.org/10.1016/j.jom.2006.05.006

Ng, I. C. L., Ding, D. X., & Yip, N. (2013). Outcome-based contracts as new business model: The role of partnership and value-driven relational assets. *Industrial Marketing Management, 42*(5), 730–743. https://doi.org/10.1016/j.indmarman.2013.05.009

Ngamvichaikit, A., & Beise-Zee, R. (2014). Customer preference for decision authority in credence services: The moderating effects of source credibility and persuasion knowledge. *Managing Service Quality, 24*(3), 274–299. https://doi.org/10.1108/MSQ-03-2013-0033

Packard, G., Moore, S. G., & McFerran, B. (2018). (I'm) happy to help (you): the impact of personal pronoun use in customer-firm interactions. *Journal of Marketing Research, 55*(4), 541–555. https://doi.org/10.1509/jmr.16.0118

Pepper, A., & Gore, J. (2014). The economic psychology of incentives: An international study of top managers. *Journal of World Business, 49*(3), 350–361. https://doi.org/10.1016/j.jwb.2013.07.002

Pepper, A., & Gore, J. (2015). Behavioral agency theory: new foundations for theorizing about executive compensation. *Journal of Management, 41*(4), 1045–1068. https://doi.org/10.1177/0149206312461054

Reim, W., Sjödin, D., & Parida, V. (2018). Mitigating adverse customer behaviour for product-service system provision: An agency theory perspective'. *Industrial Marketing Management, 74*(June 2017), 150–161. https://doi.org/10.1016/j.indmarman.2018.04.004

Rungtusanatham, M., Rabinovich, E., Ashenbaum, B., & Wallin, C. (2007). Vendor-owned inventory management arrangements in retail: An agency theory perspective. *Journal of Business, 28*(1), 111–136.

Sakao, T., Öhrwall Rönnbäck, A., & Ölundh Sandström, G. (2013). Uncovering benefits and risks of integrated product service offerings—Using a case of technology encapsulation. *Journal of Systems Science and Systems Engineering, 22*(4), 421–439. https://doi.org/10.1007/s11518-013-5233-6

Sampson, S. E., & Froehle, C. M. (2006). Foundations and implications of a proposed unified services theory. *Production and Operations Management, 15*(2), 329–343. https://doi.org/10.1111/j.1937-5956.2006.tb00248.x

Selviaridis, K., & Norrman, A. (2014). Performance-based contracting in service supply chains: A service provider risk perspective. *Supply Chain Management, 19*(2), 153–172. https://doi.org/10.1108/SCM-06-2013-0216

Shafiq, A., Johnson, P. F., Klassen, R. D., & Awaysheh, A. (2017). Exploring the implications of supply risk on sustainability performance. *International Journal of Operations and Production Management, 37*(10), 1386–1407. https://doi.org/10.1108/IJOPM-01-2016-0029

Sheffi, Y., & Rice, J. B. (2005). A supply chain view of the resilient enterprise. *MIT Sloan Management Review, 47*(1).

Tomlin, B. (2006). On the value of mitigation and contingency strategies for managing supply chain disruption risks. *Management Science, 52*(5), 639–657. https://doi.org/10.1287/mnsc.1060.0515

Van der Valk, W., & Van Iwaarden, J. (2011). Monitoring in service triads consisting of buyers, subcontractors and end customers. *Journal of Purchasing and Supply Management, 17*(3), 198–206. https://doi.org/10.1016/j.pursup.2011.05.002

Villena, V. H., Gomez-Mejia, L. R., & Revilla, E. (2009). The decision of the supply chain executive to support or impede supply chain integration: A multidisciplinary behavioral agency perspective. *Decision Sciences, 40*(4), 635–665. https://doi.org/10.1111/j.1540-5915.2009.00245.x

Walker, R., & Wilson, J. (2012). Nokia's supply chain management. *Kellogg School of Management Cases*, 1–8.

Wollenhaupt, G. (2020). *Food distributors, restaurants reinvent business models in a scramble to survive*. Supply Chain Dive.

Wu, L. W. (2011). Beyond satisfaction: The relative importance of locational convenience, interpersonal relationships, and commitment across service types. *Managing Service Quality, 21*(3), 240–263. https://doi.org/10.1108/09604521111127956

Zsidisin, G. A., Ellram, L. M., Carter, J. R., & Cavinato, J. L. (2004). An analysis of supply risk assessment techniques. *International Journal of Physical Distribution and Logistics Management, 34*(5), 397–413. https://doi.org/10.1108/09600030410545445

Mitigation of Supply Chain Vulnerability Through Collaborative Planning, Forecasting, and Replenishment (CPFR)

Leonardo de Carvalho Gomes

Abstract The chapter aims to clarify the concept of supply chain vulnerability and to show how collaborative methods such as Collaborative Planning, Forecasting, and Replenishment (CPFR) can mitigate its effects. It proposes a framework of supply chain vulnerability that includes a relation between internal and external elements. The framework also explains the relationship between resilience and robustness and supply chain vulnerability. This chapter gives an overview of the theoretical study of supply chain vulnerability. The research contains a literature review of supply chain vulnerability, supply chain resilience, supply chain robustness and correlated elements. In addition, the chapter presents a literature review on CPFR. We propose a theoretical framework to explain supply chain vulnerability and how CPFR can aim to mitigate it. The chapter provides empirical insights about supply chain vulnerability, how it is related to its internal and external components, and how collaborative methods such as CPFR can assist in mitigating vulnerability effects. Because of the chosen research approach, the research results contribute to the theoretical discussion about supply chain vulnerability. The chapter includes implications for the development of a framework regarding supply chain vulnerability and its relationship with the concepts of resilience, robustness, and other elements. We discuss the mitigation of supply chain vulnerability, as well as the contribution of CPFR to the mitigation of the effects of supply chain vulnerability. Further study of the concepts of supply chain vulnerability, resilience, and robustness is also presented.

1 Introduction

Supply Chain Management (SCM) has emerged as one of the main strategies for competitive advantage generation in companies (Zsidisin et al., 2005). The main function of SCM comprises supply chain coordination (SCC), since the adoption of a

L. d. C. Gomes (✉)
School of Engineering, Federal University of Rio Grande (FURG), Rio Grande, Brazil
e-mail: leonardo.gomes@furg.br

strategy of joint work between companies includes the planning, managing, and monitoring of information (Cooper & Ellram, 1993). In this context, companies begin to develop activities of planning, monitoring, information sharing, and decision-making, seeking higher performance in the Supply Chain (SC) and focusing on the transaction costs (external coordination) in addition to the production costs (internal coordination) (Malone, 1987).

Nevertheless, SCM has not been a simple task and, despite investments in technology and intellectual capacity, SC performance is still not satisfactory (Fabbe-Costes & Jahre, 2007; Fisher, 2007). SCs are complex systems that undergo constant turbulence, which enable the creation of potentially unpredictable disruption and make them susceptible to internal or external failures (Malone, 1987; Pettit et al., 2010). In the context of failure, Malone (1987) presented a third type of cost, in addition to transaction and production costs, which has been poorly addressed and considered in the structure of general coordination: the vulnerability cost. These three costs present differences in their proportions according to the structure of coordination, with emphasis on the vulnerability cost due to internal and external failures.

Supply Chain Vulnerability (SCV) has become a relevant theme for managers due to constant changes in the economy, and risks and disruptions that derive from situations that are unexpected in terms of the companies and the environment, which make the SC more vulnerable than before (Hendricks & Singhal, 2005a; Wagner & Neshat, 2010). Singhal and Hendricks (2002) have pointed out that significant financial losses occur due to interruptions in the SC, where the average financial return for shareholders immediately drops 7.5% when a SC rupture is announced, and, 4 months after the interruption, the total loss grows to an average of 18.5%.

Despite the theme's relevance, academic research has done little to address SCV over the years, a fact that can be verified by a simple search on the Science Direct database using the keyword "supply chain vulnerability" from 1987 to 2014. The result shows 11,817 papers retrieved. However, in another simple search of the Science Direct database, using the keyword "supply chain vulnerability" from 1987 to 2017, the result is 16,609. This shows an increase of 40.55% in only the last 3 years. Current knowledge of SCV presents limitations and most of the papers address cases and informal evidence (Hendricks & Singhal, 2005a; Wagner & Bode, 2006).

Collaborative initiatives used to construct SC are more efficient and robust that respond more quickly to market demands (Matopoulos et al., 2007). The degree of SC turbulence and complexity requires more collaboration, as well as alignment between the SC members (Ahlquist et al., 2003; Slone et al., 2007). Collaborative practices generate initiatives, methods, and systems that can identify and reduce certain SCVs.

The objective of this chapter is to contribute to the understanding of SCV through the proposition of a framework and the presentation of the hypothesis that the use of collaborative methods such as Collaborative Planning, Forecasting, and Replenishment (CPFR) can mitigate elements that promote an increase in SCV.

This chapter is based on the theoretical background of SCV. It addresses a brief conceptualization of collaboration, including CPFR and its potential to mitigate SCV. From the theoretical background, we have created a framework for understanding SCV. Afterwards, a theoretical study is executed, comparing CPFR's scope of action to the main elements and factors of SCV comprised in the proposed framework.

2 Theoretical Background

Organizations, as well as countries, communities, and individuals, are subject to several environments undergoing constant change. Thus, threats arise in these frequently turbulent environments and can vary in intensity and frequency, with internal or external origins (Bhamra et al., 2011). The effects of this turbulence are a concern for managers, mainly due to the high competitiveness of the environment to which they belong, where profit margins are limited and fines for delay and product failures are common.

Two main scenarios have raised concern for the risks of the SC: the number of catastrophes, natural disasters, and crises that have occurred throughout the years (Coleman, 2006; Wagner & Neshat, 2010), and SC complexity, which has increased due to shorter product life cycles, outsourcing, offshoring, new technologies, legislations, competitor pressure, and globalization, among other reasons (Wagner & Bode, 2009; Wagner & Neshat, 2010).

Nevertheless, risks are not only present in macro factors, but also in factors arising from the SC itself, which, due to internal weaknesses, lead to its disruption and indicate possible robustness failures of the project (Norrman & Lindroth, 2004; Ponomarov & Holcomb, 2009; Wagner & Bode, 2006). This lack of robustness results in the increase of SCV (Christopher & Lee, 2004). In addition to the phenomena mentioned, another factor observed in companies contributes to the increase in SCV: the search for greater efficiency and cost reduction. The need to reduce costs and to be efficient can make the SC more vulnerable if the risks in product and process change are not considered (Christopher & Rutherford, 2004; Cranfield, 2002; Hendricks & Singhal, 2005a; Melnyk, 2007; Pettit et al., 2010).

As the starting point of this chapter, understanding SCV comprises three phenomena that naturally promote SCV growth: SC complexity; the quantity of catastrophes, natural disasters, and crises; and the search for efficiency and cost reduction. Given the perception of these three phenomena, it is possible to begin to comprehend SCV and their interrelated elements.

SCV is related to supply chain risk management (SCRM), where companies develop both corrective and preventive actions to reduce SC risk factors. SCRM is a coordinated process between companies to identify risk sources and to reduce SCV, and it can become complex due to the great quantity of hard-to-quantify variables and the difficulty in verifying the influence of risks (Juttner et al., 2003). Wagner and Bode (2009) presented SCRM as the process of evaluating and acting on the risks based on the company's general objectives: identification, analysis, evaluation, prioritization, monitoring, and performance results. Viswanadham and Gaonkar (2008) created a SCRM framework similar to that proposed by Wagner and Bode (2009), but added two approaches: the preventive and the interceptive.

SCRM will not be explored here beyond these introductory concepts because it is not the focus of this chapter.

2.1 Vulnerability, Risk Sources, Disruptions, and Risks of the SC

The concept of vulnerability presents differences in the literature, such as the exposure to serious disturbances (Christopher & Peck, 2004; Cranfield, 2002), the tendency for risk sources and factors to prevail over risk mitigation strategies, causing adverse consequences for the SC (Juttner et al., 2003), or the unexpected deviation of regulations with negative consequences (Svensson, 2000, 2002). In this chapter, we adopt SCV as the susceptibility of SCs to damage due to a disruption, which can lead to risks to the chain—that is, negative consequences (Wagner & Bode, 2006). Although a SC disruption is the situation that leads to the occurrence of risks, the SCV determines the consequences. Therefore, three interconnected elements explain SCV: SC risk; SC disruption; and SC risk sources.

Supply Chain Risk (SCR) considers the variation of results with negative consequences, such as danger, damage, or losses (Harland et al., 2003; Juttner et al., 2003; March & Shapira, 1987; Wagner & Bode, 2006). There are two categories of SCR: risks arising from coordination problems in supply and demand; and risks arising from the interruption of normal activities, which may occur in the event of natural catastrophes, strikes and economic instabilities, political interventions or intentional acts, including terrorist attacks (Kleindorfer & Saad, 2005). Khojasteh and Abdi (2016) and Khojasteh (2018) categorized these risks into internal and external risks.

Supply Chain Disruption (SCD) is an undesired situation, which may come from external or internal sources and can lead to SCR (Wagner & Bode, 2006). For the affected companies, SCD is an exceptional situation that must be mitigated rapidly to avoid SCR. Kleindorfer and Saad (2005) presented SAM (specifying sources of risk, vulnerabilities, Assessment, and Mitigation), a methodology to specify and evaluate risk sources and mitigate SCV. The quantitative methodology focuses on the events that can cause SC ruptures and combines investments in the mitigation of identified risk sources with the probability of their occurrence.

Supply Chain Risks Sources (SCRS) address the categories or classes of possible, internal or external, events that cause SCD and are defined and classified distinctly in the literature (Chopra & Sodhi, 2004; Juttner, 2005; Svensson, 2000). Wagner and Bode (2006) organized them into three classes: demand side risks, supply side risks, and catastrophes. Juttner (2005) proposed three classes: supply, demand, and the environment. Chopra and Sodhi (2004) delineated nine classes: ruptures, delays, systems, demand forecasting, intellectual property, acquisition, receivables, inventories, and capability. Christopher and Peck (2004) defined SCRS as companies' internal risks (process and control), supply chains' internal risks (supply and demand), and supply chains' external risks (environment).

In this chapter, two SCRS types are adopted: internal (I) and external (E), which are further subdivided into five classes: demand side risks (I); supply side risks (I); infrastructure (I); regulation (E); and catastrophe (E).

Demand side risk is the result of interruptions in the supply flows to the final client, that is, downstream of the focal company. It normally appears in the form of an interruption in the physical distribution of product due to technical problems with transportation, products, and processes, among others. It can also happen due to uncertainties between the projected and the effective demand, which result in resource scarcity or excess and an increase in the bullwhip effect (Nagurney et al., 2005). Thus, it can be hypothesized here that one of the causes of such risk is an inefficient and barely collaborative SCC (Christopher & Lee, 2004).

Supply side risk is the result of interruptions in the supply flows from the suppliers to the focal company. They can occur due to problems in logistics, products, processes, and bankruptcy, among others. The opportunistic behavior of suppliers is a supply side risk (Spekman & Davis, 2004; Williamson, 1985). The bullwhip effect is also an example of supply side risk.

Product security is a risk source both in the supply side risk sources and in the demand side risk sources, with two possible approaches (Marucheck et al., 2011): product safety and product security. Product safety refers to the reduction in the probability that the use of a product will result in disease, accident, or death, or that it will have negative consequences for people, goods, or machinery. One example is the quantity of vehicle recalls due to safety issues. Problems may also arise from inadequate product storage, handling, and distribution systems. Product security involves the delivery of a product free of intentional contamination, damage, or deviations, such as sabotage, terrorism, or product misrepresentation (in the case of falsification).

Infrastructure refers to companies' infrastructure such as assets, electricity supply, water, labor, energy matrix (oil, gas, coal, and others), technology and information systems, and management systems, among other elements. The infrastructure class also comprises the SC configuration characteristic, which can have few suppliers or a larger number of suppliers.

Regulation refers to risks originating from governmental regulations, such as new laws or changes in taxation. Catastrophe refers to natural disasters, sociopolitical instability, civil movements like strikes, or attacks. We can state that in globalized supply chains, supply chain risks increase due to the catastrophe factor.

Based on this theoretical review of the four concepts, Fig. 1 illustrates the relationships between them. SCRS generate SCD, which, depending on SCV, may or may not cause SCR.

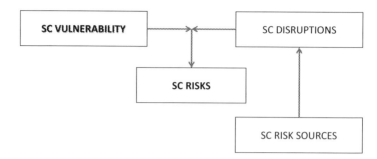

Fig. 1 Risk sources, rupture, risks, and SCV

2.2 Supply Chain Resilience and Supply Chain Robustness

Managers seek, through actions, to anticipate, absorb, and overcome SC disturbances arising from any source (Pettit et al., 2010; Pickett, 2006). Two concepts have emerged from these practices: Supply Chain Resilience (SCRes) and Supply Chain Robustness (SCRob).

While SCV is a susceptibility characteristic of the SC, resilience and robustness are characteristics of resisting or being impervious to SCRS. These three concepts are connected in the following semantic way: if a SC is more resilient, it tends to be less vulnerable and more robust in the face of interruptions and risk sources.

However, the literature presents divergences between these three concepts. Some authors understand that robustness and resilience are synonyms (Ponomarov & Holcomb, 2009; Sheffi & Rice Jr., 2005), while some others (Ivanov & Sokolov, 2013; Spiegler et al., 2012; Vlajic et al., 2012) argue that there are conceptual and strategic differences between them, even though they are complementary.

The following two subsections define SCRes and SCRob.

2.2.1 Supply Chain Resilience

Supply Chain Resilience (SCRes) is a recent scientific field that is on the rise and has been underexplored. Studies on SCRes present three drivers: SCV becomes important for business competitivity; few studies address SCV, generating limited knowledge on the subject; methodologies are necessary to manage SCV (Bhamra et al., 2011; Christopher & Peck, 2004; Pettit et al., 2010; Ponomarov & Holcomb, 2009).

SCRes has been defined in several ways by many authors: the ability to react to an unexpected interruption and to restore normal operation (Ponomarov & Holcomb, 2009), the ability of a system to return to its original state or to change to a new, more desirable state after an interruption (Christopher & Peck, 2004), and the ability of a company to survive, adapt, and grow in the face of turbulent change (Fiksel, 2006). In addition, SCRes is the capacity of a SC to deal with changes within two dimensions: agility (reactive capacity) and robustness (proactive capacity).

Table 1 Capability and vulnerability factors

Capability factors	Vulnerability factors
Flexibility in sourcing	External factor turbulences
Flexibility in order fulfillment	Deliberate threats from intentional attacks
Reserve capacity	External pressures from innovation or cultural, regulatory, or price changes
Waste elimination efficiency	SC resource limits
Waste and productivity	Sensitivity originating in product and process characteristics and conditions
Technologies and asset visibility	Connectivity originating from the external sources, such as the degree of outsourcing
Adaptability to changes and opportunities	Supplier/customer disruption originating from the susceptibility to the forces of suppliers or customers
Anticipation to predict situations	
Problem recovery	
Assets and decision-making dispersion	
Collaboration from demand forecasting to risk sharing	
Organization as skills and culture	
Market position through product differentiation and relationships	
Cyber and property security	
Financial strength to absorb certain impacts	

Adapted from Pettit et al. (2010)

Despite the concepts being similar, in this chapter, SCRes is a term more directed to the capacity of the SC to respond to disturbances and disruptions, or to adapt to a new reality. Pettit et al. (2010) presented a proactive method for SCRes management based on the principles that forces of change create SCV and management controls create capabilities. Capabilities are defined in this chapter as the attributes that enable a company to anticipate or to overcome an interruption. This method consists in an analysis of capabilities generated in the company for SCV reduction, that is, the more controls and management, the more capabilities are generated in the company and the more SCV is reduced.

Vulnerability factors and capability factors of a SC must be managed to control the SCV. Table 1 presents the factors and Fig. 2 displays the resilience zone to be sought.

In addition to the basic principle that capabilities reduce SCV, Table 1 explains the point stated by the authors that when excessive vulnerability with low capability is present, the risk is high. Excessive capability related to low vulnerability results in lower profits, as many resources are unnecessarily wasted. SC best performance occurs when vulnerability and capability are balanced.

Fig. 2 Resilience zone (Pettit et al. 2010)

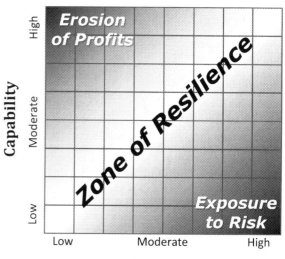

2.2.2 Supply Chain Robustness

Supply Chain Robustness (SCRob) is the characteristic of being capable of resisting a disturbance or interruption and retaining its former state, that is, without offering risks to the SC (Asbjørnslett, 2008; Christopher & Peck, 2004; Tang, 2006).

We define SCRob as the use of strategies to assist the company in reducing costs and improving customer satisfaction in normal circumstances, as well as allowing a company to sustain their operations during or after a disruption. These strategies promote the development of new contingency plans, when disruptions occur, in order to reduce risk exposition (Tang, 2006). Table 2 presents the main strategies for making a more robust SC.

SCRob differs from resilience because it does not have the capacity to adapt and reach a new stable situation (Asbjørnslett, 2008). Therefore, the robustness strategies are related to SC protection, whereas the resilience strategies are related to SC adaptation flexibility. Nevertheless, we perceive that a resilient system can generate a certain degree of robustness to the SC, and a robust system can also generate a degree of resilience. This fact is due to the observation that some strategies encompass these two characteristics, as in the case of the capability factors (Table 1) *flexibility* in sourcing and *reserve capacity*, which, besides resilience strategies, are also part of protection strategies to generate robustness in the SC. Table 2 shows that the following factors are also a part of resilience strategies: strategic postponement of the product differentiation point, flexible supply base, and flexible transportation in terms of modals and routes.

Tang (2006) presented a case where robustness strategies can also make the SC more resilient. Nokia changed the Radio-Frequency Identification (RFID) chip configuration of one of its cell phones to a modular component, enabling acquisition

Table 2 SC robustness strategies

Strategy	Main objective	Definition
Postponement	Increases product flexibility	Utilizes the concepts of product projects or processes such as standardization, modular design, and operation reversion to delay the product's differentiation point
Strategic stock	Increases product availability	Considered Just-in-case, maintains safety stocks of products to guarantee the SC functioning in the case of an interruption
Flexible supply base	Increases supply flexibility	Used to mitigate the risk of demand fluctuations or disruptions associated with a single supply source, in spite of the possible cost reduction arising from purchases of higher volumes
Make-and-buy	Increases supply flexibility	Production of some products that are subject to potential disruptions instead of buying them
Economic supply incentives	Increases product availability	Economically incentivizes certain suppliers to obtain a greater supply base
Flexible transportation	Increases flexibility in transportation	Investment in multi-modal transportation, multiple carriers, and multiple routes
Revenue Management through sales and dynamic pricing	Increases control of product demand	Manages demand when the supply of a product is interrupted or products are perishable
Dynamic assortment planning	Increases control of product demand	The set of products displayed, how they are placed on shelves, and the number of parameters for each one as a way to influence demand
Silent product rollover	Increases control of product exposure to customers	The slow, gradual, and informal insertion of new products into the market to substitute current products

from several suppliers. Thus, based on SCD, Nokia reconfigured the product and purchased the chips from other suppliers, not compromising the SC.

2.3 Vulnerability Measurement and Representation

From a strategic point of view, when the focus of studies is on SCV, the literature points to factors that represent SCV. When the focus of studies is on SCRes, the literature points to adaptation or recovery strategies after a SC's disruption. Yet, when the focus of studies is on robustness, the literature points to protection measures. Wagner and Bode (2006) represented SCV by factors or drivers as the customer dependency and supplier dependency, single source and supply concentration, and global supply.

Dependency is a property of buyer–supplier relationships, when the focal company becomes dependent on the buyer or the supplier in terms of transaction volume, technology, and monopoly, among other factors (Hallikas et al., 2005).

Supply concentration characterizes a scenario where the client company has only a small number of suppliers. Although the literature points out such benefits as reduction in the number of suppliers, and improvement in relationships (Ellram, 1991), these benefits can increase SCV (Choi & Krause, 2006). Thus, companies must reduce the number of suppliers, considering the consequences in terms of exposition to risks (Christopher & Peck, 2004; Elkins et al., 2005), or develop emergency supply sources (Sheffi, 2005).

Globalized Supply is associated with increasing uncertainty, as well as visibility and transparency reduction due to the distancing and stratification of supply sources. Parameters such as suppliers' geographical location, the product purchased, and the transportation mode characterize this driver.

Gallopín (2006) presented a holistic view of SCV as the chain's degree of sensitivity (the degree to which the chain is affected or changed because of disturbances); responsiveness (robustness and/or resilience) to threats (risks); and the SC degree of exposition to disruptive events. Through this framework, Gallopín (2006) related resilience to responsiveness.

To measure SCV, Wagner and Neshat (2010) proposed three categories of SCV drivers: demand side vulnerability, supply side vulnerability, and structure vulnerability.

Demand Side Vulnerability comprises downstream factors of the focal company, and includes, for example, client dependence (volume, financial situation, random demands), the product and its characteristics (complexity and life cycle), and distribution (physical distribution of product).

Supply Side Vulnerability comprises the upstream factors of the focal company, which includes supplier–supplier relationships, supply base complexity, suppliers' structure, financial instability, bankruptcy, and technological updating (Zsidisin & Ellram, 2003).

Structure Vulnerability refers to the degree of disintegration, complexity, and globalization of activities, including offshoring. This category lacks other factors, such as collaborative management, contingency plans and structures, process, inventory or quality controls, flexibility of the process and/or product structures, and management improvement.

Wagner and Neshat (2010) presented a method that aims to assist SC managers in managing SCV. The method takes a qualitative approach where it identifies vulnerability drivers, quantifies them and thus generates the Index of Supply Chain Vulnerability (ISCV). In addition, Wagner and Neshat (2010) presented three categories of drivers: client dependence and supplier dependence, single source and supply concentration, and globalized supply. The method consists of drawing key people to the organizations and having them identify and score such drivers, therefore generating the ISCV.

The methodology proposed by Wagner and Neshat (2010) proved to be consistent in the case study, but this methodology did not induce participants to think of external risk sources or the fragilities of SC management. The authors admitted some of the model's limitations, such as the dependence on gathering data, the dependence on experts to define the scores, and the possible lack of analysis of vulnerability over time and its consequences due to the nature of the data, among other issues. They recommended the use of initiatives, practices, and collaborative tools to mitigate the company's vulnerability, since in collaborative works, companies can preventively identify points of improvement and enhance their flexibility so as to face possible risks. Such a hypothesis is defended by some authors through the idea that managers must attenuate vulnerability proactively and in collaboration with other members of the SC (Bogataj & Bogataj, 2007; Hendricks & Singhal, 2005b; Wagner & Neshat, 2010).

2.4 SCV and Supply Chain Coordination (SCC)

A coordination structure can be defined as a pattern or model of decision-making and communication between actors performing tasks in order to achieve defined goals (Baligh & Burton, 1981; Baligh & Damon, 1980). Malone (1987) presented the concept of vulnerability as the third type of cost that affects coordination structures between companies and markets, and presented the four main structures: product hierarchy, process hierarchy, centralized markets, and decentralized markets. Regarding these structures, Malone (1987) analyzed the incidence of production costs, coordination, and vulnerability. For SCs, coordination structures present a higher complexity because of the greater number of companies, pieces of information, and products. The objectives in supply chains are also varied, which could be the effectiveness of customer service in terms of time and product availability or the efficiency in resource usage throughout the SC (Fisher, 1997).

Production costs are composed of the production capacity costs and the costs of processing delayed tasks, also known as transaction costs. Coordination costs are costs of maintaining communication channels among actors and the costs for message exchange along these links. However, vulnerability costs are the unavoidable costs of a situation change that affects the organization before it can adapt to this new situation and the costs for the company to rapidly adapt to market changes.

Malone (1987) explained that in the hierarchy structure named Product Hierarchy, the vulnerability costs are higher due to the nonexistence of similar resources to perform the same task when a failure occurs. This does not occur in the production structure named Functional Hierarchy, because when the failure occurs, it may be transferred to another resource.

This statement from Malone (1987) contradicts the idea proposed by the Toyota Production System (TPS) (Ohno, 1988), which seeks to build an efficient and flexible lean structure. To this end, the reliability of processes is developed in a preventive way and through a collaborative structure, in order to repair the failure as quickly as possible. In addition, through the TPS, due to task simplification and line/cell configuration, coordination costs are lower than in a departmentalized structure, which benefits from production efficiency through specialization and task repetition. However, there is a need for a trade-off, wherein process resources should not be reduced without a risk evaluation and a stable environment and process.

2.4.1 SCV and SC Efficiency

The search for SC efficiency has been the dominant business model in the last few decades (Cranfield, 2002). The results from improvement works led to a more competitive SC, by means of the reduction of intermediary inventory and total costs, and a faster pace of operations information and inventory (Cooper & Ellram, 1993; Poirier & Reiter, 1996). However, the improvement initiatives can lead to an increase in vulnerability and reliability if the risk factors are not considered. Inventory reductions can cause a supply interruption and have an impact on the SC. The vulnerability is increased partly due to internal SC failures, but mainly due to a rise in external phenomena, such as the Kobe earthquake that affected SCs around the world, the terrorist attack of September 11th, 2001 in the USA, and the 1997 fire at a Toyota supplier that forced the company to paralyze its operations (Cranfield, 2002). In addition, the Great East Japan Earthquake and Tsunami in 2011 is also worth mentioning, as it significantly disrupted the global supply chain.

With a greater focus on customers, after 1980, several methods were developed for a quicker response to the market, such as Just-in-time (JIT), Vendor Managed Inventory (VMI), and Continuous Replenishment (Herron, 1987; Schwarz & Weng, 2000; Waller et al., 1999; Zinn & Charnes, 2005). Later, the need for cost reduction and efficiency improvement generated a movement in the industry whereby opportunities were more focused on reducing inventories (Cranfield, 2002). The JIT concept was widely adopted and organizations became more supplier dependent. This model, despite its merit in terms of stable demands, can become less viable when demand volatility increases.

The decade of the 1990s was characterized by globalization and continuous cost reduction, with emphasis and focus on the TPS, disseminated as *lean manufacturing*: a systematic approach to identifying and eliminating waste (activities that do not add value). The "lean movement" generated different, more holistic approaches to SC mapping and evaluation to support its implementation (Jones & Womack, 2002; Rother & Shook, 1999).

Nevertheless, the search for leaner operations and more aggressive results causes companies to "create tension," reducing resources and ignoring risks. This practice results in more vulnerability in companies when it comes to not considering the process fragility. Christopher and Rutherford (2004) recommend avoiding leaning down too far, that is, going too far in the downswing, and they recommend including the cost of failure recovery in the total cost equation to identify an appropriate leanness level.

With few differences from Lean Manufacturing, the TPS focuses on strengthening bases such as process stabilization and standardization to enable safer and more sustainable resource reduction (waste) (Ohno, 1988). The TPS also acts in a resilient way, since it defends the identification of potential problems and preventive action, mainly through mistake proofing devices (Shingo, 1986), and, by using problems, it develops capabilities to repair the system as soon as possible. From the solving of problems come "lessons learned," which serve as inputs to other preventive actions.

Current literature about Lean Manufacturing does not clearly address a SC improvement by considering the risks originating from premature downsizing. Thus, the need for cost reduction and efficiency can make the SC more vulnerable if risks of a drastic change in products or processes are not taken into consideration.

2.5 SC Collaboration

Collaboration in Supply Chain Management (SCM) is a means through which companies work in an integrated way with common objectives, characterized by the significant exchange of information and technologies. Companies share the responsibilities of planning, management, execution, and monitoring the performance, as well as the risks and even the profit (Mentzer, 2001). For the development of the collaboration process, common interests among companies, transparency at work, mutual help, and defined objectives are necessary. Leadership, clear expectations, cooperation, confidence, benefits, and technology sharing are fundamental (Humphreys et al., 2001; Mentzer, 2001).

In the collaborative process, factors such as risk sharing, evaluation, monitoring, and management have rarely been approached and disseminated in the literature (Juttner et al., 2003; Mentzer, 2001). In the literature, we can find collaborative methods with distinct characteristics, applications, objectives, and limitations. Generally, collaborative methods, especially those related to SCC, have the dynamic of evaluating the global functioning of SC, its limitations, and objectives in order to make decisions to meet the demand. In this systemic and collaborative activity, it is possible to identify specific risk sources in places that can compromise the final result of the SC. The finding on which the research hypothesis is based will verify the contribution of collaborative methods to the identification of sources of risks and threats, and thus help people take actions to mitigate the SCV.

The CPFR is characterized as a collaborative method with a more holistic, structured, and collaborative approach than other methods, because it encourages the SC partners to constantly exchange information regarding their demand predictions and, through a methodical form, reach a consensus on planning considering the SC limitations (Ramanathan, 2012; Skjoett-Larsen et al., 2003).

3 Collaborative Planning, Forecasting, and Replenishment (CPFR)

CPFR is a method of collaborative management with a focus on SC integration, aiming to increase its efficiency. CPFR focuses mainly on the conjoint elaboration of sales forecasting and the replenishment of the items involved, considering existing SC limitations, either from industrial suppliers, retail clients, or the distributor (Mentzer, 2001). The improvement of CPFR happens continuously, with a necessity to experiment, analyze, innovate, and experiment again. The importance of collaboration in several aspects, both industry and retail, becomes the main CPFR advantage (Mentzer, 2001).

The CPFR functioning structure originally featured three levels. For example, in the retailing sector, the retailer acts as the buyer, a manufacturer acts as a vendor, and the customer is the final client. In industry, a manufacturer of original parts acts as the buyer and as the vendor to the final client, while suppliers act as vendors (VICS, 2004). See VICS (2004) for the framework that involves CPFR participants and activities.

The CPFR implementation process starts with the initial agreement of collaboration and intentions, then proceeds to the creation of a joint business plan in order to identify objectives. Afterwards, the CPFR team creates systematic sale forecasting and decision-making, where exceptions to the sales forecasting are created and solved by the CPFR team. After this macro planning, there is the order forecasting, identifying and solving exceptions. Orders are generated and system feedback provided to make improvements. During each stage, the coordinator defines the participants and the periodicity of interactions (Seifert, 2002).

The main point for understanding the benefits of CPFR is the internal team's commitment. By mapping the potential benefits for the company's priorities, such knowledge will aid the commitment of the parties involved. Table 3 shows the two groups that benefit from CPFR (Fliedner, 2003): Demand related and supply related.

CPFR considers four collaborative activities (phases) composed of strategy and planning; demand and supply management; execution; analysis (VICS, 2004). The strategy and planning phase refers to the establishment of rules for a collaborative relationship, as well as establishing the product mix, planning, and controls needed for a given period. The demand and supply management phase refers to the demand forecasting and the guarantee of product shipment and on-time delivery. The execution phase refers to the activities of placing orders, preparing and dispatching, receiving and stocking products at retail, recording sales transactions and making payments. The analysis phase refers to the activities of planning and the execution of exceptions. It also includes consolidating results, analyzing performance, sharing ideas, and adjusting plans for a continuous improvement of results.

The CPFR level of implementation and improvement is measured through a roadmap: Collaborative Processes; Integrated Demand Planning and Forecasting; Replenishment Processes; and SC Management (VICS, 2004).

Table 3 Benefits of CPFR

	Benefit	Definition
Demand related	Closer relationship	The existing relationship between the members of the SC is strengthened. Buyer and vendor work together through meetings from the beginning of CPFR
	Increased sales	Increased sales due to the closer collaboration needed for CPFR implementation, which results in a better business plan between buyer and vendor
	Management	Partners present and analyze the current state in view of the targets, and thus, they are able to take the appropriate actions to maintain SC functioning and to guarantee the supply to the consumer
	Better product offering	Partners perform an evaluation to assess opportunities for additional products
Supply related	Forecasting precision	Supply order forecasting is faster and presents additional information; more time available for production planning and forecasting improvement
	Inventory reduction	Forecasting uncertainty and process inefficiencies are reduced. The product can be produced via make to order instead of make to stock, based on historical forecasting, and companies can maintain additional stocks due to errors in prediction regarding lack of capability
	Improved technology	Investments in technology for internal integration can be perceived through the higher quality of forecasting information. Companies benefit from the precise and rapidly available data
	Improved return on investment	Since process flow is improved, return on investment with CPFR is significant
	Increased customer satisfaction	With more precise forecasting results and stocks results, and faster access to information, store service will be more consistent, providing consumers with greater satisfaction

Collaborative processes are composed of joint business plans, sales or new items initiatives, and result evaluation process. Integrated planning and demand forecasting is composed of information technology usage, development of the demand flag, internal integration of demand forecasting (vendor), and internal integration of demand forecasting (buyer). The replenishment process is composed of delivery reliability process (producer to client), delivery reliability process (shopkeeper's inventory), efficient receipt process, reliability and conformity of the retail process, and automated purchase order. The replenishment process consists of delivery reliability process (producer to customer), delivery reliability process (retailer's inventory), efficient receipt process, retail process reliability and compliance, automated purchase order, replenishment processes, product flow (stock/warehouse), and product flow (internal shopkeeper to shelf). SCM is composed of partnerships and trust relationships, process reengineering, operational strategy (service level and inventories), measurement/benefits, activities in the defrayal process, and product availability to consumers.

4 Methodology and Results

In order to achieve the research objectives, the first step is to create a framework for SCV representation. This framework's objective is to show elements that built on the concept of SCV and its interaction with other elements, as shown in Fig. 1. The second step is the identification of CPFR capabilities and the comparison with SCV elements and related elements. Therefore, it is possible to identify how and where CPFR can help in mitigating SCV.

4.1 Supply Chain Vulnerability (SCV) Framework

Aiming to better understand and represent SCV, a framework was developed that included some elements, factors, and interdependencies. In addition, risk management and the deployment of the concept of SCV were added. Figure 3 shows the developed framework. Dashed arrows represent actions and continuous arrows represent the effects.

Following the Gallopín's research (2006), where vulnerability is understood as the degree of sensitivity of the chain (the degree to which the chain is affected or altered by disturbances), its responsiveness (robustness and/or resilience) to threats

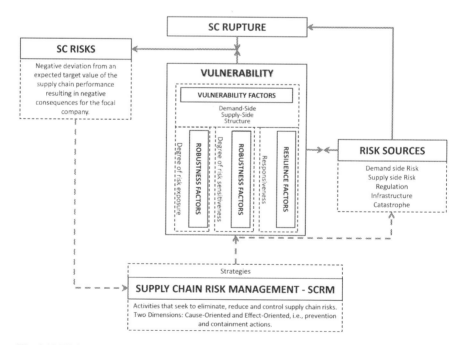

Fig. 3 SCV framework

(risks) and degree of exposition to disturbing events, these three elements were adopted as factors representing SCV. The responsiveness factor was directed to the resilience condition of the SC. The factors Degree of SC Sensitivity and Degree of Exposition to disturbing events of the SC were directed to the condition of SC robustness.

These three factors form SCV and are used to identify vulnerability drivers in three SC categories—Demand Side Vulnerability, Supply Side Vulnerability, and Structure Vulnerability, as defined in Sect. 2.3 of this chapter. The drivers of each category depend on each SC, as well as on the vulnerability factors.

The SCRM was inserted into the proposed framework (Fig. 3) because it is an element that is interrelated with SCV and the other aspects. The SCRM receives inputs from risks and problems occurring in the SC, and preventively receives inputs from the internal activities of identifying potential risk sources. By receiving these inputs, actions are generated to reduce SCV, increasing SCRes or SCRob and mitigating possible risk sources. An example of actions or strategies arising from risk management can be given from a SCD where some stores failed to receive the products for sale on a commemorative day. The interruption was caused by a delay in leather processing, due to an imported chemical that was out of stock, and the emergency import was delayed due to port inspections. Corrective actions could be product regularization and an agreement with the supplier for a contingency inventory. Preventive actions for this problem could be the development of alternative suppliers and raw materials. Other preventive actions could arise, such as anticipating the production of the collection and anticipating orders, among others.

SCRS generate disturbances that can create SCD. These disruptions can generate SC risk, depending on its vulnerability. SCRS that generate SCD, such as for example a natural disaster of any kind, may not compromise the result of the SC if it is robust or if it is able to rapidly adapt, thus avoiding SC risk.

4.2 Analysis of CPFR's Contribution to SCV Mitigation

Besides contributing to a better comprehension of SCV, this chapter's objective is to verify the hypothesis of using collaborative methods to mitigate SCV. In other words, the aim is to establish which elements of SCV can be reduced by collaborative methods. CPFR was adopted as the collaborative method to test the hypothesis of its contribution to SCV mitigation. CPFR was chosen because it presents higher and more comprehensive SCC and collaboration characteristics than other methods (Fliedner, 2003).

The first analysis was the theoretical comparison of CPFR (capabilities) acting scope against the SCV factors in the framework of Fig. 3. The CPFR capabilities (Sect. 3) were theoretically analyzed and the capabilities that could mitigate SCV factors were identified. Table 4 presents this analysis.

As seen in Table 4, some CPFR capabilities or activities can contribute to the mitigation of SCV factors by being addressed at certain stages of CPFR.

Table 4 CPFR Capabilities × SC Vulnerability Factors

CPFR Capabilities \ SC Vulnerability Factors	Responsiveness	Degree of risk sensitivity	Degree of risk exposure
Collaborative Processes	x	x	x
Integrated Planning and Demand Forecasting	x		
Replenishment Processes		x	
SC Management	x		

Collaborative processes are a CPFR capability that consists of the development of joint business and operational plans and the process of results evaluation (the business plan commonly includes the risks). Therefore, this capability contributes to mitigating the three vulnerability factors. Integrated Demand Planning and Forecasting Capability can contribute to the responsiveness factor, for example, by using IT and demand flags in addition to the demand forecasting integration between buyers and sellers. The Replenishment Processes capability can contribute to the vulnerability factor of Degree of Risk Sensitivity, once it includes processes focused on the delivery reliability. SC management can identify the responsiveness factor once it encompasses the relationship of trust and commitment between companies, the improvement of processes, and the management of the operational process focusing on the level of service and inventories.

The second theoretical analysis was the verification of compatibility between CPFR and capability factors (shown in Table 1) presented by Pettit et al. (2010) as SCV mitigating factors, improving SC robustness and resilience. Table 5 shows the second theoretical analysis, pointing out the capability factors that can be part of the CPFR acting scope. This analysis highlights the compatibility of CPFR with the capability factors of Collaboration, Flexibility, Integration, Speed, and Redundancy.

From the analysis shown in Table 5, a schematic representation of the CPFR contribution to mitigating SCV was added into the framework of Fig. 3. Figure 4 shows the new framework with the CPFR contribution, that is, the topics in italics are the factors to which CPFR can contribute, according to the analysis.

Table 5 Capability factors × CPFR compatibility

Capability factor	Description	CPFR compatibility
Collaboration	Capability of working effectively with suppliers and clients in the search for mutual benefits to the SC (Collaborative forecasting, client management, communications, order postponement, product life cycle management, risk sharing)	X
Culture	Creation of an organizational culture for risk and vulnerability management; decision autonomy and problem solving in operational and strategic levels	
Density	SC density denotes the geographical space within the chain. Density is inversely proportional to the geographical spacing	
Flexibility	"Being able to easily bend without breaking." Flexibility ensures that the alterations caused by the event of risk can be absorbed by the supply chain through effective responses	X
Integration	Emphasis on the importance of the interaction of logistics aspects, upstream and downstream of the SC. It includes order interaction, inventories, transportation, and distribution to facilitate SC transparency	X
Agility	It means "speed of movement, action or operation, fastness and agility" and it is defined as the capability of the SC to respond to unpredicted changes from the environment (risks)	X
Visibility	It comprises the awareness of the SC's operational assets (identity, structure, and localization) and the environmental assets (identification of the risk and vulnerability sources)	
Redundancy	It is the concept of maintaining spare resources to be used in the case of an interruption, focusing on reliability and delivery conformity	X

5 Discussion of the Results

It can be verified that the proposed hypothesis can be accepted, that is, CPFR can contribute, preventively or correctively, to the activities of risk management, to increasing the capabilities of resilience and robustness and to mitigating sources of risk in SCs.

CPFR was found to assist in the mitigation of SCV from two perspectives. The first is through collaborative activities, which allow for discussion, identification, generation of action plans, and monitoring of possible sources of risks as well as the possible appearance of preventive actions. The second is through activities that increase the SC's capabilities, that is, by strengthening the SC's characteristic of resilience.

The contribution to the SCV reduction is restricted to the CPFR extension in the SC: if the collaboration is between the focal company and the client, the contribution to mitigating SCV and sources of risk is restricted to this scope. Because the collaborative methods have different characteristics, it was not possible to study the contribution of other collaborative methods, such as VMI, to the mitigation of SCV.

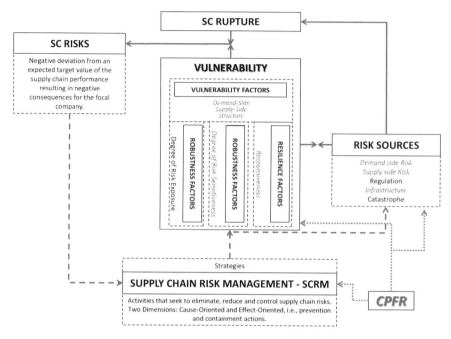

Fig. 4 Framework of the CPFR contribution to mitigating SCV

Increasing the capabilities of a SC to mitigate SCV should be evaluated through a cost–benefit trade-off. Just as with the search for increased efficiency and cost reduction, whether these actions can or cannot make the SC more vulnerable should also be assessed. According to Tang (2006), there is a challenge regarding the costs and the benefits related to the robustness of a SC, that is, companies invest in robustness strategies as a "guarantee" for SC competitiveness, but there exists an associated cost that is hard to measure.

The present research neither analyzed or detailed in which stages of the CPFR the activities of SCV mitigation could be inserted nor how could they be executed.

6 Conclusions and Recommendations for Future Research

The objective of this chapter was to contribute to the understanding of SCV through proposing a framework and to present the hypothesis that the use of collaborative methods such as CPFR can mitigate elements that increase SCV. For the development of the research, a SCV framework was created in order to understand its components and their relationships with other elements. This framework enabled the consolidation of several concepts and elements that constitute SCV. Through a literature review and the subsequent creation of the framework, it was possible to understand the complexity and to clarify the concept of SCV, as well as to better comprehend the elements that constitute and are related to SCV. From this framework, it was possible to evaluate, according to the methodological procedures used, the contribution of CPFR to the mitigation of SCV. The SCV framework created can contribute, in practical and theoretical terms, to future discussions or case studies on the subject.

The research results showed the contribution of CPFR from two perspectives: discussion method, identification, generation of action plans and monitoring of possible sources of risk, and the creation of capabilities to increase SCRes. These results showed that the SCC developed through CPFR incorporates, in a structured and systemic way, elements that can lead to the mitigation of SCV. However, it is understood that this research must be applied in a real environment to validate the theoretical evidence found.

Another topic of discussion raised by this research was the possibility of increasing SCV caused by continuous improvement activities. The continuous identification and reduction of waste through Lean Manufacturing can increase sources of risks if these issues are not considered.

The use of collaborative methods, such as CPFR, are not practices for managing vulnerability or risks sources in the SC, but rather they are practices that can help in mitigating SCV. CPFR does not include specific activities for identifying risks sources, despite the possibility of assisting in the mitigation of SCV. Therefore, a possibility and suggestion for future research is the inclusion of risk analysis activities in the SCC, through CPFR. This may allow for more efficient coordination in terms of results.

The results of SCV mitigation depend on the extent to which CPFR is applied in the SC. Also, it is not possible to conclude that the verified contribution of CPFR will be the same for other SCC methods, and thus analysis of the specific contribution of each method is required, which can be done through the method used in the present research.

A relevant point in the theoretical background was the necessity for a trade-off between the increase in robustness, resilience, and SCV, because when too many preventive and corrective actions are generated, the cost of the SC rises so that it may no longer to be profitable and competitive. This topic could be further explored in future research. Another subject that may be explored in further research is the need for a risk assessment when improvements for efficiency increase are made. In other words, the degree of SC "Lean downsizing" should be analyzed in terms of its future sustainability and failure to generate new SCVs.

References

Ahlquist, G., Irwin, G., Knott, D., & Allen, K. (2003). Enterprise resilience. *Best's Review, 104*(3), 88.

Asbjørnslett, B. (2008). Assessing the vulnerability of supply chains. In G. A. Zsidisin & B. Ritchie (Eds.), *Supply chain risk: A handbook of assessment, management and performance* (pp. 15–33). Springer.

Baligh, H. H., & Burton, R. M. (1981). Describing and designing organizational structures and processes. *International Journal of Policy Analysis and Information Systems, 5*, 251–266.

Baligh, H. H., & Damon, W. W. (1980). Foundations for a systematic process of organization structure design. *Journal of Information and Optimization Sciences, 1*, 133–165.

Bhamra, R., Dani, S., & Burnard, K. (2011). Resilience: the concept, a literature review and future directions. *International Journal of Production Research, 49*(18), 5375–5393.

Bogataj, D., & Bogataj, M. (2007). Measuring the supply chain risk and vulnerability in frequency space. *International Journal of Production Economics, 108*(1–2), 291–301.

Choi, T. Y., & Krause, D. R. (2006). The supply base and its complexity: implications for transaction costs, risks, responsiveness, and innovation. *Journal of Operations Management, 24*(5), 637–652.

Chopra, S., & Sodhi, M. S. (2004). Managing risk to avoid supply-chain breakdown. *Sloan Management Review, 46*(1), 53–61.

Christopher, M., & Lee, H. L. (2004). Mitigating supply chain risk through improved confidence. *International Journal of Physical Distribution & Logistics Management, 34*(5), 388–396.

Christopher, M., & Peck, H. (2004). Building the resilient supply chain. *The International Journal of Logistics Management, 15*(2), 1–13.

Christopher, M., & Rutherford, C. (2004). Creating supply chain resilience through agile six sigma. *Critical Eye*, Jun–Aug, 24–28.

Coleman, L. (2006). Frequency of man-made disasters in the 20th century. *Journal of Contingencies and Crisis Management, 14*(1), 3–11.

Cooper, M., & Ellram, L. M. (1993). Characteristics of supply chain management and the implications for purchasing and logistics strategy. *The International Journal of Logistics Management, 4*(2), 13–24.

Cranfield. (2002). *Supply chain vulnerability: executive report, school of business*. Cranfield University.

Elkins, D., Handfield, R. B., Blackhurst, J., & Craighead, C. W. (2005). 18 ways to guard against disruption. *Supply Chain Management Review, 9*(1), 46–53.

Ellram, L. M. (1991). A managerial guideline for the development and implementation of purchasing partnerships. *International Journal of Purchasing and Materials Management, 27*(3), 2–8.

Fabbe-Costes, N., & Jahre, M. (2007). Supply chain integration improves performance: the Emperor's new suit? *International Journal of Physical Distribution & Logistics Management, 37*(10), 835–855.

Fiksel, J. (2006). Sustainability and resilience: toward a systems approach. *Sustainability: Science, Practice, & Policy, 2*(2), 1–8.

Fisher, M. L. (1997). What is the right supply chain for your product? *Harvard Business Review*, March-April, 105–116.

Fisher, M. L. (2007). Strengthening the empirical base of operations management. *Manufacturing & Service Operations Management, Hanover, 9*(4), 368–382.

Fliedner, G. (2003). CPFR: An emerging supply chain tool. *Industrial Management & Data Systems, Bingley*, 14–21.

Gallopín, G. C. (2006). Linkages between vulnerability, resilience, and adaptive capacity. *Global Environmental Change, 16*, 293–303.

Hallikas, J., Puumalainen, K., Vesterinen, T., & Virolainen, V. M. (2005). Risk-based classification of supplier relationships. *Journal of Purchasing and Supply Management, 11*(2–3), 72–82.

Harland, C., Brenchley, R., & Walker, H. (2003). Risk in supply networks. *Journal of Purchasing and Supply Management, 9*(2), 51–62.

Hendricks, K. B., & Singhal, V. R. (2005a). An empirical analysis of the effect of supply chain disruption on long-run stock price performance and equity risk of the firm. *Production and Operations Management, 14*(1), 35–52.

Hendricks, K. B., & Singhal, V. R. (2005b). Association between supply chain glitches and operating performance. *Management Science, 51*(5), 695–711.

Herron, D. P. (1987). Integrated inventory management. *Journal of Business Logistics, 8*(1), 96–116.
Humphreys, P. K., Lai, M. K., & Sculli, D. (2001). An inter-organizational information system for supply chain management. *International Journal of Production Economics, 70*(3), 245–255.
Ivanov, D., & Sokolov, B. (2013). Control and system-theoretic identification of the supply chain dynamics for planning, analysis and adaptation of performance under uncertainty. *European Journal of Operational Research, 224*(2), 313–323.
Jones, D. T., & Womack, J. (2002). *Seeing the whole—Mapping the extended value stream* (pp. 1–100). The Lean Enterprise Institute.
Juttner, U. (2005). Supply chain risk management: understanding the business requirements from a practitioner perspective. *The International Journal of Logistics Management, 16*(1), 120–141.
Juttner, U., Peck, H., & Christopher, M. (2003). Supply chain risk management: outlining an agenda for future research. *International Journal of Logistics: Research and Applications, 6*(4), 197–210.
Khojasteh, Y. (2018). developing supply chain risk mitigation strategies. In Y. Khojasteh (Ed.), *Supply chain risk management*. Springer. https://doi.org/10.1007/978-981-10-4106-8_6
Khojasteh, Y., & Abdi, M. R. (2016). Japanese supply chain management. In P. Haghirian (Ed.), *Routledge handbook of Japanese business and management*. Routledge.
Kleindorfer, P. R., & Saad, G. H. (2005). Managing disruption risks in supply chains. *Production and Operations Management, 14*(1), 53–68.
Malone, T. W. (1987). Modeling coordination in organizations and markets. *Management Science, 33*(10), 1317–1332.
March, J. G., & Shapira, Z. (1987). Managerial perspectives on risk and risk taking. *Management Science, 33*(11), 1404–1418.
Marucheck, A. S., Greis, N., Mena, C., & Cai, L. (2011). Product safety and security in the global supply chain: issues, challenges and research opportunities. *Journal of Operations Management, 27*(7), 707–720.
Matopoulos, A., Vlachopoulou, M., Manthou, V., & Manos, B. (2007). A conceptual framework for supply chain collaboration: empirical evidence from the agri-food industry. *Supply Chain Management: An International Journal, 12*(3), 177–186.
Melnyk, S. A. (2007). Lean to a fault? *Council of Supply Chain Management Professionals Supply Chain Quarterly, 3*, 29–33.
Mentzer, J. (2001). Managing supply chain collaboration. In J. Mentzer (Ed.), *Supply chain management* (pp. 83–84). Sage.
Nagurney, A., Cruz, J., Dong, J., & Zhang, D. (2005). Supply chain networks, electronic commerce, and supply side and demand side risk. *European Journal of Operational Research, 164*(1), 120–142.
Norrman, A., & Lindroth, R. (2004). Categorization of supply chain risk and risk management. In C. Brindley (Ed.), *Supply chain risk* (pp. 14–27). Ashgate.
Ohno, T. (1988). *Toyota production system: beyond large-scale production*. Productivity Press.
Pettit, T. J., Fiksel, J., & Croxton, K. L. (2010). Ensuring supply chain resilience: development of a conceptual framework. *Journal of Business Logistics, 31*(1), 1–21.
Pickett, C. (2006). Prepare for supply chain disruptions before they hit. *Logistics Today, 47*(6), 22–25.
Poirier, C. C., & Reiter, S. E. (1996). *Supply chain optimization*. Berrett-Koehler.
Ponomarov, S. Y., & Holcomb, M. C. (2009). Understanding the concept of supply chain resilience. *The International Journal of Logistics Management, 20*(1), 124–143.
Ramanathan, U. (2012). Supply chain collaboration for improved forecast accuracy of promotional sales. *International Journal of Operations & Production Management, 32*(6), 676–695.
Rother, M., & Shook, J. (1999). *Learning to see*. The Lean Enterprise Institute.
Schwarz, L. B., & Weng, Z. K. (2000). The design of a JIT supply chain: the effect of lead-time uncertainty on safety stock. *Journal of Business Logistics, 21*(2), 231–253.

Seifert, D. (2002). *Collaborative Planning Forecasting and Replenishment: How to create a supply chain advantage.* Galileo Business.

Sheffi, Y. (2005). *The resilient enterprise: Overcoming vulnerability for competitive advantage.* MIT Press.

Sheffi, Y., & Rice, J. B., Jr. (2005). A supply chain view of the resilient enterprise. *MIT Sloan Management Review, 47*(1), 40–49.

Shingo, S. (1986). *Zero quality control: source inspection and the poka-yoke system.* Productivity Press.

Singhal, V. R., & Hendricks, K. B. (2002). How supply chain glitches torpedo shareholder value. *Supply Chain Management Review, 6*(1), 18–24.

Skjoett-Larsen, T., Thernoe, C., & Andresen, C. (2003). Supply chain collaboration: theoretical perspectives and empirical evidence. *International Journal of Physical Distribution & Logistics Management, 33*(6), 531–549.

Slone, R. E., Mentzer, J. T., & Dittmann, J. P. (2007). Are you the weakest link in your company's supply chain? *Harvard Business Review, 85*(9), 116–127.

Spekman, R. E., & Davis, E. W. (2004). Risky business: expanding the discussion on risk and the extended enterprise. *International Journal of Physical Distribution & Logistics Management, 34*(5), 414–433.

Spiegler, V. L. M., Naim, M., & Wikner, J. (2012). A control engineering approach to the assessment of supply chain resilience. *International Journal of Production Research, 50*(21), 6162–6187.

Svensson, G. (2000). A conceptual framework for the analysis of vulnerability in supply chains. *International Journal of Physical Distribution & Logistics Management, 30*(9), 731–749.

Svensson, G. (2002). A conceptual framework of vulnerability in firms' inbound and outbound logistics flows. *International Journal of Physical Distribution & Logistics Management, 32*(2), 110–134.

Tang, C. S. (2006). Robust strategies for mitigating supply chain disruptions. *International Journal of Logistics: Research and Applications, 9*(1), 33–45.

VICS. (2004). CPFR: an overview. VICS. Available at: https://www.gs1us.org/DesktopModules/Bring2mind/DMX/Download.aspx?Command=Core_Download&EntryId=492#:~:text=Collaborative%20Planning%2C%20Forecasting%20and%20Replenishment%20(CPFR%C2%AE)%20is%20a,and%20fulfillment%20of%20customer%20demand.&text=CPFR%20has%20also%20influenced%20industry,consumer%20packaged%20goods%20(CPG). Accessed 27 Dec 2020.

Viswanadham, N., & Gaonkar, R. S. (2008). Risk management in global supply chain networks. In C. S. Tang (Ed.), *Supply chain analysis.* Springer.

Vlajic, J., Van Der Vorst, J. G. A. J., & Haijema, R. (2012). A framework for designing robust food supply chains. *International Journal of Production Economics, 137*(1), 176–189.

Wagner, S. M., & Bode, C. (2006). An empirical investigation into supply chain vulnerability. *Journal of Purchasing & Supply Management, 12*(6), 301–312.

Wagner, S. M., & Bode, C. (2009). Dominant risks and risk management practices in supply chains. In G. A. Zsidisin & B. Ritchie (Eds.), *Supply chain risk: A handbook of assessment, management and performance* (pp. 271–290). Springer.

Wagner, S. M., & Neshat, N. (2010). Assessing the vulnerability of supply chains using graph theory. *International Journal of Production Economics, 126*, 121–129.

Waller, M., Johnson, M. E., & Davis, T. (1999). Vendor-managed inventory in the retail supply chain. *Journal of Business Logistics, 20*(1), 183–203.

Williamson, O. E. (1985). *The economic institutions of capitalism: Firms, markets, relational contracting*. Free Press.

Zinn, W., & Charnes, J. M. (2005). A comparison of the economic order quantity and quick response inventory replenishment methods. *Journal of Business Logistics, 26*(2), 119–141.

Zsidisin, G. A., & Ellram, L. M. (2003). An agency theory investigation of supply risk management. *Journal of Supply Chain Management, 39*(3), 15–27.

Zsidisin, G. A., Melnyk, S. A., & Ragatz, G. L. (2005). An institutional theory perspective of business continuity planning for purchasing and supply management. *International Journal of Production Research, 43*(16), 3401–3420.

The Boom and Bust of Medical Supplies During the COVID-19 Pandemic

Tung Nhu Nguyen

Abstract Hoarding and phantom orders reflect negative buying behavior in times of crisis, such as after a natural disaster. The COVID-19 outbreak, which began in late 2019 and spread worldwide, witnessed a boom in essential products. This chapter aims to analyze the unexpected surge in demand for medical supplies through the lens of behavioral supply chain management. Using the face mask for illustration, this study illustrates its demand and supply during the COVID-19 outbreak. Despite different opinions on the face-mask mandate, their demand has soared during the pandemic in many countries. We review how hoarding behavior influences the stock management model and proposes collaboration among all actors in a supply chain to address the hoarding problem. The chapter contributes to a comprehensive and insightful analysis of strategies to resolve the boom and the bust of medical supplies during the pandemic.

1 Introduction

Sometimes customers buy more than they need in times of crisis, such as in wartime or after disasters (Sterman & Dogan, 2015). For example, people bought too much gasoline after the Sandy SuperStorm in the USA in 2012. People also rushed for the antibiotic ciprofloxacin during the anthrax attack in 2001 or flu vaccines in 2004 due to the flu outbreak (Sterman & Dogan, 2015).

A similar soaring demand phenomenon occurred during the COVID-19 outbreak, which began in late 2019 and spread worldwide. During a disease outbreak, products in shortage include disinfecting products (e.g., sanitizer, bleach), personal hygiene products (e.g., toilet paper, paper towels), food (e.g., eggs, canned foods), and those products necessary during lockdowns (e.g., flour, yeast) (Meyersohn, 2020). For instance, during the pandemic in India in 2021, the demand for liquid medical

T. N. Nguyen (✉)
International University – Vietnam National University, Ho Chi Minh City, Vietnam
e-mail: nntung@hcmiu.edu.vn

© The Author(s), under exclusive license to Springer Nature Switzerland AG 2022
Y. Khojasteh et al. (eds.), *Supply Chain Risk Mitigation*, International Series in Operations Research & Management Science 332,
https://doi.org/10.1007/978-3-031-09183-4_6

oxygen jumped up to 800% in this Asian country, causing a severe shortage (BusinessToday, 2021).

As a consequence of supply shortages and even price gouging, the selling prices of essential products abruptly increased. For example, third-party sellers of hand sanitizers charged approximately six times higher than the regular price of this disinfecting product on the Amazon website (Terlep, 2020). Worse, when health products (e.g., face masks, ventilators, oxygen) are out of stock, people in need cannot acquire them for personal health protection. This fathom demand also causes economic consequences for manufacturers. Suppliers increase production capacity for the product in shortage to match supply with demand. Alternatively, they repurpose their routine production to the needed product line, as in the case of General Electrics for ventilator manufacture under the Defense Production Act (Wayland, 2020).

The above phenomenon creates a puzzle for manufacturers who have to choose between increasing or not increasing the production rate. In the immediate run, expanding the production of an essential product would meet its soaring demand during an emergency scenario. On the other hand, if the shortage is phantom due to the bulk purchase or hoarding behavior, it conveys a wrong message on the soaring demand for the product to the manufacturer. However, when the outbreak is over, the need for the product falls, and there is the phenomenon of oversupply.

This chapter investigates the causes of hoarding behavior toward essential products during the COVID-19 pandemic and explores how this behavior affects organizational decisions on the production rate. Analyses in this chapter are helpful for suppliers in identifying behavioral elements that may inflate demand and how they influence rational decisions on the production rate. Integrating behavioral factors into supply chain decision-making and performance evaluation is one of the current and future research directions because this contributes to the clarification of concepts behind the realignment of supply chains (Cohen & Kouvelis, 2021).

The chapter contributes to a comprehensive and insightful analysis of strategies to resolve the boom and the bust of medical supplies during the pandemic. The rest of this chapter is organized as follows. Section 2 reviews the "bullwhip effect" as a supply chain phenomenon in the pandemic and psychological factors affecting demand. Then, we focus on hoarding behavior and its effects on the conventional stock management structure. In Sect. 3, we explain why the high demand for the face mask, a personal protective product, is likely to be a demand bubble. Finally, we analyze some practical solutions to reduce the impact of hoarding behavior.

2 Theoretical Perspectives

2.1 Boom and Bust: The Bullwhip Effect

The bullwhip effect is defined as the phenomenon where orders to the supplier have larger variances than sales to the buyer (Cohen & Kouvelis, 2021; Lee, 2002;

Metters, 1997). Under bullwhip effect theory, psychological elements distort the demand information (Chen et al., 2000). Behavioral influence in supply chains has been extensively discussed in the literature on the bullwhip effect. For example, an analysis of industrial production data from 1950 to 2013 revealed greater variances in material production than consumer goods production (Sterman, 2015). These variances propagate upstream in supply chains in the form of amplification, and the booms and busts in material production lag behind those in manufacturing. For instance, oil and gas exploration fluctuates approximately three times more than oil and gas production in the oil industry. Similar amplifications have been witnessed upstream in supply chains in other sectors, such as electronic equipment and machine tools (Anderson Jr et al., 2000).

Causes of the bullwhip effect are the issues of a supply chain infrastructure or related processes, including demand forecast updating, order batching, price fluctuation, rationing, and shortage gaming (Metters, 1997). In addition, irrational human behavior explains this demand amplification (Sterman, 2015). From the behavioral perspective, major causes of the bullwhip effect include demand updating based on erratic orders, order batching, speculating, and hoarding (Chen et al., 2000; Sterman & Dogan, 2015).

The bullwhip effect reportedly occurs in many products, including pasta, soup, and soft drinks (Sheffi, 2022). Rather than duplicating this analysis of the bullwhip effect for these products, we scrutinize this phenomenon for a particular product category, i.e., medical supplies, which experienced a sudden surge in demand in the COVID-19 pandemic. During the pandemic outbreaks from 2000 to 2021, their supply shortages and supply chain disruptions delayed delivery, causing retailers to order more than they need "just in case" (Sheffi, 2022). The bullwhip phenomenon occurred during the COVID-19 pandemic, when retailers ordered too much to replenish stock to meet the soaring demand for essential products (e.g., the high need for face masks in 2020). In turn, distributors and manufacturers amplify their order size (Sheffi, 2022). Suppliers update their demand forecast based on the current increasing demand of the product in need, while they may know or do not want to know that this soaring demand is temporary (NYTimes.com, 2020).

Too many orders have triggered an increase in production scale. Given the global scope of the increasing demand for medical supplies (e.g., face masks) during the COVID-19 pandemic (or the boom scenario), manufacturers have ramped up production to reduce stockout. Garment companies have expanded their product lines to face masks. Some manufacturers have repurposed their production lines to face masks. Global manufacturers of medical supplies scaled up their production to meet the soaring demand during the 2020–2021 period. For example, in February 2020, China, the world's largest exporter of surgical masks, increased production by approximately 12 times. Another example is that Foxconn has changed its production of Apple devices over to face masks (Mcgregor, 2020).

Nevertheless, the characteristics of the bullwhip effect reveal that after the surge is the purge in demand. Then oversupply occurs when the demand for face masks, for example, is stable. This bust scenario is likely to happen when the pandemic is under control, the vaccination rate is high, and masks are no longer mandatory. In this purge, their demand will plunge, causing oversupply and operational inefficiency, including inventory overinvestment, lost revenues, lower service levels, and ineffective transportation (Metters, 1997). Financially, suppliers and retailers have to pay higher inventory-holding costs, salvage it at a deep discount or even discard it.

For instance, making too many cloth masks in Japan during the peak of the COVID-19 pandemic (known as the "Abenomasks" Program) has caused such high inventory-holding costs that the government of Japan has planned to dispose of unused cloth masks. It costs approximately 600 million yens (or US$5.3 million) to stock up to 81 million undistributed masks (Kyodo News, 2021). Therefore, this event can trigger a reduction in the production of face masks, leading to a bullwhip-amplified economic downturn, especially when consumer demand drops due to restricted money supply or raised interest rates (Sheffi, 2022). Witnessing a demand drop, retailers and manufacturers will most likely cut inventory. Considering such a production reduction by manufacturers of medical supplies on a broader global scale, economic losses will be huge. This scenario may be analogous to the bullwhip effect in the 2008 financial crisis. Due to its broad scale, the bullwhip effect was severe in this downturn. For example, the U.S. manufacturers cut inventories by 15%, retailers lost their sales by approximately 30%, and imports dropped by 30%. As a consequence of this effect in the 2008 financial downturn, in the USA, they reported the loss of 100,000 manufacturing jobs and more supplier bankruptcies in the automotive industry than in previous years (Sheffi, 2022).

2.2 Psychological Factors Affecting Demand

Hoarding is a psychological phenomenon of storing up supplies, such as speculation, causing the scarcity of products, from the essential (e.g., food, gasoline) to the nonessential (e.g., technological products) (Sterman, 2015). Psychologists relate this phenomenon to behavioral and emotional factors. Hoarding behavior, which causes difficult-to-predict variability in demand, is one of the critical causes of the well-known bullwhip effect in supply chains (Sterman & Dogan, 2015). Hoarding behavior by retailers and consumers is one of the illustrative examples of the impact of psychological elements on purchase behavior. For example, retailers and consumers, for instance, in a pandemic outbreak, for fear of the shortages of essential products, ordered too much. The false increase in demand creates an untrue supply/demand imbalance. Supply shortages and demand amplifications that hoarders cause are temporary (Bendoly et al., 2015). The number of orders will sharply fall when hoarding behavior disappears, especially when the pandemic crisis is over. When the number of orders from retailers declines, suppliers have to reduce their production rate. Due to the demand boom and bust, suppliers find it hard to predict the variability in demand (Cohen & Kouvelis, 2021).

Hoarding behavior causes a false supply/demand imbalance and prolongs and disrupts supply chains. For illustration, Fig. 1 demonstrates how the phantom orders,

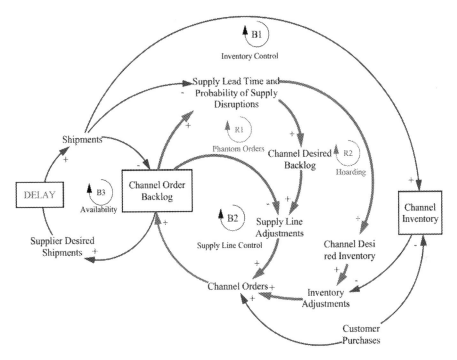

Fig. 1 Stock management structure with phantom orders (Sterman & Dogan, 2015).

due to hoarding, cause longer supply lead time and supply disruption in a typical stock management structure (Sterman & Dogan, 2015).

According to Fig. 1, in the "supply chain control" loop, order backlogs are subject to desired channel orders, anchored to expected customer purchases, and adjusted to inventory level and supply line. In the loop of "phantom orders," an increase in channel order backlog prolongs supply lead time and possible supply disruptions, which would increase the designed backlog and the supply line adjustment (Sterman & Dogan, 2015). The false boom in demand due to phantom orders may create a bullwhip effect. Retailers order more to replenish the empty shelves and update their sales forecasts based on a false need. In turn, distributors and manufacturers also follow the same pattern of amplifying order size. When demand drops, those upstream actors in the supply chain suffer a "painful sting," as depicted by Sheffi (2022) in his review on MIT Sloan Review on the bullwhip effect during the pandemic:

> In response to lower demand—and to work off their bloated inventories—retailers slash new orders. Distributors likewise stop ordering more product from manufacturers and even cancel any outstanding orders until both the retailers and the distributors have sold their excess inventory. Each player up the chain suffers a worse and longer fall-off in demand, leading to reduced manufacturing and layoffs, which in turn exacerbate the contraction in demand, leading to and then exacerbating an economic downturn. Thus, the crack of the

whip comes to the supply chain when boom flips to bust and intensifies the recession. (Sheffi, 2022).

2.3 Factors Affecting Supply

2.3.1 Inventory Models

We assume that any retailer wants to maximize profit, so he computes the optimal order quantity for this optimization objective. The single-period inventory model or newsvendor model sets an order quantity for a single-period product to maximize the expected profit. While classical inventory models such as economic order quantity (EOQ) aim to minimize total inventory-related costs, the single-period inventory model deals with maximizing expected profit. The model is constructed on the marginal analysis where marginal profit is canceled off by the marginal cost (Khouja, 1999).

On the one hand, ordering too much will provide the retailer with more than enough supplies without any stockout concern. However, on the other hand, a retailer ordering too much may incur additional inventory-holding costs if the demand is too low. Balancing the marginal costs of shortage and excess is the foundation for the single-period inventory model.

The undermentioned inventory model has an objective for profit maximization. Let Q be the order quantity for the decision, x is the variable demand; $f(x)$ is the probability density function of x; p *is* the selling price per unit; and c *is* the unit purchase cost. Then, we express the profit function for Q and x:

$$\text{Profit}(Q, x) = \begin{cases} px - cQ & \text{if } x \leq Q \\ pQ - cQ & \text{if } x \geq Q \end{cases}$$

We establish the following expected profit function:

$$E[\text{Profit}(Q)] = \int_0^\infty P(Q, x_0) f_x(x_0) dx_0$$

$$E[\text{Profit}(Q)] = \int_0^Q (px - cQ) f_x(x_0) dx_0 + \int_Q^\infty (pQ - cQ) f_x(x_0) dx_0$$

$$E[\text{Profit}(Q)] = p \int_0^\infty (x) f_x x_0 dx_0 - cQ - p \int_Q^\infty (x - Q) f_x x_0 dx_0$$

According to the above equation, the expected profit equals the expected profit minus the cost of buying Q units (cQ) minus the expected lost sales revenue due to shortage, which is equivalent to the number of units short times the unit selling price. The procedures to maximize profit include determining an optimal order quantity by marginal analysis (i.e., balancing shortage cost and excess cost)[1].

However, in practice, a human decision-maker may not use the calculated optimal order quantity for profit maximization in the single-period inventory model (Becker-Peth et al., 2020). For example, during the COVID-19 pandemic, a supplier may choose a push production strategy. In a push strategy, the producer manufactures as many goods as he wants to push the finished goods through the distribution system to the stores. In a push strategy, a seller uses sales promotion, personal selling, advertisements, and other promotional tools to convince buyers to stock their supplies (Nickels et al., 2008).

Although manufacturers of face masks are most likely to be aware that overproduction would lead to oversupply in the supply chain, they may increase production capacity. To reach the new target, they either maximize the utilization of the existing production equipment or add more capacity.

What are the justifications for not employing the optimal order quantity?

There are some justifications for the push production strategy. First, the supplier does not want to miss the chance of increased sales when the demand is at a peak. Another reason is that the manufacturer is overconfident in selling their outputs through different distribution channels. Nevertheless, disruptions to their supply chains would risk on-time delivery. For example, a retailer on its distribution channel may have multiple suppliers other than a single supplier, and he may refuse to order large quantities from any single supplier. To modify the single-period inventory models to reflect practice, Khouja (1999) compiled a list of extensions to the model as follows: (1) Extensions to different objectives and utility functions; (2) different supplier pricing policies; (3) different newsvendor pricing policies and discounting structures; (4) random yields; (5) different states of information about demand; (6) constrained multi-products; (7) multi-product models with substitution; (8) multi-echelon systems; (9) multilocation models; and (10) multiple periods in the selling season. Recently, Becker-Peth et al. (2020) proposed a multiperiod inventory model for cases with many budget cycles.

---

[1] A critical ratio or service level using marginal analysis uses the following equation:

$$P[x \leq Q] = \frac{c_s}{(c_e + c_s)}$$

where x is the varying demand; Q is the order quantity; c_s is the shortage cost; c_e is the excess cost. Then, the calculated service level can be inverted into a safety factor to determine a safety stock given standard deviation in a normal distribution pattern, based on which an optimal order quantity can be computed.

Would a human decision-maker compute the optimal order quantity in a newsvendor model for a rational inventory decision? Maybe not. The following section explains irrational decisions.

2.3.2 Psychological Elements of Suppliers

As discussed above, phantom demand may cause soaring demand. What makes a retailer order too much or a manufacturer ramps up production, even though they may be aware of phantom orders or uncertain future demand? Psychological elements may cause their actions. We can analyze this behavior through the lens of risk behavior, irrational decisions, utility, and prospect theories.

Theories used to explain how psychological elements influence rational decisions include the Quantal Response Equilibrium (QRE) Theory and Prospect Theory.

The QRE Theory helps explain why a decision is irrational. Under the QRE theory, the newsvendor's decisions are influenced by noise, whose level is determined by the degree of rationality of the decision-maker (Donohue et al., 2019). The probability distribution of selecting an order quantity, q_i, is determined by

$$\Pr(q_i) = \frac{e^{\lambda E[\pi(q_i)]}}{\sum_q e^{\lambda E[\pi(q)]}}$$

The above equation generates a probability distribution according to the level of rationality, λ. When the level of rationality approaches infinity, the decider has unlimited rationality. The newsvendor's optimal order quantity is a special case of the QRE model where the order quantity, q, shrinks to a point, q^*, or the optimal order quantity (Donohue et al., 2019). However, a decider whose rationality level is low cannot reach the optimal order point.

Moreover, a man's decision is influenced by his prospect. According to the Prospect Theory, the decision-maker becomes risk-seeking when he foresees that the cost of lost sales would be high in case of stockout, and he does not want to miss a chance to win significant revenues, even though that chance is small (Kahneman & Tversky, 1979; Redelmeier & Tversky, 1992).

2.3.3 Issues with Stock Management Structure

Classical operations management theories assume that actors in a supply chain are rational decision-makers linked in a physical stock–flow structure (Sterman, 2015). Thereby, instability in supply chains results from the interaction of the rational actors within this physical structure. The stock management structure is one of the models assuming rationality in actors in supply chains. A supplier popularly uses the stock management structure as a tool for inventory control using systems dynamics

(Donohue et al., 2019). This structure reflects the dynamic relationships between many variables that affect the production rate.

The stock management structure has three feedback loops: order fulfillment, work-in-process (WIP), and inventory control loops (Sterman, 2015). In the order fulfillment loop, shipments cannot meet orders if the inventory level is too low. The WIP control and inventory control loops aim to adjust the production rate to the desired levels of WIP and inventory. When the demand is excessive, retailers tend to place multiple orders or order large quantities from numerous suppliers because they do not want the supplies to be out of stock, resulting in the cost of lost sales. That is, retailer purchase behavior follows a continuous review pattern, i.e., continually checking the demand and making adjustments to inventory to ensure the adequacy of stock for sales. According to this stock management model, the desired production starts are fixed to the desired production rate and adjusted to bring the WIP level in line with the desired WIP level (Sterman, 2015).

Decision Biases of Suppliers

As presented above, in a typical stock management structure, the feature of stock adjustment to order rates aims to increase inventory levels to maintain inventory coverage and customer service (Sterman, 2015). However, the stock management model ignores several factors that make decisions irrational (Metters, 1997). On the buyer's side, their purchase behavior may not come from actual demand but local profit optimization. Perhaps they buy in large quantities to take advantage of quantity discounts. The stock management model also fails to include capacity and materials constraints, limiting production starts. Another assumption of the stock management structure is that each supplier has only one customer and each customer has only one supplier to avoid phantom orders (Bendoly, 2013). This assumption is challenged when hoarders place multiple orders on many sellers. In practice, phantom orders happen as a consequence of hoarding. Therefore, the stock adjustment of the production rate and inventory level to the order rate is based on inaccurate demand information. Although the number of orders for a specific essential product is inflated due to hoarding behavior, the supplier's stock management structure is not designed to distinguish between true and false demand due to its characteristics (Sterman, 2015). Sterman (2015) pointed out that the consequence of the stock management structure is the creation of demand amplification, which is temporary.

Anchoring Bias

The stock management structure suggests that production decisions are based on anchoring and judgment heuristics (Bolton & Katok, 2008; Katok & Wu, 2009; Schweitzer & Cachon, 2000). Looking back at the supply line in the stock management structure, we find that the production rate is anchored on the expected order rate and then adjusted to the adequacy of finished goods and work-in-process inventory.

The anchoring factor quantifies the degree of anchoring on the expected demand. If α is an anchoring factor, an adjustment factor is labeled $(1 - \alpha)$. Initially, managers anchor on the expected demand, or μ, using the anchoring factor α (Schweitzer & Cachon, 2000). The adjustment process is modeled as the following equation:

$$q = \alpha \times \mu + (1 - \alpha) \times q^*$$

where q is the production quantity, α is the anchoring factor, $(1 - \alpha)$ is the adjustment factor, and q^* is the nominal quantity. On one extreme, when the adjustment factor $(1 - \alpha) = 0$, managers ignore the nominal quantity but anchor on expected demand. On the other extreme, when the anchoring factor $\alpha = 0$, the manager does not anchor on the expected demand but completely adjusts the order quantity optimally.

Like most heuristics for organizational decisions, demand anchoring cannot avoid flaws (Hammond et al., 2006). Demand anchoring is an example of a pernicious mental phenomenon that causes decision bias (Hammond et al., 2006). If a manager anchors on past demand to decide how much and how fast to produce, the old figures of the previous need become anchors, which the manager adjusts based on other factors (Hammond et al., 2006). Unfortunately, the past demand may be false due to phantom orders. Demand anchoring based on unrealistically high demand causes oversupply. To remedy decision bias due to demand anchoring, a manager can use the following techniques: viewing a problem from different perspectives, seeking information from a wide variety of people, and avoiding preconceptions (Hammond et al., 2006).

Demand Chasing

Demand chasing refers to using a recent demand realization to determine the production level to maximize sales (Bolton & Katok, 2008; Katok & Wu, 2009; Schweitzer & Cachon, 2000). The manufacturer looks at the most recent sales peak and decides to increase the production level. The literature on behavioral inventory decisions indicates that demand chasing is one of the heuristics techniques a manager follows in inventory-related decisions (Donohue et al., 2019). Following this heuristic, a manager increases order quantities after high-demand realization and decreases order quantities after low demand realizations.

Chasing demand may be an easy way for planners to estimate future demand. However, it demonstrates limited cognitive ability or willingness to use a more rational method for demand estimation (Donohue et al., 2019). Managers should make decisions based on complete information. They need to know "comprehensive information regarding all possible order quantities" (Donohue et al., 2019). Additional valuable information for informed decisions is about "the profit distribution, the probability to sell all units ordered, the break-even sales level, and the likelihood to incur losses" (Donohue et al., 2019).

2.3.4 Other Supply Issues

In addition to decision biases due to the human factors described above, some obstacles, such as material shortages and logistics disruptions, may cause delays in supply. Ivanov (2020) categorized the features of supply chain risks during epidemic outbreaks into three components: (1) long-term interruption and its unpredictability, the simultaneous propagation of disruptions in the supply chain; (2) pandemic propagation; and (3) simultaneous disruptions in supply and logistics infrastructure. For example, raw material shortages delay product delivery downstream in a supply chain, exacerbating operational inefficiency and reducing the service level (Pinto, 2020). The COVID-19 pandemic has also caused many logistical disruptions. Ivanov (2020) argued that an epidemic outbreak, such as SARS, MERS, Ebola, Swine Flu, or COVID-19, began on a small scale but quickly spread over many geographical locations, causing large-scale disruptions in many supply chains. Up to 94% of companies listed in the Fortune 1000 had experienced supply disruptions due to the coronavirus (Fortune, 2020).

2.4 Strategies to Respond to Supply–Demand Imbalance

To reduce the bullwhip effect, managers can employ technical and management solutions. Technically, managers can modify their existing supply chain infrastructure and processes to mitigate the potential bullwhip effect concerning demand forecast updating, order batching, price fluctuation, rationing, and shortage gaming (Metters, 1997). Specifically, mitigation strategies include coordinated demand forecasting through information-sharing technologies such as point-of-sale data, electronic data interchanges, and vendor-managed inventory. If the bullwhip effect is caused by order batching, mitigation solutions may be Internet ordering, consolidation, and logistics outsourcing to reduce ordering costs. In addition, everyday low price or activity-based costing can work if the bullwhip effect is due to price fluctuations. Finally, allocation based on past sales alleviates the shortage of gaming or rationing (Geary et al., 2006; Metters, 1997). These strategies can reduce information distortion and increase operational efficiency by reducing lead time and echelon-based inventory control (Metters, 1997). Regarding vaccine supply, Bamakan et al. (2021) identified some of the most critical factors influencing the bullwhip effect reduction (CBER) of the COVID-19 vaccine supply chain (CVSC). They are information sharing, logistics transportation centers, lead time reduction, optimal resource allocation, internal processes, reliability, flexibility, training of employees, accuracy in order quantity, quick responsiveness, inventory management, and management skills (Bamakan et al., 2021).

In addition to mitigation strategies, managers should employ an avoidance strategy to systematically eliminate any avoidable causes of the bullwhip effect, including business systems engineering to integrate technical, cultural, organizational, and financial aspects (Geary et al., 2006). It is also important to educate managers about "example good practices" in reducing the bullwhip effect, encouraging them to share data to minimize its impact and eliminate the "functional silos" within organizations (Geary et al., 2006).

Technologies used for sharing information to deal with the bullwhip effect include Geographical Information System (GIS), the Internet of Things (IoT), and cloud computing, which enable the collection, storage, and transfer of information about logistical flow for better collaboration among actors in a supply chain (Pinto, 2020).

In addition, removing production and supply chain bottlenecks for medical supplies requires the engagement and coordination of various stakeholders, including the government. In practice, some governments intervene in the production of face masks. For example, Chinese manufacturers have received generous support (e.g., preferential land policies, subsidies) from their government for expanding production capacity for medical supplies. Winner Medical, a mask manufacturer in China, received up to $4 million per year as a subsidy from the government of China (NYTimes.com, 2020). In many countries, manufacturers have invested a great deal in building new factories and setting up new equipment to be more independent from China's exports and boost producing locally made medical supplies. For example, France wanted to make its homemade masks by the end of 2021. In the USA, President Trump's industrial policies included a push for the federal government to buy American-made medical supplies for self-reliance and production. As a result, some U.S. manufacturers shifted their conventional products to partial-trapping materials required to produce medical masks, such as the case of QYK Brands, a California-based factory (NYTimes.com, 2020). In addition, supplying medical equipment quickly to the neediest places (e.g., hospitals, infection clusters, badly affected regions, and nations) requires support from many stakeholders (the local government, the public, humanitarian groups, volunteer groups). For instance, local groups share demand information on the most demanding locations for timely supply delivery (Ye et al., 2020).

Businesses also implement risk response strategies in the dynamic supply chain environment. For instance, global supply disruptions may be due to export-restricting policies imposed by exporting nations. During the pandemic, China sometimes prioritized the exports of medical masks to just a few recipient countries, restricting their supplies to many nations (NYTimes.com, 2020). Some relevant supply chain management can explain those actions. Charpin et al. (2021) found that a company's supply chain was affected by perceived political risks. From the resource dependency perspective, Darby et al. (2020) discovered that companies increased the inventory level to cope with policy changes. Fifarek et al. (2008) found that offshore manufacturing may negatively affect home innovation in the long run and suggested that the government should support home manufacturing. This argument explains why the governments of many nations subsidize manufacturers of medical masks.

In addition, bridging the gap in accessing medical supplies between richer and poorer countries is a social objective. Therefore, nonprofit organizations can help resolve the bottlenecks in supplying medical products to poorer countries. The shortages of medical supplies in developing countries during the pandemic were more severe than in developed countries due to the lack of commitment from manufacturers, who find these markets unattractive due to their lack of ability to pay (Martin et al., 2020). Therefore, wealthier nations have had quicker access to vaccines, for example, than poorer countries. Researchers have proposed some models for the effective delivery of medical equipment for developing countries, including the concept of theoretical maximum capacity, which "refers to the highest level of capacity committed by the manufacturer and serves as a benchmark to evaluate the relative performance of all contracts" (Martin et al., 2020). The contract designs use optimal parameters at various times with available budgets (e.g., from donors, self-financing). In practice, the Global Health Organization (GHO) has used different optimal contract designs subject to budget availability to encourage manufacturers' capacity commitment to supply vaccines to developing countries (Martin et al., 2020)

3 The Case of Face Masks

At the peak of the COVID-19 pandemic in 2020, the demand for face masks was estimated to be ten times greater than the world production capacity (OECD, 2020). People wear masks as an effective way to prevent the spread of the novel coronavirus in their community.

In 2020, the estimated monthly demand for N95 masks, which could block 95% of small particles, was 290 million units for healthcare workers, while monthly production capacity by U.S. producers, such as 3M, Honeywell, Moldex-Metrix, and Prestige Ameritech, was only 80 million units (Gereffi, 2020). The undersupply created a severe shortage in retail stores. There has been a point of time people purchase face masks in bulk and in panic, as reported by CNN:

> U.S. drugstores, retailers and suppliers are racing to keep up with the surging interest in cleaning products as fears over coronavirus intensify. Demand for products such as hand sanitizers, face masks and cleaning wipes has spiked, according to CVS, Walgreens and others. CVS warned it may cause supply shortages (Meyersohn, 2020)

3.1 Supplier Behavior

A sharp increase in demand for face masks during the outbreak of the COVID-19 pandemic urges manufacturers to increase their production rate. The following case of 3M is an illustrative example.

3M Corporation, a Minnesota-based corporation with face-mask production factories in the USA and China, is one of the largest N95-mask suppliers for the U.S. market. Before the pandemic outbreak in 2020, the company supplied approximately 5 million N95 masks to U.S. healthcare workers (Gereffi, 2020). However, the pressure to match face mask supply with its demand in the USA pressured this corporation to increase the monthly production rate to 95 million units per month in May 2020 (Gereffi, 2020).

Indeed, increasing the production rate for face masks to meet its emergency demand is purely an economic decision to create a new supply/demand balance and demonstrate social responsibility (Mcgregor, 2020). However, our point of interest here is to examine this organizational decision through the lens of behavioral operations. Concretely, we want to investigate how demand chasing and anchoring explain a supplier's decision on the production rate.

3.1.1 Demand Chasing

A manufacturer wants to adjust the production rate to the desired order rate. The increasing production of face masks reflects the behavior of chasing high demand. Simply put, demand-chasing behavior means that face-mask manufacturers increase their production rate to meet their increasing demand. The demand-chasing strategy is risky for some reasons. First, during the pandemic outbreak, face-mask shortages may be overstated. By chasing these shortages for demand forecasts, suppliers may overestimate the demand. As discussed before, these shortages may be due to hoarding or panic purchases. Second, when the pandemic ends and wearing masks is no longer mandatory, there will be a decline in the number of their orders. If a manufacturer invests in expanding the production capacity of his factories just to meet a temporary high demand, this decline will cause underutilization. Third, oversupply risk is also due to too many suppliers in the face-mask market. Face masks are not difficult to imitate and are not rare products, and there are substitutes (e.g., homemade cloth masks, reusable face masks), so it is easy for many new suppliers to enter the market. For example, in April 2020, when COVID-19 broke out, IRIS USA, a Japan-headquartered manufacturer of household products, decided to invest $10 million to make face masks (Naczek, 2020).

3.1.2 Demand Anchoring and Stock Adjustment

According to the physical stock management structure, overproduction may also be attributed to inventory adjustment (Fig. 1). According to the stock management structure, the inventory control and WIP control loops adjust production to move the inventory and WIP to desired inventory levels.

To express the dependence of the order rate on the expected demand, the Adjustment for Stock and the Adjustment for Supply Lines, we use the equation proposed by Sterman (2015):

$$\text{OR} = \text{MAX}(0, D^e + A_S + A_{SL}) \tag{1}$$

where D^e is the expected demand, which is anchored on the manufacturer's belief about the incoming orders; A_S is the Adjustment for Stock and A_{SL} is the Adjustment for Supply Chain.

Let S' be the desired stock, S be on-hand stock and S.L. be an on-order inventory, we rewrite Eq. (1):

$$\text{OR}_t = \text{MAX}(0, D^e + \alpha(S' - (S_t + \beta \text{SL}_t)) + \varepsilon_t \tag{2}$$

$$D_t^e = \theta D_t + (1 - \theta) D_{t-1}^e \tag{3}$$

where α is the fraction of the perceived inventory discrepancy each period t; θ is the smoothing parameter of demand, β is the fraction of the supply line of unfilled orders; and ε is the error term.

The above equations show that the decision rule for orders depends on only three factors: expected demand, inventory, and the supply line. Manufacturers increase the production rate if they have seen the actual incoming orders, D_t, and the shortage of stock. This decision rule appears to be followed by face-mask producers who raise their production to meet the latest demand and adjust the falling inventory and supply line level to the desired stock level. However, these equations fail to capture how people place orders (Sterman, 2015). For example, buyers may make bulk purchases of face masks as hoarding behavior, leading to a temporary drop in inventory. Suppose the producer increases the production rate or even expands capacity to match this false and temporary demand. In that case, there is a risk of excess cost when the demand returns to its low normal level.

3.2 Buyer Behavior

When buyers order too much in bulk for several reasons, such as price gouging or simply fear of future shortages, it conveys the wrong signal of high demand to the supplier, who then increases production to meet this phantom demand. When there is some delay in supply lines due to production change-over or capacity increase, people may speculate on a supply shortage. As a consequence of temporary needs, there is an increase in retail prices (Harwell, 2020). In its 2020 report, the selling price of surgical masks increased by 20 times in the COVID-19 crisis compared to before the crisis (OECD, 2020). In just 2 months, from January to February 2020, the price of N95 respirator masks increased by approximately four times, from US$17 to US$70 (OECD, 2020).

Hoarding does not always mean price gouging (Sterman & Dogan, 2015). Bulk purchase is also attributed to panic or fear appeal (Witte & Allen, 2000). In February

2020, CNN reported bulk and panic purchases of cleaning products such as hand sanitizers, face masks, and cleaning wipes in retail stores, including CVS and Walgreens, during the COVID-19 pandemic (Meyersohn, 2020). Some descriptive words used to illustrate this unusual behavior are "madness of crowds" (Merton, 1948) and "irrational exuberance" (Shiller, 2015). According to the self-fulfilling prophecy theory proposed by Merton (1948), the rumor of supply shortages triggers hoarding behavior. As a consequence, people purchase face masks in bulk and panic.

Consumers also make bulk purchases for fear of shortage. A demand bubble for a product occurs when people buy large quantities of the product because they fear its scarcity. During the COVID-19 pandemic, the demand for face masks bubbles because of this mentality. Retailers witnessing this increasing demand would place orders of large volume for fear of stockout. In countries where the shortages of products used to happen, buyers who fear their experience of product shortage would come again to them and their family, so they want to stock as many products as possible in case of no supply. Buyers tend to base their predictions for product shortages on their experience as a recallability trap (Hammond et al., 2006).

3.3 Remedies

In practice, actors in supply chains have responded to supply shortages. Table 1 indicates the major countermeasures to prevent or mitigate the impacts of hoarding. Among the countermeasures listed in Table 1, collaboration in information sharing between retailers and suppliers is critical to reducing the impact of hoarding behavior. Parties in supply chains should know that the shortage of essential products in a crisis period is not chronic. Therefore, they need to collaborate to respond to this momentary shortage. For example, retailers and supermarkets should be alert to hoarding behavior and send the right message to manufacturers. In addition, when the high-demand product falls under the essential category (e.g., food, grocery, medical supplies), the retailer needs greater buffer stock or safety stock in case of any abrupt increase in demand, depending on their corporate culture and other competitive factors (Sheffi, 2021). Safety stock avoids the shortage of essential products, so people in need can buy them even when bulk purchases occur.

Furthermore, with retailer safety stock, manufacturers do not have to increase their production capacity for a seasonal demand peak in the immediate and short term. For example, Toyota Corporation successfully avoided chip supply shortages after learning some critical lessons on this shortage after the 2011 earthquake in Japan. Concretely, rather than following an immediate or short-term inventory adjustment, the company carefully reviewed long-term market dynamics to build up stockpiles of chips in anticipation of future shortages (Sheffi, 2021).

Table 1 Countermeasures for hoarding behavior

Party involved	Countermeasures	Examples
Supplier	Not chasing sudden demand	Allocate based on past sales, not a sudden surge in demand (Saturn, H.P.)
	Scheduling	Shared capacity and supply information (H.P., Motorola)
	Substitute products	Reusable face masks (e.g., textile face masks)
	Greater inventory	Toyota with great stockpiles in case of future shortages
	Flexibility in production	Easy change-over to different products
Distributor	Redistribution	Prioritize product distribution to the people who need the product the most. For example, prioritize surgical masks for health professionals
Retailer	Rationing	Walmart imposing purchase limits Grocery stores restricting open hours The first item is sold at a regular price, and the second item at a prohibitive price
	Banning sales that can increase the price	Online marketplace eBay banned U.S. listing for hand sanitizers and surgical masks to stop the surge in price gouging
Government	Managing demand	Restricting exports of medical supplies, including face masks
	Consumer education	Communication to members on the consequences of bulk purchase
	Penalty on price gouging	Law enforcement (e.g., a violation of price gouging law in California can be penalized with up to 1 year in prison and/or a fine up to $10,000)

4 Conclusions

A sharp increase in demand for face masks during the outbreak of the COVID-19 pandemic motivates manufacturers to increase face-mask production as they want to demonstrate their social responsibility in jointly combatting the pandemic. Beyond that, they produce face masks and other personal protective equipment to avoid stockout costs. A rational production planner should adjust the production rate to the desired inventory levels. According to the stock management structure, the chosen production level is dependent on the order rate. However, this physical stock management structure does not capture how and why retailers and end consumers order. Buyers may acquire more face masks than they need due to psychological elements. The face-mask shortage in pandemic outbreaks is a typical example of phantom orders due to hoarding and panic purchases. These psychological elements distort demand information during the COVID-19 pandemic (Cohen & Kouvelis, 2021). To be knowledgeable about phantom orders, retailers should collaborate with manufacturers in sharing demand information and applying demand control measures such as rationing and purchase limits. In addition, large stockpiles should be prepared to prepare for any sudden future shortages. A production planner should also consider the probability of a looming decline in demand for face masks. This scenario is most likely to occur when the COVID-19 pandemic is over, vaccination coverage is perfectly high, and universal face masks are not mandatory. Therefore, surplus stock probability results in extra holding costs (Kyodo News, 2021). Thus, too much investment capital in medical face mask production may not yield a positive cash inflow.

References

Anderson, E. G., Jr., Fine, C. H., & Parker, G. G. (2000). Upstream volatility in the supply chain: The machine tool industry as a case study. *Production and Operations Management, 9*(3), 239–261.

Bamakan, S. M. H., Malekinejad, P., Ziaeian, M., & Motavali, A. (2021). Bullwhip effect reduction map for COVID-19 vaccine supply chain. *Sustainable Operations and Computers, 2*(July), 139–148. https://doi.org/10.1016/j.susoc.2021.07.001

Becker-Peth, M., Hoberg, K., & Protopappa-Sieke, M. (2020). Multiperiod inventory management with budget cycles: rational and behavioral decision-making. *Production and Operations Management, 29*(3), 643–663. https://doi.org/10.1111/poms.13123

Bendoly, E. (2013). *The handbook of behavioral operations managment* (Vol. 53, Issue 9, pp. 1689–1699). McGraw-Hill. https://doi.org/10.1017/CBO9781107415324.004

Bendoly, E., van Wezel, W., & Bachrach, D. G. (2015). *The handbook of behavioral operations management: Social and psychological dynamics in production and service settings.* Oxford University Press.

Bolton, G. E., & Katok, E. (2008). Learning by doing in the newsvendor problem: A laboratory investigation of the role of experience and feedback. *Manufacturing & Service Operations Management, 10*(3), 519–538.

BusinessToday. (2021, April 21). 600% jump in oxygen demand from hospitals. *BusinessToday.In.* https://www.businesstoday.in/latest/economy-politics/story/600-jump-in-oxygen-demand-from-hospitals-293981-2021-04-21

Charpin, R., Powell, E. E., & Roth, A. V. (2021). The influence of perceived host country political risk on foreign subunits' supplier development strategies. *Journal of Operations Management, 67*(3), 329–359.

Chen, F., Drezner, Z., Ryan, J. K., & Simchi-Levi, D. (2000). Quantifying the bullwhip effect in a simple supply chain: the impact of forecasting, lead times, and information. *Management Science, 46*(3), 436–443. https://doi.org/10.1287/mnsc.46.3.436.12069

Cohen, M. A., & Kouvelis, P. (2021). Revisit of AAA excellence of global value chains: robustness, resilience, and realignment. *Production and Operations Management, 30*(3), 633–643. https://doi.org/10.1111/poms.13305

Darby, J. L., Ketchen, D. J., Jr., Williams, B. D., & Tokar, T. (2020). The implications of firm-specific policy risk, policy uncertainty, and industry factors for inventory: a resource dependence perspective. *Journal of Supply Chain Management, 56*(4), 3–24. https://doi.org/10.1111/jscm.12229

Donohue, K., Katok, E., & Leider, S. (2019). *The handbook of behavioral operations.* Wiley. https://doi.org/10.1002/9781119138341

Fifarek, B. J., Veloso, F. M., & Davidson, C. I. (2008). Offshoring technology innovation: A case study of rare-earth technology. *Journal of Operations Management, 26*(2), 222–238.

Fortune. (2020). *Finance. Fortune 1000. 94% of the Fortune 1000 are seeing coronavirus supply chain disruptions.* https://fortune.com/2020/02/21/fortune-1000-coronavirus-china-supplychain-impact

Geary, S., Disney, S. M., & Towill, D. R. (2006). On bullwhip in supply chains—historical review, present practice and expected future impact. *International Journal of Production Economics, 101*(1), 2–18. https://doi.org/10.1016/j.ijpe.2005.05.009

Gereffi, G. (2020). What does the COVID-19 pandemic teach us about global value chains? The case of medical supplies. *Journal of International Business Policy*, 1–15. https://doi.org/10.1057/s42214-020-00062-w

Hammond, J. S., Keeney, R. L., & Raiffa, H. (2006). *The hidden traps in decision making*. Harvard Business Review. Harvard Business School Publishing.

Harwell, D. (2020, March 28). Gouged prices, middlemen and medical supply chaos: Why governors are so upset with Trump. *Washington Post*. https://www.washingtonpost.com/business/2020/03/26/gouged-prices-middlemen-medical-supply-chaos-why-governors-are-so-upset-with-trump/

Ivanov, D. (2020). Predicting the impacts of epidemic outbreaks on global supply chains: A simulation-based analysis on the coronavirus outbreak (COVID-19/SARS-CoV-2) case. *Transportation Research Part E: Logistics and Transportation Review, 136*, 101922.

Kahneman, D., & Tversky, A. (1979). Prospect theory—An analysis of decision under risk. *Econometrica*. https://doi.org/10.2307/1914185

Katok, E., & Wu, D. Y. (2009). Contracting in supply chains: A laboratory investigation. *Management Science, 55*(12), 1953–1968.

Khouja, M. (1999). The single-period (news-vendor) problem: Literature review and suggestions for future research. *Omega, 27*(5), 537–553. https://doi.org/10.1016/S0305-0483(99)00017-1

Kyodo News. (2021, December 21). *Japan gov't to dispose of unused "Abenomasks" amid growing costs.* https://english.kyodonews.net/news/2021/12/f35c100e6736-japan-govt-to-dispose-of-unused-abenomasks-amid-growing-costs.html

Lee, H. L. (2002). Aligning supply chain strategies with product uncertainties. *California Management Review*. https://doi.org/10.2307/41166135

Martin, P., Gupta, D., & Natarajan, K. V. (2020). Vaccine procurement contracts for developing countries. *Production and Operations Management, 29*(11), 2601–2620. https://doi.org/10.1111/poms.13229

Mcgregor, G. (2020). Masks and hand sanitizer replace iPhones and perfume. Firms redeploy factories to make coronavirus supplies. *Fortune.com*. https://fortune.com/2020/03/17/coronavirus-mask-hand-sanitizer-factory/

Merton, R. K. (1948). The self-fulfilling prophecy. *Antioch Review, 8*, 193–210.

Metters, R. (1997). Quantifying the bullwhip effect in supply chains. *Journal of Operations Management, 15*(2), 89–100. https://doi.org/10.1016/S0272-6963(96)00098-8

Meyersohn, N. (2020). CVS and Walgreens warn there could be a shortage of hand sanitizer. *CNN*. https://edition.cnn.com/2020/02/28/business/hand-sanitizers-wipes-cvs-walgreens-coronavirus/index.html

Naczek, M. (2020, April 9). IRIS USA to invest $10 million into face mask production in Pleasant Prairie. *Milwaukee Business Journal*. https://www.bizjournals.com/milwaukee/news/2020/04/09/iris-usa-to-invest-10-million-into-face-mask.html

Nickels, W. G., McHugh, J. M., & McHugh, S. M. (2008). *Understanding Business* (9th ed.). McGraw-Hill Irwin.

NYTimes.com. (2020, July 5). *China dominates medical supplies, in this outbreak and the next.* https://www.nytimes.com/2020/07/05/business/china-medical-supplies.html?smid=url-share

OECD. (2020). *The face mask global value chain in the COVID-19 outbreak: Evidence and policy lessons.* https://www.oecd.org/coronavirus/policy-responses/the-face-mask-global-value-chain-in-the-COVID-19-outbreak-evidence-and-policy-lessons-a4df866d/

Pinto, C. A. S. (2020). Knowledge management as a support for supply chain logistics planning in pandemic cases. *Brazilian Journal of Operations & Production Management, 17*(3) https://doi.org/10.14488/bjopm.2020.033

Redelmeier, D. A., & Tversky, A. (1992). On the framing of multiple prospects. *Psychological Science*. https://doi.org/10.1111/j.1467-9280.1992.tb00025.x

Schweitzer, M. E., & Cachon, G. P. (2000). Decision bias in the newsvendor problem with a known demand distribution: Experimental evidence. *Management Science, 46*(3), 404–420.

Sheffi, Y. (2021). How companies can break the hoarding habit. *Medium.com*. https://medium.com/mitsupplychain/how-companies-can-break-the-hoarding-habit-b43747a5a3b7

Sheffi, Y. (2022). Prepare for the bullwhip's sting. *MIT Sloan Management Review*. https://sloanreview.mit.edu/article/prepare-for-the-bullwhips-sting/?use_credit=10a08cdd741bb5dd841466abff753a3c

Shiller, R. J. (2015). *Irrational exuberance: Revised and expanded third edition*. Princeton University Press.

Sterman, J. D. (2015). Booms, busts, and beer. In *The handbook of behavioral operations management: Social and psychological dynamics in production and service settings* (pp. 203–210). Oxford University Press.

Sterman, J. D., & Dogan, G. (2015). "I'm not hoarding, I'm just stocking up before the hoarders get here.": Behavioral causes of phantom ordering in supply chains. *Journal of Operations Management, 39–40*(1), 6–22. https://doi.org/10.1016/j.jom.2015.07.002

Terlep, S. (2020, March 5). Amazon dogged by price gouging as coronavirus fears grow. *Wall Street Journal*. https://www.wsj.com/articles/amazon-dogged-by-price-gouging-as-coronavirus-fears-grow-11583417920

Wayland, M. (2020, April 16). GE, Ford sign $336 million federal contract to make ventilators for coronavirus outbreak. *CNBC*. https://www.cnbc.com/2020/04/16/ge-ford-sign-336-million-federal-contract-for-ventilator-production.html

Witte, K., & Allen, M. (2000). A meta-analysis of fear appeals: Implications for effective public health campaigns. *Health Education & Behavior, 27*(5), 591–615.

Ye, Y., Jiao, W., & Yan, H. (2020). Managing relief inventories responding to natural disasters: Gaps between practice and literature. *Production and Operations Management, 29*(4), 807–832.

Part III
Supply Chain Sustainability and Resilience

Sustainability Practices for Enhancing Supply Chain Resilience

Alejandro Ortiz-Perez, Elena Mellado-Garcia, and Natalia Ortiz-de-Mandojana

Abstract In this chapter, we analyze how sustainability practices can improve the resilience of the supply chain against different types of disruptions. Firstly, we delimit the concept of resilience and its different dimensions in social sciences (pre-adversity capabilities, in-crisis organizing and adjusting, and post-crisis resilience responding and recovery). Secondly, we describe and classify the disruptions firms face related to their supply chain by distinguishing between internal, social, environmental, and global disruptions. Thirdly, we propose resilient practices through which sustainability commitment can foster relationships with suppliers and customers, by creating mechanisms to deal with these four types of disruptions in the supply chain, and by increasing in the three different dimensions of resilience. Finally, we conclude the chapter by analyzing some managerial and academic implications and proposing some topics and ideas for future research.

1 Introduction

Recently, the complex context and dynamic markets, changing regulations, and the appearance of global disruptions like economic crises or the current COVID-19 pandemic have created much uncertainty and risk in organizations (Sarkis, 2020; Shih, 2020). Organizations must operate in contexts that are abnormal, exceptional, or extreme (Hällgren et al., 2018), and their abilities to survive or avoid bankruptcy have been seriously affected. Owing to these problems, the concept of resilience has been obtaining increasing attention from scholars and practitioners in the "Business" and "Management" fields. Specific research fields in which resiliency has been studied include human resources (e.g., Cooke et al., 2021; Santoro et al., 2021), innovation and new technologies (e.g., Bertschek et al., 2019; Korhonen et al.,

A. Ortiz-Perez (✉) · E. Mellado-Garcia · N. Ortiz-de-Mandojana
Department of Business Management II, Faculty of Economics and Business Management, University of Granada, Granada, Spain
e-mail: aleortiz@ugr.es

2021), the supply chain's operations (e.g., Negri et al., 2021), and the implementation of green practices and sustainability (e.g., Çop et al., 2021). This chapter focuses on these two last domains.

There are now more connections among suppliers than ever before. Most firms subcontract many of their production processes, creating a huge dependency on their supply chain. Thus, this dependence increases the uncertainty and difficulties related to the external environment, including additional factors and the possibility of more complicated external disruptions. As a result, firms' abilities to predict and accurately manage disruptions related to the supply chain are an important part of their resilience.

Ortiz-de-Mandojana and Bansal (2016) have examined how firms' sustainability practices can foster their resilience and have found evidence supporting the idea that a long-term vision and a social and environmental practices implementation can help firms mitigate threats and an environmental shocks adaptation or avoidance. Supply chain collaboration in times of globalization and increasing demands on sustainability is extraordinarily complex but could result in substantial performance improvements if pursued in a thorough and strategic manner (Blome et al., 2014). Following this idea, previous studies have pointed out the potential of sustainability commitments among firms and their suppliers to create more resilient supply chains (e.g., Fahimnia & Jabbarzadeh, 2016; Jabbarzadeh et al., 2018; Shashi et al., 2020). However, deeper study is still needed to understand and clarify how sustainability can improve the resilience of the supply chain against different types of disruptions.

To address this issue, we first delimit the concept of resilience and its different dimensions in the social sciences. Secondly, we list different supply chain disruptions described in previous literature (i.e., internal, social, and environmental) and include a new category (i.e., global) to adapt the previous classification of disruptions to the current complex environment. We then propose a combined classification (resilience dimensions and supply chain disruptions) to better categorize and understand the sustainability practices that can be used to increase resilience in the supply chain.

This chapter contributes to the literature by offering a better understanding of how sustainability practices can foster relationships with suppliers and customers to create the three dimensions of resilience. Additionally, inspired by this previous literature, we have proposed specific sustainability practices that could be adequate to each disruption and dimension of resilience. This identification opens the door to future studies for further investigating the mechanisms underlying these relationships and test them empirically. By analyzing these practices focusing on the different types of disruptions in the supply chain that firms may face (internal, social, environmental, and global), researchers and managers can obtain more concrete guides to increase resilience.

2 Definition of Resilience

Resilience is an interdisciplinary concept; each discipline focuses on different aspects of resilience, creating diverse but related definitions (Linnenluecke, 2017). For example, ecological resilience has been defined as the ability of a system to absorb disturbances without modifying its primary function and structure (Holling, 1996), while in the field of engineering, resilience emphasizes the speed of recovery after a disturbance (Pimm, 1991). However, these definitions are not the most appropriate for social and dynamic systems like supply chains.

As applied to social systems, resilience incorporates the ideas of adaptation, learning, and self-organization in addition to the ability to persist after a disturbance (Folke, 2006). Therefore, resilience not only appears in social systems in response to moments of crisis, but it is also continually applied as systems anticipate and adjust to changes in the environment (Gittell et al., 2006; Hamel & Välikangas, 2003; Ortiz-de-Mandojana & Bansal, 2016).

According to Williams et al. (2017), resilience is an interactive process of adaptation related to a firm's ability to understand, respond to, and absorb variations and to maintain, recover, or build new resources. This emphasis on anticipation and adaptation explains why research on organizational resilience has been largely explored separately from crisis management, under the assumption that resilient organizations avoid crises (Williams et al., 2017). Thus, while research on crisis management focuses on organizations and systems' ability to return to their normal functioning after a crisis, resilience focuses on their ability to maintain reliable performance despite adversity.

Based on the findings of previous literature, organizational resilience can be analyzed in two different situations. On the one hand, resilience can be studied as an outcome; coupled with the crisis-as-event perspective, resilience would naturally be situated after the event and would consist of the ability to recover. On the other hand, resilience can be studied as a process; coupled with the crisis-as-process perspective, resilience would naturally be situated earlier. This is coincident with previous analyses of resilience in the supply chain, such as that of Ponomarov and Holcomb (2009, 131), who define supply chain resilience as "the adaptive capability of the supply chain to prepare for unexpected events, respond to disruptions, and recover from them by maintaining continuity of operations at the desired level of connectedness and control over structure and function."

Following this line of research that proposes the possibility of distinguishing organizational resilience in different situations, Williams et al. (2017) define "resilience" as the process by which an actor (i.e., individual, organization, or community) builds and uses its capability endowments to interact with the environment in a way that positively adjusts and maintains functioning in three different situations: before, during, and following adversity. We can thus consider resilience to be a three-dimension process with pre-adversity capabilities, in-crisis organizing and adjusting, and post-crisis resilience responding and recovery. Based on the proposal of

Table 1 Dimensions of resilience

Dimension	Explanation
Pre-adversity capabilities	Processes aimed at anticipating, preventing, or mitigating potential dangers or disruptions before damage is done. Any adjustments made to deal with unexpected or unknown contingencies to prevent them from arising into a triggering event. These additionally include any resource or capacity that helps the organization to prevent any danger or crisis
In-crisis organizing and adjusting	Maintaining positive functioning in the aftermath of a disaster or disruption. When exposed to a disturbance, organizations try to generate the most effective responses that can preserve the best performance. These include intra- and inter-organizational responses because the supply chain can be affected
Post-crisis resilience responses and recovery	If a disruption or disaster could not be avoided, organizations need to recover as fast as possible. This dimension of resilience focuses on organizations' capacity to return to their original level of performance before the disruption or disaster

Source: Author's elaboration based on Williams et al. (2017)

Williams et al. (2017), Table 1 offers a short description of these three dimensions of organizational resilience.

3 Categorizing Disruptions in the Supply Chain

In this section, we classify the different types of disruptions that organizations may experience within the supply chain. In this sense, we follow the work of Miller and Engemann (2019), who consider three types of disruptions: internal, social, and environmental disruptions.

Miller and Engemann (2019) state that internal disruptions are related to the supply chain itself, as it comprises the infrastructure and resources that they need to operate. Some examples they give are supply-chain design, its capacity to deliver and its flexibility, the incompetent infrastructure (air, rail, water, electrical), and the managerial deficiencies between demand and supply, among other aspects. Whereas internal disruptions are those that occur inside a firm, social disruptions are related to social and organizational systems that are external to a firm. Some examples include adverse responses from local communities or political risks, such as changes in trade agreements, government and international organization regulations and guidelines, nongovernmental organization (NGO) campaigns, wars, or mistreatment of workers in factories. Lastly, they define environmental disruptions as incidents related to disasters such as tsunamis, earthquakes, the depletion of raw material resources, fires, or any catastrophe that affects the natural environment and their aftermath in the community and economy.

In addition to these three typologies, we propose a fourth type of disruption that we name "global disruptions." This new type of disruption includes situations like

Table 2 Examples of studies about the different types of disruptions

Type of disruption	Authors	Event studied
Internal disruptions	Pavlov et al. (2019)	They analyzed internal disruptions of supply chains and they found that establishing a contingency plan is vital for a firm
Environmental disruptions	Eweje and Sakaki (2015)	They mentioned the Tsunami of Japan in 2011 in which Japanese firms pulled their resources together and were involved in helping the affected victims and communities
	Williams and Shepherd (2016)	They studied the Haiti Earthquake, considering that the creation of local ventures was important for the success of the recovery
	Dwivedi et al. (2018)	They centered their work on natural disasters that occurred in Bangladesh (e.g., Bhola cyclone in 1970, Daulatpur-Saturia whirlwind in 1989, inundations in 1998)
Social disruptions	Gittell et al. (2006)	They analyzed airlines' responses to the September 11th crisis and found that airlines that did not fire employees during the crisis recovered faster than others
Global disruptions	Rao and Greve (2018)	They studied the effect of Spanish flu in Norway and how different communities reacted to the pandemic
	Ivanov (2020); Kovács and Sigala (2021); Shih (2020) Siagian et al. (2021)	These scholars analyzed the actual situation of COVID-19 from different perspectives: the impact of this disruption on firms, the used policies to deal with the pandemic crisis; and the lessons to prevent difficulties in future similar disruptions

pandemics (COVID-19, Spanish flu), financial crises (financial crisis of 2007–2008), and situations that exceed the previous categories (for example, a war conflict).

Using this classification, we can categorize previous literature on supply chain disruptions according to the type of disruption they analyzed. Table 2 includes examples of previous studies on these four types of disruptions.

4 Sustainability Practices for Dealing with Supply Chain Disruptions

In this section, we discuss how specific sustainability practices can address the four types of supply chain disruptions (i.e., internal, social, environmental, and global), respectively. By reviewing the literature about disruptions and the different practices to deal with them, we summarize all the ideas and propositions in a table. Table 3 can help to clarify which practices can be better to deal with the different disruptions facing the different dimensions of resilience.

4.1 Sustainability Practices Based on Firms' Internal Planning, Organization, and Internal Stakeholders

To reach a proper functioning of the firm, clear objectives and practices to achieve them must be defined. Strategic planning determines the direction of the actions that firms will implement to achieve their objectives and overcome disruptions. Therefore, internal organization and planning are necessary for firms to make successful decisions and increase their resilience and sustainability. Pavlov et al. (2019) propose developing a contingency plan with procedures and instructions for the proper functioning of a firm under the possibility of adverse situations as a sustainable practice. This contingency plan should assess risks and identify measures and techniques to respond to such risks. In fact, it represents a recovery practice that is created at the pre-disruption stage and executed at the post-disruption stage. This plan must consider sustainability issues such as reducing resource consumption and nonutilized resource reservations.

One of the most important tools in firms' strategic planning is internal communication and internal decision-making (de Vries et al., 2022). Thus, the preparation and implementation of a contingency plan require communication mechanisms between all internal stakeholders in the firm (employees, managers, and owners). Providing a global approach to the current and upcoming situation and evaluating all the scenarios of the firm's actions for all involved and thus informing, and involving them in the conversation help internal stakeholders feel more comfortable because they know what to expect. Thereby, through information exchanges, all members contribute different points of view to develop a contingency plan that is completely efficient in moments of disruption. Employees may think that developing a contingency plan is a primary function of the firm's top management; however, the firm can motivate employees through incentives (extra remuneration, bonuses, time off, or others) as a way of expressing its appreciation for their collaboration in the development of the plan. In this way, job satisfaction and employee performance increase. In situations with heavy disruptions, employees are one of the most vulnerable members of the firm, and it is important that they feel that they are being supported by the firm itself. As an example, Gittell et al. (2006) show that firms that do not fire employees during a crisis recover more quickly than firms that bet on dismissal. Another example is the situation created by the COVID-19 pandemic, in which many employees have been fired or have suffered a reduction in their salaries (Siagian et al., 2021). However, some firms have decided to maintain jobs even when they cannot continue with the production process, and this decision has improved employees' motivation and popular opinion across such firms. These

Table 3 Examples of sustainability practices that can be used to deal with disruptions in the supply chain

		Dimensions of resilience			
		Pre-adversity		In-crisis	Post-crisis and recovery
		Common practice	Specific practices		
Typology of disruption	Internal disruption	Design communication mechanisms between internal and external stakeholders and ensure high coordination. Foster long-term relationships, motivation, and collaboration between them	Establish a contingency plan that identifies response measures and techniques Promote job satisfaction through incentives that motivate employees Provide internal stakeholders with a global approach to the situation	Activate the contingency plan and control its implementation Implement collaboration and communication mechanisms Organize events and meetings to exchange information between the different agents and find a conjunct solution Teamwork and constant communication with employees (feedback process)	Measure the impact of the plan and extract useful information for future contingency plans Exchange information between employees regarding the results of the plan and modify the original with new ideas by organizing events and group meetings Implement a feedback process and evaluate whether the employees have carried out the transmission of the information correctly
	Social disruption		Participate in training courses on possible social disruptions Prioritize the stakeholders that have a direct impact on the firm	Increase employees' motivation by establishing guidelines for flexibility at work Foster relationships with interest groups such as workers' unions or local NGOs	Strengthen relationships with stakeholders Determine what tactics and ideas for improvement have been achieved through these relationships
	Environmental disruption		Operate in more than one place, taking care of environmental concerns Comply with global regulations that guarantee	Pull resources together with other members of the supply chain and get involved in helping the affected victims and communities	Create local ventures for the success of recovery in the affected area Acquire knowledge and learning about caring for the

(continued)

Table 3 (continued)

		Dimensions of resilience			
		Pre-adversity		In-crisis	Post-crisis and recovery
		Common practice	Specific practices		
			elements of responsible management Understand the regulations of the countries in which they operate and be proactive		environment to avoid future natural catastrophes
	Global disruption		Collaborate with members of other supply chains, thanks to internationalization Have trusted partners and help them in other disruptions increasing the payment period or preventing them from going bankrupt to obtain the same behavior in a disruption of its own	Establish a clear chain of commands among versatile groups. Foster information sharing and coordination within the supply chain	Have a good communication and marketing program to show the efforts of the organization

elements that can improve the recovery of a firm are aligned with the results of Gittell et al. (2006).

4.2 Sustainability Practices Based on Relationships and Collaborations with External Stakeholders

Considering that many disruptions are related to aspects external to firms, the sustainability practices we propose are based on relationships with stakeholders (suppliers, customers, society, and governments). To proactively manage these relationships, on the one hand, the firm must identify the external stakeholders that have a direct impact on it, such as those that are affected by its activity and its current relationship with each of them (Vizcaino et al., 2021). It must also foster relationships with interest groups such as workers' unions or local NGOs, always prioritizing those interested in the best practices that can benefit the firm (Yilmaz-Börekçi et al., 2021).

Hence, when firms assume that collective action is needed to deal with disruptions and acknowledge that external stakeholders should find a solution to a collective action problem together, especially when their reputations are interdependent and when there may be negative legitimacy spillover effects (De Bakker et al., 2019), both are more resilient. Firms should consequently define a stakeholder relationship plan to ensure that their corporate behavior meets the expectations of their stakeholders, strengthen relationships with stakeholders that generate value for the firm, and ensure that their collective actions are sustainable (Dev et al., 2021). However, the information exchange between the supply chain and the firm is necessary to face any type of disruption. This sustainability practice is supported by Ivanov (2020), who proposes that collaboration with other members of the supply chain can help reduce outbreaks and provide faster recovery. In this sense, firms can supervise partners and establish procedural guidelines for good coordination and management when an adverse situation occurs.

In addition, we argue that a stable relationship with external stakeholders is essential to the contribution of resources in any type of disruption (e.g., reconstruction of houses and property, food, health resources, employment, and economic resources, among others) (Eweje & Sakaki, 2015). Supply chains have the advantage of being located in different areas (vulnerable or not), making it easier for them to help victims and the affected communities by providing resources and assistance from other parts of the world. However, governments and customers located in the same country as the firm can also provide the necessary resources.

4.3 Sustainability Practices Based on Business, Financial, and Regulatory Orientation

Each country has a series of regulations and requirements that must be met to avoid conflicts. It is important to know these requirements in advance, since in some cases, their implementation may take some time. Firms must know the regulations of the countries in which they and their supply chain partners operate to be proactive in the face of any disaster in addition to know the characteristics of the area and the decisions to be made when a disruption has occurred (e.g., natural disaster, a war conflict, or financial crisis). Once a disaster or disruption has occurred in a specific geographic area, firms and supply chains located in other geographic areas should acquire knowledge and learn about caring for the environment to avoid future similar disruptions.

Additionally, most of the disruptions have such a high impact on the economy that they can generate a global crisis, especially global disruptions. Several of the works that we have cited as examples have focused on the study of global health crises, such as the Spanish flu and the current COVID-19 pandemic (e.g., Ivanov, 2020; Rao & Greve, 2018). They argue that contagious disease outbreaks that are transmitted from person to person lower within-group social integration, foster distrust, and create cooperation deficits that lower resilience. In addition, Ivanov (2020) considers that epidemic outbreaks represent a special case of supply chain disruptions that are characterized by three components: (a) long-term disruption and unpredictable scaling, (b) simultaneous disruption propagation in the supply chain (i.e., the ripple effect) and epidemic outbreak propagation in the population (i.e., pandemic propagation), and (c) simultaneous disruptions in supply, demand, and logistics infrastructure. Unlike other disruptions, epidemic outbreaks start small but scale quickly and disperse over many geographic regions. In line with these arguments, Kovács and Sigala (2021) mention that COVID-19 has impacted production lines and manufacturing capacities. The movement of people and materials has been blocked, causing supply chain disruptions. Mainstream supply chain management has been at a loss in responding to these disruptions, mostly due to a dominant focus on minimizing costs for stable operations while following lean, just-in-time, and zero-inventory approaches.

Therefore, firms affected by such serious disruptions prioritize their payments to seek resources and mechanisms to survive, which represents a great economic loss in the firm itself. Due to the importance of collaboration between different agents, we propose that partner firms help affected firms by increasing payment terms. The objective is that, collaboratively, this would prevent the affected firms from going bankrupt and, thus, obtain the same behavior in their own disruption. Another sustainability practice that we suggest is based on the creation of local ventures (e.g., Williams and Shepherd 2016) within the affected areas to alleviate victims' suffering of the victims. The examples taken from the humanitarian response to the COVID-19 pandemic analyzed by Kovács and Sigala (2021) are also relevant here.

These scholars show how standardization, innovation, and collaboration from social sustainability within the supply chain are valid features to improve resilience.

4.4 Sustainability Practices Based on an Environmental Orientation

Environmental disruptions do not affect all geographic areas equally. Some areas are more vulnerable to certain types of natural disasters than others. We propose to develop multiple connections among firms and supply chains that are located in different geographical areas where natural disasters differ from each other and, therefore, the damage caused, differ. Thereby, when a natural disaster has affected an area, other firms and supply chains commit to help and supply the necessary resources to support victims to be accorded the same treatment during a disruption of its own. Therefore, having good relationships with international supply chains is a key factor in this type of disruption.

However, supply chains must ensure that they comply with global regulations that guarantee elements of responsible management, such as ISO 14001 certificate, or those that certify a source of raw material with the least environmental impact, in addition to committing to reducing emissions, discharges, and energy consumption besides ensuring the sustainable management of products and mobility (eco-mobility). A sustainable supply chain can prevent many natural disasters internationally.

Finally, we summarize the different practices outlined in Table 3. In this table, practices are categorized by the different types of disruptions (internal, social, environmental, and global disruptions) and, in addition, by the different dimensions of resilience that have been explained in the previous sections of this chapter (pre-adversity moment, in-crisis, post-crisis). We now specify in what type of disruption and in what moment of a crisis each proposed practice should be applied.

Before any type of disruption, we suggest identifying the firm's direct stakeholders and designing communication mechanisms and information sharing between all members (internal and external stakeholders). There are two important functions that firms should develop to make collective actions successful. On the one hand, firms must maintain the motivation and satisfaction of their own employees; on the other hand, firms must maintain long-term relationships with external stakeholders, such as supply chains, customers, and governments, among others (Dev et al., 2021). Holding events and meetings with the aim of information exchange is a sustainability practice that can be used to find conjoint solutions. Further, firms must prioritize stakeholders that have a direct impact on the firm and foster relationships with interest groups such as workers' unions or local NGOs, always prioritizing those interested in the best practices that can benefit the firm. The objective is to maintain long-term relationships with the stakeholders with the greatest impact on the firm to get support during possible disruptions.

In the case of internal disruption, we find that firms must establish guidelines for high coordination and management to create a contingency plan that identifies measures and techniques that can be used to respond to any disruption. In the time of a crisis, motivated employees can improve a firm's performance. Once the situation of adversity has been overcome, a feedback process should be implemented to evaluate whether the employees have correctly transmitted the information. Furthermore, it should be evaluated whether it was possible to communicate with and motivate employees. At the end of a crisis, the firm must measure the plan's impact: how, when, and where the contingency plan is made and the acquired benefits and the limitations and emphasize the objective criteria for assessing impact, since this may provide useful information for future contingency plans.

Preparing employees to face possible social disruptions is a challenge that requires holding training courses on possible social disorder, workers' rights, the operation of NGOs, and international regulations, among others. During disruptions or crises, employees must be highly motivated to form a resilient firm; this can be encouraged through flexible guidelines at work. Scholars who have studied changes in working conditions and flexibility in times of crises (e.g., Lin et al., 2021; Sorribes et al., 2021) have found that employees' performance and their engagement with firms increase when motivation is strengthened through adaptative practices focused on employee well-being and satisfaction. As an example, due to the COVID-19 pandemic crisis, many firms have offered their employees a choice between working at the office or telecommuting (Lin et al., 2021), adapting the employee's working hours to their personal conditions. Other adaptative practices that firms can implement to increase motivation and resilience include extending deadlines for the delivery of tasks, granting rewards, and fostering relationships with interest groups such as labor unions or local NGOs. Additionally, firms should maintain constant communication with employees and exchange information on disruptions, allowing employees to understand their participation in the firm and increasing their commitment to return to stability.

Although communication practices and relationships with other members are necessary for all types of disruptions, we propose several specific practices for environmental disruptions. Before any natural catastrophe occurs which has serious impacts on a community, we suggest that firms develop relationships with supply chains in different geographical areas so that the affected areas can easily obtain resources. To this end, an understanding of the environmental regulations of other countries is an important factor to anticipate possible natural disasters, besides complying with all global regulations that guarantee responsible management (sustainable materials and low levels of pollution in business operations, among others). During and after a natural disaster in a geographic area, we propose that strong relationships with other trusted partners could lead to jointly contributing resources to the affected areas (food, financing, and labor) and helping to create local ventures for the victims.

Lastly, global disruptions are more complex and can create a global crisis, leaving communities without economic resources in most cases. Once again, collaborative work between different members (firms, governments, and suppliers) acquires a significant value in the face of global disruptions (Grant & Wunder, 2021). In addition, we propose that extending the payment periods and conditions between firms could prevent the affected organizations from going bankrupt. By the same token, we propose that firms and supply chains establish a good communication and marketing program to show the efforts the organization is making.

5 Conclusions

This chapter provides a better understanding of how sustainability practices can improve the management of the supply chain to create resilience. Based on the findings of previous literature, we first analyze the different dimensions of resilience (pre-adversity, in-crisis organizing, and post-crisis resilience) and the typology of disruptions in the supply chain (internal, environmental, social, and global disruptions). We then classify sustainability practices that can increase the different dimensions of resilience in supply chains during different types of disruptions.

Firms do not survive simply due to their internal resources; rather, they survive because of their ability to adapt to and/or interact dynamically with their environment. Thus, it is important to consider how they manage difficulties related to the supply chain. In this work, we have shown how the role of sustainability practices can determine a firm's survival or recovery from supply chain disruptions.

Previous studies have analyzed different practices that can help improve resilience to different types of disruptions (e.g., Miller & Engemann, 2019). The contribution of this chapter is to deepen this research and propose a classification of sustainability practices that can foster resilience by combining the classification of disruptions and the three dimensions of resilience. This classification can provide a better understanding of how these sustainable practices work to increase resilience. Although we have presented some examples of disruptions and the practices that have been followed to solve them, many more could be useful when adapted to different contexts; some could even be used interchangeably for different types of disruptions. However, we have focused on practices that are related to sustainability and are useful for resilience in the three possible dimensions and to prepare for different kinds of disruptions.

This work can provide practitioners with a better comprehension of how sustainability commitment can foster relationships with suppliers and customers. For example, in the ISO created in 2017, 22316:2017 focuses on the resilience of organizations. This ISO suggests the improvement of the risk management of all the aspects of a firm, specifically the relationship with suppliers. The concepts laid out in this chapter provide practitioners with some clues for managing firms that seem to have or are trying to recover from financial problems. This may help them understand that establishing relationships with suppliers or customers focused on sustainability commitment can improve their ability to avoid disruptions or their success during disruptions and increase their recovery rates after a disruption. Therefore, the division of resilience into three dimensions is important. We have also highlighted some practices that include collaborations with other partners in the supply chain to improve resilience.

The issues highlighted in this chapter can be used to propose some ideas for future research. We believe that more empirical research is necessary to determine how firms' sustainability practices or their relationships with suppliers or customers can improve resilience. Our proposals included on Table 3 open the door to future studies that could deepen the mechanisms underlying these relationships and empirically test them. Further research could elaborate upon other ways in which sustainability can help avoid or at least mitigate disruptions (pre-adversity capabilities), as resilience is usually studied during or after a disruption. Future research should also address the issue of determining the differences between the role of suppliers and customers in the fostering of resilience.

This work is especially important in situations of crises or disruptions like the global pandemic of 2020, which demonstrated the necessity of resilient supply chains in resisting the problems generated by the spread of COVID-19. Our work also provides information for improving the pre-adversity system to avoid risks and mitigate the problems associated with future crises and disruptions.

Acknowledgment This work was supported by the Spanish State Research Agency grant number PID2019-107767GA-I00/SRA, State Research Agency/10.13039/501100011033

References

Bertschek, I., Polder, M., & Schulte, P. (2019). ICT and resilience in times of crisis: evidence from cross-country micro moments data. *Economics of Innovation and New Technology, 28*(8), 759–774.

Blome, C., Paulraj, A., & Schuetz, K. (2014). Supply chain collaboration and sustainability: A profile deviation analysis. *International Journal of Operations & Production Management, 34*(4), 639–663.

Cooke, F. L., Wood, G., Wang, M., & Li, A. S. (2021). Riding the tides of mergers and acquisitions by building a resilient workforce: A framework for studying the role of human resource management. *Human Resource Management Review, 31*(3), 100747.

Çop, S., Olorunsola, V. O., & Alola, U. V. (2021). Achieving environmental sustainability through green transformational leadership policy: Can green team resilience help? *Business Strategy and the Environment, 30*(1), 671–682.

De Bakker, F. G., Rasche, A., & Ponte, S. (2019). Multi-stakeholder initiatives on sustainability: A cross-disciplinary review and research agenda for business ethics. *Business Ethics Quarterly, 29*(3), 343–383.

Dev, N. K., Shankar, R., Zacharia, Z. G., & Swami, S. (2021). Supply chain resilience for managing the ripple effect in Industry 4.0 for green product diffusion. *International Journal of Physical Distribution & Logistics Management, 51*(8), 897–930.

De Vries, T. A., van der Vegt, S. G., Scholten, K., & van Donk, D. P. (2022). Heeding supply chain disruption warnings: When and how do cross-functional teams ensure firm robustness? *Journal of Supply Chain Management, 58*(1), 31–50.

Dwivedi, Y. K., Shareef, M. A., Mukerji, B., Rana, N. P., & Kapoor, K. K. (2018). Involvement in emergency supply chain for disaster management: a cognitive dissonance perspective. *International Journal of Production Research, 56*(21), 6758–6773.

Eweje, G., & Sakaki, M. (2015). CSR in Japanese companies: Perspectives from managers. *Business Strategy and the Environment, 24*(7), 678–687.

Fahimnia, B., & Jabbarzadeh, A. (2016). Marrying supply chain sustainability and resilience: A match made in heaven. *Transportation Research Part E: Logistics and Transportation Review, 91*, 306–324.

Folke, C. (2006). Resilience: The emergence of a perspective for social–ecological systems analyses. *Global Environmental Change, 16*(3), 253–267.

Gittell, J. H., Cameron, K., Lim, S., & Rivas, V. (2006). Relationships, layoffs and organizational resilience: Airline responses to crisis of September 11th. *Journal of Applied Behavioral Science, 42*(3), 300–329.

Grant, J., & Wunder, T. (2021). Strategic transformation to sustilience: learning from COVID-19. *Journal of Strategy and Management, 14*(3), 331–351.

Hällgren, M., Rouleau, L., & de Rond, M. (2018). A matter of life or death: how extreme context research matters for management and organization studies. *Academy of Management Annals, 12*(1), 111–153.

Hamel, G., & Välikangas, L. (2003). The quest for resilience. *Harvard Business Review, 81*(9), 52–62.

Holling, C. S. (1996). Engineering resilience versus ecological resilience. In P. Schulze (Ed.), *Engineering within ecological constraints*. National Academy Press.

Ivanov, D. (2020). Predicting the impacts of epidemic outbreaks on global supply chains: A simulation-based analysis on the coronavirus outbreak (COVID-19/SARS-CoV-2) case. *Transportation Research Part E: Logistics and Transportation Review, 136*, 101922.

Jabbarzadeh, A., Fahimnia, B., & Sabouhi, F. (2018). Resilient and sustainable supply chain design: sustainability analysis under disruption risks. *International Journal of Production Research, 56*(17), 5945–5968.

Korhonen, J. E., Koskivaara, A., Makkonen, T., Yakusheva, N., & Malkamäki, A. (2021). Resilient cross-border regional innovation systems for sustainability? A systematic review of drivers and constraints. *Innovation: The European Journal of Social Science Research, 34*(2), 1–20.

Kovács, G., & Sigala, I. F. (2021). Lessons learned from humanitarian logistics to manage supply chain disruptions. *Journal of Supply Chain Management, 57*(1), 41–49.

Lin, W., Shao, Y., Li, G., Guo, Y., & Zhan, X. (2021). The psychological implications of COVID-19 on employee job insecurity and its consequences: The mitigating role of organization adaptive practices. *Journal of Applied Psychology, 106*(3), 317–329.

Linnenluecke, M. (2017). Resilience in business and management research: a review of influential publications and a research agenda. *International Journal of Management Reviews, 19*(1), 4–30.

Miller, H. E., & Engemann, K. J. (2019). Resilience and sustainability in supply chains. In G. A. Zsidisin & M. Henke (Eds.), *Revisiting supply chain risk*. Springer Series in Supply Chain Management.

Negri, M., Cagno, E., Colicchia, C., & Sarkis, J. (2021). Integrating sustainability and resilience in the supply chain: A systematic literature review and a research agenda. *Business Strategy and the Environment, 30*(7), 2858–2886.

Ortiz-de-Mandojana, N., & Bansal, P. (2016). The long-term benefits of organizational resilience through sustainable business practices. *Strategic Management Journal, 37*, 1615–1631.

Pavlov, A., Ivanov, D., Pavlov, D., & Slinko, A. (2019). Optimization of network redundancy and contingency planning in sustainable and resilient supply chain resource management under conditions of structural dynamics. *Annals of Operations Research*, 1–30.

Pimm, S. L. (1991). *The balance of nature? Ecological issues in the conservation of species and communities*. The University of Chicago Press.

Ponomarov, S. Y., & Holcomb, M. C. (2009). Understanding the concept of supply chain resilience. *International Journal of Logistics Management, 20*(1), 124–143.

Rao, H., & Greve, H. R. (2018). Disasters and community resilience: Spanish flu and the formation of retail cooperatives in Norway. *Academy of Management Journal, 61*(1), 5–25.

Santoro, G., Messeni-Petruzzelli, A., & Del Giudice, M. (2021). Searching for resilience: The impact of employee-level and entrepreneur-level resilience on firm performance in small family firms. *Small Business Economics, 57*(1), 455–471.

Sarkis, J. (2020). Supply chain sustainability: Learning from the COVID-19 pandemic. *International Journal of Operations & Production Management, 41*(1), 63–73.

Shashi, Centobelli, P., Cerchione, R., & Ertz, M. (2020). Managing supply chain resilience to pursue business and environmental strategies. *Business Strategy and the Environment, 29*(3), 1215–1246.

Shih, W. C. (2020). Global supply chains in a post-pandemic world. *Harvard Business Review, 98*(5), 82–89.

Siagian, H., Zeplin Jiwa, H. T., & Ferry, J. (2021). Supply chain integration enables resilience, flexibility, and innovation to improve business performance in COVID-19 era. *Sustainability, 13*(9), 4669–4688.

Sorribes, J., Celma, D., & Martínez-Garcia, E. (2021). Sustainable human resources management in crisis contexts: Interaction of socially responsible labour practices for the wellbeing of employees. *Corporate Social Responsibility and Environmental Management, 28*(2), 936–952.

Vizcaino, F. V., Cardenas, J. J., & Cardenas, M. (2021). A look at the social entrepreneur: the effects of resilience and power distance personality traits on consumers' perceptions of corporate social sustainability. *International Entrepreneurship and Management Journal, 17*(1), 83–103.

Williams, T. A., Gruber, D. A., Sutcliffe, K. M., Shepherd, D. A., & Zhao, E. Y. (2017). Organizational response to adversity: Fusing crisis management and resilience research streams. *Academy of Management Annals, 11*(2), 733–769.

Williams, T. A., & Shepherd, D. A. (2016). Building resilience or providing sustenance: Different paths of emergent ventures in the aftermath of the Haiti earthquake. *Academy of Management Journal, 59*(6), 2069–2102.

Yilmaz-Börekçi, D., Rofcanin, Y., Heras, M. L., & Berber, A. (2021). Deconstructing organizational resilience: A multiple-case study. *Journal of Management & Organization, 27*(3), 422–441.

A Theoretical Framework for Supply Chain Resilience Planning

Jennifer F. Helgeson and Alfredo Roa-Henriquez

Abstract The current literature on resilience planning, especially related to supply chains, rarely considers the difference between intended and actualized behaviors toward mitigation and adaptation actions. However, a potential contributor to taking on supply chain pre-disaster planning tends to be individual and institutional risk perceptions encountered in the Theory of Planned Behavior (TPB) (Ajzen I Organizational Behavior and Human Decision Processes 50:179–211, 1991). The theory suggests that an agent's intentions to implement mitigation and/or adaptation actions inform their resilience capacity toward a given disaster event that is comparable to data on interruption and recovery post-event.

The gist of this chapter hinges on the idea that supply chain management (SCM) has been largely successful in providing a normative framework supporting decisions involved in the design (e.g., plant location, sourcing and procurement, transportation), planning (e.g., demand forecasting, aggregate planning), coordination (e.g., organizational talent, collaboration, and integration), and risk management (e.g., excess capacity, inventory buffers, suppliers diversification); however, the field has largely not addressed the underlying behavioral mechanism that drives an agent's decision-making process in the specific context of pre-disaster planning (i.e., mitigation) and adaptation decisions. Failing to understand why intention and actualized behavior toward mitigation and adaptation differ is an obstacle to effectively coping with disruption risks and may pose a threat to the resilience of the supply chain given that some mitigation actions aim at building or increasing a firm's inherent resilience capacity and at improving its ability to adapt to disruptions potentially affecting business continuity. This topic is particularly relevant for

J. F. Helgeson (✉)
Engineering Laboratory, National Institute of Standards and Technology, Gaithersburg, MD, USA
e-mail: Jennifer.helgeson@nist.gov

A. Roa-Henriquez
College of Business and Challey Institute for Global Innovation and Growth, North Dakota State University, Fargo, ND, USA
e-mail: alfredo.roa@ndsu.edu

small- and medium-sized enterprises (SMEs) that play a critical role within their communities and do not have resources to incorporate sophisticated business continuity plans or emergency management plans within their risk management frameworks.

1 Introduction

In much of the resilience planning literature, especially related to supply chains, the difference between intended and actualized behaviors toward mitigation and adaptation actions is not recognized nor discussed. One possible reason behind the lack of studies in this specific topic is that the nature of natural disasters is so complex that the tasks of (1) obtaining a pre-event snapshot of a given agent's behavioral intention around mitigation and adaptation and (2) comparing intention and actual behavior immediately before and after a disruptive event are almost unattainable. Another possible reason may be related to previous findings suggesting that prior disaster experience is positively related to pre-disaster planning (see, e.g., Dahlhamer & D'Souza, 1997), which may lead to the belief that firms just replicate those actions that have been useful in the past in coping with future disruptions. However, it has been found a non-significant statistical influence of prior disaster experience and pre-disaster planning on long-term recovery (e.g., Webb et al., 2002), indicating that if a true relationship exists, this might be mediated by another variable or influenced by different factors that have not currently being empirically explored in the literature (e.g., behavioral). For instance, past research indicates that individuals do not always engage in pre-disaster planning or mitigation—even when they have sufficient resources, preparedness training, or a history of disaster exposure (National Research Council, 2006; Kunreuther et al., 2013).

A potential and significant contributor to taking on mitigation and adaptation in a context of business resilience to disasters tends to be individual and institutional risk perceptions. The Theory of Planned Behavior (TPB) (Ajzen, 1991) offers additional considerations that mediate across intended versus actualized behaviors, namely attitude toward the behavior (ATT), subjective norm (SN), and perceived behavioral control (PBC). The TPB can be directly applied in the domain of business resilience planning—where a category of mitigation and adaptation behaviors may be considered. Yet, it is well worth mentioning that intended and actualized behaviors also depend upon complex interactions among agents who are internal and external to a given firm, as well as the option sets available due to the firm size and geographic reach. Furthermore, there are complex interactions between businesses and the communities in which they are situated. Much of the research on the role of businesses in community resilience is dominated by larger businesses due to the use of biased metrics such as press releases (McKnight & Linnenluecke, 2019); however, the critical role that small- and medium-sized enterprises (SMEs) play within a community is significant. SMEs contribute significantly to local economic development and job creation (Morrison et al., 2003; Schaer et al., 2018). They are invested in their communities and improve local median household income, reduce poverty, and decrease income inequality (Blanchard et al., 2012; Lyson et al., 2001; Tolbert et al., 1998). Grimm (2013) integrated all businesses into a theoretical community continuity planning model, indicating that local resilience could be increased through comprehensive business and community planning. However, the mechanisms for such planning are not fully understood (Howe, 2011; Spillan & Hough, 2003; Yoshida & Deyle, 2005). In this chapter, we treat the level of analysis as a business represented by an owner or manager of the firm[1] and suggest that the model presented based on TPB can be applied across firms, controlling for size and type.[2]

In this chapter, we also discuss risk management at the firm-level and review the need for research that addresses planning for resilience in addition to risk management in the context of the need to plan ahead and recovery post-disaster. To date, behavioral supply chain management (BSCM) has largely addressed issues such as inventory management and the "bullwhip" effect[3] procurement, forecasting, buyer–supplier relationships, and information sharing (for a literature review, see, e.g., Fahimnia et al., 2019). All of these research areas are applicable to the specific domain of managing supply chain risk, but have largely not been addressed in the specific context of firm-level resilience planning for a disaster event. Schorsch et al. (2017) provided a meta-theory of BSCM and indicated that two areas for continued research are (1) broadening the scope of inventory and capacity decision-making as well as (2) incorporating more research that addresses integration of cognitive and social research on the decision-making process.

The rest of the chapter is organized as follows. Section 2 provides background and a review of literature relevant to supply chain resilience planning and disaster management processes. Section 3 provides an overview of relevant theoretical frameworks that deal with mental models, iterative learning, and prediction of protective actions in addition to the TPB. Section 4 introduces a framework for application of the TPB into a firm's risk-related planning decisions ex ante and ex post given a shock. Lastly, Sect. 5 summarizes areas for future research and provides conclusions.

---

[1] Although a firm is usually referred in the literature as a corporation or large enterprise with multiple business, in this chapter, we use the words "firm," "business," and "organization" as synonyms.
[2] However, the few studies that have focused on SMEs have found that there is a lack of planning due to the cost and complexity of creating a plan (Runyan, 2006). Finances and size contribute to the relatively low planning and resilience of SMEs (Alesch et al., 2001; Zhang et al., 2009).
[3] See Lee et al. (1997) to explore a detailed explanation on the "bullwhip" effect.

2 Background and Relevant Literature

2.1 *Value Chain vs Supply Chain*

Analyzing an operational activity or set of operational activities is key for a firm to maintain competitiveness. This analysis is usually framed within the firm's strategy, the ultimate goal of which is to create value[4] for buyers that exceeds the cost of creating the product so the firm can achieve competitive advantage (Porter, 1985). *Value chain* analysis refers to the process by which a firm identifies primary and support activities that add value to its final product and analyze how these may be altered to enhance customer value and/or increase the competitive strength of the organization (i.e., its ability to increase revenue). The approach rests upon the need to separately analyze the contribution of organizational activities and independently assess their value added.

Primary activities involve inbound logistics (i.e., receiving, storage, and distribution of inputs), operations (i.e., transformation of inputs into outputs), outbound logistics (i.e., storage, distribution, and delivery of products), marketing and sales (i.e., consumers are made aware of and can purchase products), and service (i.e., enhance product's value such as customer support). The analysis also includes the examination of support activities such as procurement, human resource management, technology, and infrastructure.

Among all the activities undertaken by firms, the ones related to inbound logistics, operations, and outbound logistics frequently take place within a wider system—the *supply chain*—that comprises a network of firms, each of them adding value to the chain (Porter, 1985). In a system like this, a single organization is not forced to perform all the primary activities because each individual firm relies upon others in their supply network. In this context, value creation occurs at the level of the supply system and one source of competitive advantage for an individual firm is determined by the linkages with its network partners (Lowson, 2002). Fundamentally, the supply chain encompasses the flow and storage of the raw material, semi-finished goods, and the finished goods from point of origin to its final destination (i.e., consumer consumption), disposal, and links across suppliers, manufacturers, wholesalers, distributors, retailers, and the customer. Managing a supply chain, therefore, implies delivery of superior customer value at lesser cost by considering the set of upstream and downstream linkages between the different processes and activities executed by those firms involved in the network (Christopher, 1998).

In the current business environment, firms that have all their primary activities vertically integrated within their operations are not common, which implies that

---

[4] Porter (1985) defines value as the amount buyers are willing to pay for what a firm provides them. Value is measured by total revenue, a reflection of the price a firms' product commands and the units it can sell. A firm is profitable if the value it commands exceed the costs involved in creating the product.

most firms are interconnected with others and cannot avoid[5] disruptions derived from any risk or vulnerability embedded in its supply network. This is a reason of why a given firm needs to remain aware of its wider value system or supply chain in which it is situated; however, there are many constituent elements that may influence a firm's performance, but are beyond its direct control (e.g., transportation networks and utility availability). By default, planning at the value chain scope is more easily addressed and controlled by a given firm since it covers the process in which a company adds value to its raw materials to produce products eventually sold to consumers. In this process, the firm aims at achieving competitive advantage since it understands the activities that are performed more cheaply or differently from other firms within the same industry (i.e., competitors). Managing the supply chain, on the other hand, requires coordination with all participants in the chain. In this case, value creation is analyzed at the system level and the ultimate goal of the supply network is aimed at enhancing overall customer satisfaction. This chapter is predicated around how firms cope with disruptions in their supply chains and continue with their operations after an external event occurs.

2.2 Supply Chain Risks and Vulnerabilities

Research on supply chain risk[6] and vulnerability[7] is relatively recent. Although the term *supply chain* dates to the early 1980s (Oliver & Webber, 1992), in the early twenty-first century, there was a surge of studies on the importance of management to minimize firms' risk exposure to global supply chains (e.g., Johnson, 2001; Sheffi, 2001; Jüttner et al., 2003; Christopher & Lee, 2004; Chopra & Sodhi, 2004; Tang, 2006). This research trend was the result of global competition and external events that propelled the decision of firms to redesign their strategies, analyze a broad set of risk categories (e.g., terrorism, political, disruptions, delays, systems risk, forecast risk, procurement risk), exploit growth opportunities, and mitigate potential threats in the operations and performance of their supply chains (see, e.g., Johnson, 2001; Chopra & Sodhi, 2004). Currently, advances in *Supply Chain Risk Management*

---

[5]Technically, the firm cannot avoid any disruption from occurring within its supply network, but it can minimize the effect that any disruption may have on its normal operations.

[6]Although the definition of "risk" can be viewed from a positive and a negative perspective (e.g., business growth opportunity vs business continuity threat), the supply chain literature has emphasized on risks from a downside perspective from which disruptions can emerge. Additionally, the field extends beyond the boundaries of the single firm to include the flows exchange derived from interactions with other firms (Jüttner, 2005; Peck, 2006).

[7]In the supply chain literature, vulnerability is defined as "the existence of random disturbances that lead to deviations in the supply chain of components and material from normal, expected, or planned schedules or activities, all of which cause negative effects or consequences for the involved manufacturer and its sub-contractors" (Svensson, 2000, p. 732).

(SCRM)[8] have moved beyond the well-known "bullwhip" effect as the result of emerging risks and experiences from large-scale disruptions such as the 9/11 World Trade Center attacks in 2001, the outbreak of SARS in 2003, and more recently, the COVID-19 pandemic. All these factors have influenced corporate policies and business processes, including, but not limited to relationships with suppliers, risk pooling, and collaboration in the chain (e.g., Tomlin, 2006; Cao & Zhang, 2011). The COVID-19 pandemic will likely have long-lasting effects on the relationship between government and suppliers of critical goods (see, e.g., Sodhi & Tang, 2021).

The emergence of new risks and the increasingly global nature of supply chains have led the field of SCRM to shift focus toward the analysis of risks derived from short- and long-term impacts of climate change and changing weather patterns. This reconsideration in the global landscape of risks factors is exemplified by the *Global Risk Report 2019* (World Economic Forum, 2019). A recent version of this report considers several environmental factors (e.g., extreme weather events, natural disasters, failure of climate-change mitigation and adaptation efforts) among the top five concerns threatening global economic development.[9] Yet, the 2019 Report does not include the risk of a pandemic or spread of infectious diseases on supply chains among its five highest concerns, highlighting how underprepared the world was to cope with an event like COVID-19. These types of risk had been last included in a World Economic Forum (WEF) Report in 2015 among the five most impactful or disruptive events on supply chains. Its inclusion in the ranking likely arose from experiences with the Ebola epidemic, which occurred from 2014 to 2016, given that leaders have increased preparedness capacity after major outbreaks but, "as the effects of the outbreak fade, neglect sets in again until a new outbreak erupts" (WEF, 2019, p. 48). A similar case has occurred in the past with businesses in the implementation of mitigation measures aimed at reducing physical damage from extreme weather events (EWE). It is important to consider that firms are often reluctant to make major investments, as they are unable to realize financial benefits in the short run; however, previous research indicates that the ability of businesses to recover from EWE disasters appears to be influenced by the amount of property and/or physical damage experienced (e.g., Webb et al., 2002). However, certain types of protective actions may not be available to all firm types, especially SMEs, that are highly sensitive to the cost of implementation, technology, mix of resources, resource availability, among other factors (Dormady et al., 2019). In the case of infrastructure systems, however, developed economies have made large investments that aim to avoid deaths and supply chain disruptions derived from climate change risks, particularly in places where coastal flood risks are highly concentrated (WEF, , 2019).

Current global supply chain networks are vulnerable to a variety of risks and these have been broadly documented in the literature (e.g., Chopra & Sodhi, 2004; Tang,

---

[8] Later in the section, we consider the constitutive elements of a SCRM framework.

[9] Specifically, the report identifies three environmental risk factors among the top five by likelihood of occurrence and three environmental risk factors among the top 5 by magnitude of impact.

2006). To cope with these supply chain risks, the consensus in the field has been the consolidation of the firm's potential threatening events into a "SCRM framework," which pertains to "the identification of potential sources of risk and implementation of appropriate strategies through a coordinated approach among supply chain risk members, to reduce supply chain vulnerability" (Jüttner et al., 2003, p. 201). The process also involves analysis of short-run cost savings and long-run profitability of the supply chain in the context of each risk (Manuj & Mentzer, 2008).

The SCRM framework incorporates four key steps toward risk management, namely: *(1) identification of risk sources, (2) analysis of risk drivers, (3) measurement of risk consequences,* and *(4) implementation of risk mitigating strategies* (Jüttner et al., 2003).

1. *The identification of risk sources* provides a dimension of the extent of potential disruptions (ibid.). The potential risks are far ranging; for example, the WEF defines risk categories as: (1) economic, (2) environmental, (3) geopolitical, (4) societal, or (5) technological. Some SCRM literature highlights three overarching risk source types: (1) environmental risk sources, (2) network-related risk sources, and (3) organizational risk sources (ibid.). Environmental involves risks exogenous to the supply chain (e.g., natural hazards), network-related risks deal with interactions between organizations that make-up a given supply chain (e.g., outsourcing), and organizational involves risks that are internal to the organization (e.g., machine failure) and likely easier for the firm to control.
2. The second element, *risk drivers*, refers to those factors that are prone to exacerbate the underperformance of a supply chain because of the level of organizational exposure (e.g., a business heavily dependent on a supplier located in a country with high political instability).
3. The third element, *measurement of consequences*, is related to outcomes and effects on the supply chain (e.g., a natural hazard that turns into a disaster causes measurable economic impacts).
4. The last element, *mitigating strategies,* are actions aimed to neutralize potential negative effects on the supply chain (ibid.; Chopra & Sodhi, 2004). The option set of potential mitigation strategies depends on the type of risk involved. Disruptions caused by natural disasters (e.g., hurricanes, earthquakes) may be mitigated by having redundant suppliers and a "Strategic Emergency Stock," which should not be used to buffer day-to day fluctuations (Sheffi, 2001; Chopra & Sodhi, 2004). Delays caused by high capacity utilization, inflexibility, or poor quality at the supply source can be mitigated by adding capacity and inventory and increasing responsiveness, flexibility, and source capability (ibid.). Procurement risk caused by dependence on suppliers is mitigated by adding capacity and inventory, having redundant suppliers and increasing flexibility (ibid.). Forecast risk derived from longer supply lines, system uncertainties, and long lead times is mitigated when there are higher levels of collaboration and sharing of information (i.e., visibility) among supply chain participants (Sheffi, 2001; Chopra & Sodhi, 2004; Christopher & Lee, 2004). Some mitigating strategies have the potential to address multiple hazards simultaneously (e.g., expanding capacity and inventory

to cope with extreme weather and procurement risks by bringing on-line physical assets not previously in use and adding inventories before the events) and are preferred to others that may address only one type of risk. However, the implementation of mitigation strategies should be complemented with an analysis of short-run cost savings and long-run profitability. For instance, adding inventory may reduce some risks but may also increase inventory holding costs and the rate of product obsolescence, among others (Chopra & Sodhi, 2004; Manuj & Mentzer, 2008).

In the past, some large firms have reduced risks of supply chain disruptions by storing inventory at *strategic locations* (e.g., distribution centers, logistic hubs) so that all relevant businesses in the region can share in the case of an acute need.[10] Public organizations and communities-at-large may mirror similar strategies by setting up a central Emergency Management Center with clearly defined participant roles and the type of permissible interactions among participants (i.e., which entities get priority in supply provision). For instance, the U.S. Centers for Disease Control and Prevention (CDC) has historically kept large quantities of medicine and medical supplies, known as the strategic national stockpile (SNS), at certain strategic locations in the USA intended to protect the American public against any national emergency (Tang, 2006). Other strategies involve reviewing capacity versus available inventory and planning for operations focused on cash flow opposed to profits (e.g., increase days payable outstanding, reduce days sales outstanding) (Sheffi, 2001, 2020).

2.3 *Supply Chain Resilience*

A *crisis* is defined as the state of a system whose viability is threatened by a low-probability, high-impact event (Pearson & Clair, 1998; Van der Vegt et al., 2015). The identification and analysis of these risks requires collecting historical data, adjusting the data to probability distributions, and estimating statistical models with the objective of predicting future negative consequences. Large corporations frequently depend on these estimations and predictions to design their mitigation strategies (e.g., insurance subscriptions or other types of hedging instruments). Usually, all these risks (i.e., environmental), the ones identified along the supply chain (i.e., network-related), and those internal to the business (i.e., organizational) are integrated in a general Enterprise Risk Management (ERM) framework, which is a more detailed SCRM framework used by large firms to manage their risks (Fiksel, 2015). However, evidence has indicated that an ERM framework may be inadequate to help some organizations rebound effectively from business interruption losses[11]

---

[10] See, e.g., Tang (2006), for a case on Toyota.

[11] Business Interruption losses, which are usually measured by *Gross Sales Revenue*, begin at the point when the event happens but continues until the system has recovered (Dormady et al., 2019).

and insufficient to provide protection against a variety of events that are low-risk, high-consequence, and idiosyncratic in nature.

There are three high-level reasons that have been provided as to why an ERM approach is insufficient to help businesses and their supply chains to completely recover after an external shock (e.g., a natural disaster or a pandemic). One potential explanation is that the whole ERM approach depends on a thorough identification of a given firm's exposure to different hazards as well as the mapping of these risks onto probabilities of occurrence and magnitude of impact. Even if risks can be anticipated, the firm needs reliable information and credible assumptions to estimate associated probabilities and predict the magnitude of their consequences, otherwise the organization runs the risk of misallocating the use of its resources (Fiksel, 2015). A second explanation considers that an ERM approach and the notion of "optimization" pre-event may be unrealistic in a constantly changing business environment; risks faced by businesses are not static and are adjusted by managers as they see opportunities (ibid.). A third explanation suggests that outcomes are often products of compound events that coincide in space and time and their combined consequences are difficult to predict and anticipate (Fiksel, 2015; Van der Vegt et al., 2015). To cope with events that cannot be fully anticipated, are hard to quantify, and have the potential of disrupting global supply chains, some researchers and practitioners have shifted their attention over the last years from a focus of risk management to a focus of increasing supply chain *resilience*[12] (Fiksel, 2015; Van der Vegt et al., 2015).

It is recognized that in today's environment of large interconnected supply chains, disruptions leading to business interruptions and associated losses are inevitable and businesses need to adapt and not merely react (Christopher & Peck, 2004; Skipper & Hanna, 2009). However, it is possible that in the short-term some businesses neglect to adopt adaptive measures and their effects on supply chain resilience because they have incorporated sophisticated business continuity plans or emergency management plans within their ERM frameworks. Nonetheless, firms are at an advantage if they think of resilience as a strategic process that hastens business recovery while also creating competitive advantage, which can have co-benefits on a business' daily activities, and enhance bottom-line objectives.

Research on supply chain resilience has proliferated in recent years and the Supply Chain Management (SCM) field has benefited from definitions and research from other fields.[13] For instance, the U.S. Federal Emergency Management Agency (FEMA) defines *supply chain resilience* as "the ability of a preexisting network of

---

In fact, Business Interruption losses are not necessarily triggered by natural disasters, pandemics, or other catastrophic events. Small disturbances can also have large effects on the organization and threaten its survival (Fiksel, 2015).

[12] The etymology of the word resilience is "resilire" meaning to rebound (Van der Vegt et al., 2015). An ERM approach continues to be a valuable tool when disruptions are predictable and should be complemented with a resilience approach when disruptions are less predictable (Fiksel, 2015).

[13] See, e.g., Rose (2004) for a definition of economic resilience and Adger (2000) for a definition of ecological resilience.

demand and supply to deploy surviving capacity, and/or introduce new capacity, under severe duress. It is the ability of a network, or portion of a network, to continue moving (directing, redirecting, flowing) goods and services even when important elements of the network are no longer operating. For example, the continued flow of water, food, and fuel while the electric power grid is not operating would be an expression of supply chain resilience" (p. 3). In the supply chain literature, supply chain resilience is defined as "the adaptive capability of the supply chain to prepare for unexpected events, respond to disruptions, and recover from them by maintaining continuity of operations at the desired level of connectedness and control over structure and function" (Ponomarov & Holcomb, 2009, p. 131). The commonalities in both definitions indicate the need of the supply chain system to secure adaptive capabilities to maintain operations. Although there is consensus around the need to secure adaptive capabilities along the disruption stages in the supply chain (i.e., readiness, responsiveness, and recovery), there continues to be discrepancies around: (1) how to secure adaptive capabilities (i.e., formative elements), (2) whether the formative elements are "antecedents" or are integrated in the supply chain resilience, and (3) whether the formative elements are captured at the capability or at the resource level (Jüttner & Maklan, 2011). Some authors agree on the formative resilience elements that capture and integrate the adaptive capacity of supply chain resilience as flexibility, velocity, visibility, and collaboration (see, e.g., Christopher & Peck, 2004; Ponomarov & Holcomb, 2009; Jüttner & Maklan, 2011; Scholten & Schilder, 2015), but disagree on the capture level of the formative elements, which can be obtained at an individual and detailed resource level (Sheffi, 2005), as product of a system-level re-engineering of the supply chain (Christopher & Peck, 2004), or by integrating and coordinating resources that usually span functional areas that are present in the supply chain (Ponomarov & Holcomb, 2009).

Some scholars do not consider redundancy to be part of the formative elements, but rather a prerequisite to resilience because adaptive capabilities are more related to the integration and coordination of resources employed by supply chain processes and less related to the balancing requirements between redundant resources and efficiency (see, e.g., Christopher & Peck, 2004; Jüttner & Maklan, 2011; Scholten & Schilder, 2015). Other scholars, however, consider redundancy as part of the formative elements that ensure adaptive capabilities (e.g., Sheffi & Rice, 2005). Nonetheless, the SCM literature still needs to move beyond theory to offer additional management guidance on the implementation and operationalization of the concept of supply chain resilience (Scholten et al., 2014).

2.4 Supply Chain Resilience and Disaster Management

In contrast to the SCM literature that still fails in providing guidance on the implementation and operationalization of the concept of supply chain resilience, some organizations provide managers with recommendations on how to analyze supply chains in disaster contexts and on how to enhance system resilience. For instance, FEMA offers practical guidance to deal with disasters and indicates how to develop logistics planning based on the results of supply chain resilience (FEMA, 2019). Additionally, anecdotal evidence suggests that firms' responses to disasters depends on numerous factors, including sector and type of disaster (e.g., hurricane, tsunami, earthquake), although the consensus is to maintain operations, or at least return operations to normal functioning levels, as quickly as possible. For instance, after being hit by Superstorm Sandy, restaurants in New York altered menus to make them more oriented to family-style dining and created emergency funds aimed at financially sustaining the business to keep it afloat during recovery. Other businesses such as retailers altered their logistic plans and restocked those products that customers needed the most (Ilyashov, 2020).

In general, the lack of a theoretical framework that informs how resilience is implemented and measured in the SCM field has been compensated for by other theoretical approaches in some cases (e.g., economic resilience) or complemented by the emergence of new academic fields. This is the case with the research field of *humanitarian logistics*, which is an umbrella term that covers different type of operations, including disaster relief and support for developing regions constantly exposed to disasters (Kovács & Spens, 2007). The relevant set of assumptions differentiates humanitarian logistics from more traditional applications of SCM, which frequently deal with a predetermined set of suppliers, manufacturing sites, and a predictable demand. Yet, these factors are assumed to be unknown in humanitarian logistics (as well as the demand associated to relief goods and all activities involved), which are characterized by constraints that hinder the performance of the operations in large-scale emergencies because they occur in chaotic environments (Kovács & Spens, 2007).

Since disaster management deals with the organization and management of resources and responsibilities aimed at relieving and facilitating humanitarian aspects in disaster contexts (Scholten et al., 2014), humanitarian logistics has received increasing interest both from supply chain scholars and practitioners. The reason is that different organization types may potentially be involved in humanitarian efforts (e.g., local businesses, governments, military, NGOs, aid agencies, logistics providers) with the objective of alleviating the suffering of affected individuals. These circumstances show that increasing the resilience of the supply chain can contribute to disaster impact reduction, as almost 80% of the activities involved in the chain (i.e., inbound logistics, operations, and outbound logistics) can be tweaked to improve the planning, delivery, and distribution of relief goods (van Wassenhove, 2006).

3 Introducing Theoretical Frameworks

There has been movement in the supply chain literature to incorporate some intentional proxy in the measurement of resilience—*risk awareness* (Christopher & Peck, 2004), which "develops the adaptive capabilities to prepare for, respond to and recover from a disruption/disaster" (Scholten et al., 2014, p. 225). However, it is not only awareness, but also intentions arising from attitudes, subjective norms, and perceived behavioral control, and actualized behavior (i.e., why or why not actual behavior is in concert with intention) that determines true preparedness and in turn, the recovery of a business and its associated supply chain in the aftermath of a disaster.

Previous work on multi-hazard planning has generally focused on the existence of a plan, not on the result of planning. This focus has led to limited understanding of how hazard planning and past experience may influence outcomes of a new hazard experience. This need is becoming increasingly apparent, as there is recognition of the need to address compound and cascading natural and technological hazards with effects or causes originating from an initial hazard (Cutter, 2018; Phillips et al., 2020). There may also be multiple overlapping stressors that do not originate from the same hazard (Clarke et al., 2018) and concurrent risks. Adapting from experience and addressing the persistent stressors relevant to a given firm is critical to improved performance.

This section provides a summary of frameworks applicable to considering learning and adaptation of a firm's resilience planning. To date these frameworks have largely not been applied to supply chain management generally nor at the individual firm-level.

3.1 Mental Models and Triple-Loop Learning

Argyris and Schön (1978) introduced the concept of organizational cognition when confronted with a challenge, such as climate change. This forms the foundation for social learning, where the scale of learning increases from the individual to a group, business, or society (Tuler et al., 2017; Webler et al., 1995). Tuler et al. (2017, p. 63) defines social learning as, "changes in how social groups, such as organizations or communities (geographical or of practice) engage in sense-making activities, such as problem solving and decision making." Learning is often associated with the subject's perception of risk which in turn is influenced by cognitive, subconscious, socio-cultural, and other factors (Helgeson et al., 2012).

Mental models are a fundamental way by which to understand organizational learning and builds on Craik's (1943) description of the mind constructing "small-scale models of reality" that help shape behavior and set an approach to acting in response to a stimulus. Jones et al. (2011) provide an interdisciplinary synthesis of the theoretical development and practical applications of the mental model construct.

The process of learning can be conceptualized as a back-loop process through which mental models are shifted over time and through experience. Helgeson et al. (2012) highlighted the potential use of mental modeling to better understand participants' perceptions which could then be applied to track learning. Similar to an individual or governmental organization, companies have unique patterns of beliefs and histories that determine the decision-making criteria before, during, and after a

A Theoretical Framework for Supply Chain Resilience Planning

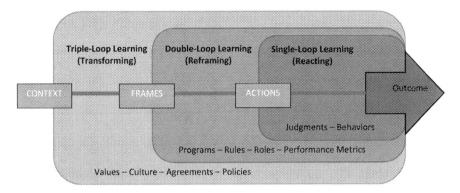

Fig. 1 Looped learning

crisis (Currah & Wrigley, 2004; Grinyer & Spender, 1979). This can make assessment of learning particularly difficult as there are multiple possible levels of inquiry and analysis.

Single- and double-loop learning-concepts were developed by Argyris and Schön (1978) and are based upon "a theory of action" perspective designed by Argyris (1982). Yet, Tosey et al. (2011) highlight that conceptualizations of triple-loop learning are diverse, often have little theoretical rooting, are sometimes driven by normative considerations, and lack support from empirical research. Single-loop learning is characterized by "following the rules," whereas double-loop learning for an organization is characterized by "changing the rules." As presented in Fig. 1, in single-loop learning, the organization addresses inconsistencies and impractical planning by adopting actions to mitigate and improve the situation accordingly in the present and future. In double-loop learning, post-event, members of the organization are able to reflect on whether the rules (i.e., plans) should be fundamentally changed, not only on whether deviations have occurred but also how to correct them. Triple-loop learning is characterized by "learning how to learn" and is typically evoked when members of the organization reflect on how they think about (i.e., conceptualize) the rules, not only whether the rules need to be altered.

Triple-loop learning has been assessed in the context of supply chain adaptative behaviors toward greater resilience to a limited extent (e.g., Ramish & Aslam, 2016). The third loop of learning enables continuity of supply chain operation in an uncertain environment, and thus the ability to adapt and perform specific functions in an environment full of challenges and tensions (Ponomarov & Holcomb, 2009), as presented in Fig. 2. The adaptive behavior consistent with the triple-loop learning principles has been referred to as preadaptation (Ameli & Kayes, 2011). Though we focus on the individual firm level, it should be noted that a resilient supply chain comprises many firms and enterprise functions that can be described as agents. Agents influence the course of events in the supply chain as they learn, i.e. decide on the inclusion of a specific link in the implementation of specific tasks, strive for expansion into new markets, and implement integrated processes of product development with suppliers and customers (Choi et al., 2001). Thus, learning by a given

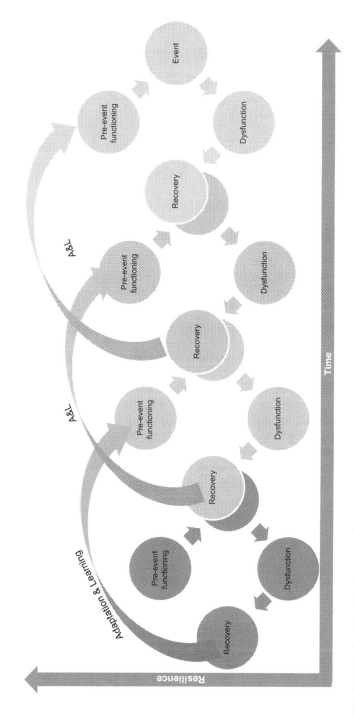

Fig. 2 Triple-loop learning for resilience at the firm-level over time. Based on Pierel (2020)

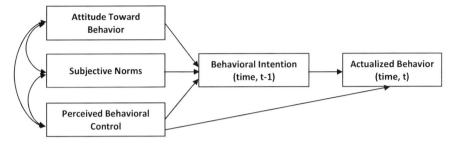

Fig. 3 Schematic of the Theory of Planned Behavior

agent (i.e., a given firm) along the supply chain may not improve the circumstances of other firms along the supply chain and may even make circumstances worse for others. Previous research suggests that interviewing owners and managers provides the appropriate level of analysis to understand business culture; however, selecting this group of respondents relies upon the belief that there is reasonable interorganizational communication (e.g., knowledge of employees missing work due to flood impacts) (Augier & Teece, 2009; Grinyer & Spender, 1979; Schindehutte & Morris, 2001).

3.2 Theory of Planned Behavior (TPB)

The relationship between attitudes and behavior has been a major topic of investigation in social psychology (e.g., Eagly & Chaiken, 1993) and the best-known model of the relationship is the Theory of Reasoned Action (Ajzen & Fishbein, 1977) and its elaboration in the Theory of Planned Behavior (Ajzen, 1991). In this framework, attitudes and subjective norms about a behavior (as well as perceived behavioral control) influence behavioral intentions which, in turn, determine the likelihood of the behavior occurring (Fig. 3). Although it has been used extensively in other fields, it has been infrequently used in geographic or disaster research (Ajzen, 1991; Najafi et al., 2017; Wang & Ritchie, 2012) and not at all in value chain or supply chain research, to the best of our knowledge.

According to the TPB, a specific behavior is determined by a combination of intention and perception of control over performing the behavior. Furthermore, the TPB identifies three global components (i.e., attitude toward the behavior, subjective norm, and perceived behavioral control) that together contribute to the creation of the intention; moreover, behavioral beliefs, normative beliefs, and control beliefs tend to be reliable predictors. The TPB has been used to understand planned disaster preparation and real behaviors for businesses and individuals in a limited number of studies to date (Daellenbach et al., 2018; Najafi et al., 2017; Passafaro et al., 2019; Wang & Ritchie, 2012). There is also a handful of papers that look at the TPB from the consumer-side (e.g., Giampietri et al., 2018; Al-Swidi et al., 2014). To the best of our knowledge there is no research connecting supply chain resilience at the firm level to the TPB explicitly. However, the TPB is a framework that fits the context of firm-level preparedness.

The Theory of Reasoned Action (Fishbein & Ajzen, 1975; Ajzen & Fishbein, 1980) provides a schematic linkage of behavioral intention with the elemental attitudes and subjective norms with which the actor is faced.

The underlying concept is summarized by:
$$BI = (ATT)W_1 + (SN)W_2 + (PBC)W_3,$$
where

BI: behavioral intention
ATT: attitude toward performing behavior
SN: subjective norms
PBC: perceived behavioral control
W_{1-3}: empirically derived weights

In past applications of the model (e.g., consumer behavior) psychologists have obtained relative values for W_1, W_2, and W_3 through complex self-reported scales over specific detailed scenarios (e.g., Hale et al., 2003).

Section 4 of this chapter presents the application of the TPB within the context of a firm planning for supply chain resilience.

3.3 Additional Considerations

Given that the TPB focuses on the relationship between behavior and attitudes, it largely assumes that the information to make an appraisal (via attitude) of the behavior is available (Ajzen, 2006). Thus, acknowledging deep uncertainty in the space of making an appraisal requires broader investigation of what influences bear on the planned and actualized decisions. When confronted with uncertain and potentially high stakes decisions, literature suggests people rely on several approaches for decision-making, including past experience with related events such as natural hazards, level of perceived risk, and other aspects of System 1 (Intuitive) and System 2 (Reasoning) thinking (Kahneman, 2011). It is also important to consider that for many firm owners/operators, there may be intangible goals, such as autonomy, personal satisfaction, and lifestyle that rank above traditional business performance and subsequently motivate their strategic planning (or lack thereof) (Wang et al., 2007).

Triandis' (1977) Model of Interpersonal Behavior provides a reasonable schematic linking social aspects to influencing conditions, attitudes, experience, and emotions. Though this model has been implemented less than the Ajzen–Fishbein formulation, where it has been applied, it appears to have additional explanatory value, particularly by including the concepts of beliefs and habits. Triandis offers an explicit role for affective factors and social factors that extend beyond norms to include self-concept on behavioral intentions. This formulation may be applicable to

A Theoretical Framework for Supply Chain Resilience Planning

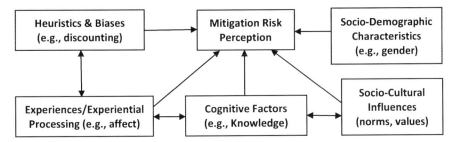

Fig. 4 Staged model that highlights relevant factors around the perception of a given mitigation risk. Adapted from the "Climate Change Risk Perception Model" (CCRPM) (Helgeson et al., 2012)

firm-level mitigation decisions in future work, as it captures many of the criticisms leveled at rational choice theory.[14]

Staged models from the social psychology literature (Fig. 4) offer structural relationships between both System 1 and System 2 elements and the pathway to an individual's behavioral preference, such as the TPB. Such a model makes it possible to trace the development of a risk preference through isolating factors which are responsible for fast and frugal methods versus those which are semi-conscious interactions with the option set of potential decisions and finally, well-reasoned and conscious considerations.

4 Proposed Theoretical Model: Informing Evolving Supply Chain Resilience

Christopher (2005) indicates that resilience implies being flexible and agile, as well as being able to adapt quickly. Following the definition of Ponomarov and Holcomb (2009), a resilient supply chain must integrate adaptive capabilities to prepare, respond, and recover from interruptions. This definition suggests that after a disruptive event, the level of supply chain function depends on the embedded adaptive capabilities and to which the firm has access. Additionally, the ability of the firm to learn from past disruptions and use those experiences to enhance and/or build capacities is a key property of resilience to future disruptions (Ramish & Aslam, 2016; Dormady et al., 2019). Sheffi (2005) provides examples on the ability of firms to continue to operate in disruptive environments after incorporating learning from past disruptions.[15]

---

[14] However, valuing constituent factors requires deep analysis of an owner/operator's preferences and decision-making methods, as well as understanding details of the decision space surrounding mitigation factors available to the firm. Gathering this level of data is likely not be operationalizable at this juncture for a reasonable sample size.

[15] This is the reason why some supply chain resilience definitions include a more systemic approach and look at the organization beyond a set of capabilities that can be utilized to cope with shocks and

Following, we propose a theoretical multi-period or dynamic structural model framed in a disaster context (e.g., EWEs), that allows the assessment of post-event actualized behaviors and intended behaviors toward future potential disruptions. In the model, behavior is represented by the decisions to implement mitigation and adaptation actions before and after the event, respectively. In the context of disasters, the resilience of the supply chain is characterized by when the system is able to continue its operations after the event occurs. As previously indicated, there are some actions that allow businesses to build capacity and strengthen the resilience of the supply chain before the occurrence of the event. The model assumes a learning pattern in which the business leverages past experiences of mitigation and adaptation and their influence on recovery to prepare the firm for future disasters. In this regard, building pre-event capacity implies learning from past disruptions and improving the firm's adaptive capabilities, which allow the supply chain to recover after disruption, returning to its original state or achieving a more desirable state of supply chain operations (Ponomarov & Holcomb, 2009).

4.1 Disaster Preparedness Behavior in a Supply Chain Context

Disasters have the potential to severely interrupt the operations of businesses and supply chains. Research routinely suggests that disruptions tend to break down most supply chains, affect their financial performance, and compromise the survival of those businesses participating in the chain (Hendricks & Singhal, 2005). In order to minimize the effects of potential disruptions and to enhance supply chain resilience, firms need to implement actions that enable them to sustain operations during and after disasters (Tang, 2006). These mitigation strategies are usually characterized as pre-disaster proactive planning actions that aim to build resilience capacity so that business interruption be avoided or minimized (Chopra & Sodhi, 2004; Tang, 2006). An example of these mitigation strategies is provided in Table 1. The lack of a general theoretical framework in the SCM literature informing how to implement these strategies has been replaced with guidance through the use of operations research techniques, decision processes, and case studies. For instance, Hale and Moberg (2005) propose a pre-disaster decision process based on a mathematical approach for establishing an efficient network of secure storage facilities that can effectively support multiple supply chain facilities.

Businesses can also employ other mitigation action types before a disruptive event (e.g., changes in physical infrastructure such as dry-proofing and flood gate installation, developing a written emergency action plan, conducting disaster drills, learning and training employees on first aid, lifting inventories off the ground,

---

instead consider it as a learning system with the ability to "grow in the face of turbulent change" (Fiksel, 2006).

Table 1 Select mitigation strategy types by risk type. Adapted from Chopra and Sodhi (2004)

	Disruptions	Delays	Forecast risk	Procurement risk	Receivables risk	Capacity risk	Inventory risk
Add capacity	↓					⇑	↑
Add inventory	⇓						⇑
Redundancy of suppliers		⇓					↑
Increase responsiveness		⇓	⇓				⇑
Increase flexibility		↓	⇓	⇓		⇑	↑
Aggregate or pool demand			⇓			⇑	⇑
Increase capability				↓			
Increase customer accounts					↓		

↓ reduces risk; ⇓ greatly reduces risk; ↑ increase risk; ⇑ greatly increases risk purchase of increased insurance) as well as post-event (e.g., asking for government assistance, or filing claims on business interruption or property insurance) in their recovery with an eye toward developing greater mitigation toward future events (see, e.g., Dahlhamer & D'Souza, 1997; Webb et al., 2002; Jahre et al., 2016). The main difference among these actions is that some of them are difficult to adopt in the future, but others are more easily adoptable pre-disaster actions. This suggests that these mitigation planning activities are constrained by a temporal element; some actions can be completed temporally close to the onset of an event and are highly adjustable (e.g., elevating equipment before a flood) and others are possible to enact close to an event's onset, but require planning well ahead of the potential event (e.g., enacting an emergency plan or completing an off-site data backup). Additionally, implementing mitigation actions can be seen as an organizational learning process where different activities may be adopted ahead of the occurrence of catastrophic future events after considering previously acquired knowledge about past mitigation effects on business recovery (Ponomarov & Holcomb, 2009).

Some of these actions have aimed at increasing awareness and creating a supportive management culture toward building supply chain resilience. However, these actions have been out of the range of most SCM literature because implementation is not associated with resources involved in the primary operational activities or in the firm's production function. Some of these actions (e.g., learning first aid, conducting disaster drills, developing emergency plans) have not been connected to resilience from a SCM perspective because they do not directly aim at building pre-disaster capabilities for the firm to continue and sustain business operations during and after disruptions (Tang, 2006). This is a reason of why these mitigation activities have been traditionally related to disaster management, but not to supply chain resilience. Recent supply chain resilience research has started considering resilience as a fundamental aspect of preparedness and emergency management with positive effects on business recovery (Scholten et al., 2014).

The literature provides empirical evidence intended to explain the influence of these types of mitigating actions on long-term business recovery. The evidence indicates no significant statistical effect, even after controlling for business size and sector (Webb et al., 2002). Among the reasons clarifying this result, Webb et al. (2002) have suggested that businesses do too little to prepare for disasters, that mitigation actions are more focused on employee life safety than on avoiding business interruption, and that businesses focus on site-specific and not off-site disruptions. Rose (2007) considers that, in the absence of property damage, those pre- and post-event resilience tactics aimed at sustaining some level of operation during and after disasters will likely restore operations completely following the occurrence of the event. In case of property damage, building resilience capacity prior to the disaster may not be sufficient and full recovery can be achieved after the business efficiently invests over time in repair and reconstruction of its capital stock.

In terms of whether businesses prepare for disasters, the evidence is mixed; recent research indicates that the decision to prepare ahead of disasters varies across groups widely. Josephson et al. (2017) suggest that there are some owner and firm's characteristics that make a small business more likely to mitigate and prepare ahead of hurricanes. For example, they find that female business owners and those with a college education or higher, businesses with prior disaster experience, or those who own the property are more likely to have a written emergency plan and thus are less likely to be low mitigators. This is not the case for sole proprietors and businesses in the construction and wholesale industries, which are prone to mitigate less.

4.2 Approach

Recently, an alternative modeling framework has explored the factors that may influence the decisions to implement mitigation and adaptation actions to cope with disasters. This behavioral approach is based on the TPB (Ajzen, 1991). For instance, Zhang et al. (2020) empirically examine the determinants of rice producers' intentions toward mitigation and adaptation by estimating a structural equation model (SEM) that relies on TPB assumptions. Najafi et al. (2017) employ a SEM with data collected from a cross-sectional survey implemented in Tehran that studied the determining factors affecting disaster preparedness behavior of individuals. In a similar approach Daellenbach et al. (2018) combined SEM with cluster analysis and used survey data collected from individuals living in Australia and New Zealand to identify different segments of the population in terms of disaster preparation. In this case, the authors found four different segments: (1) unprepared and uninterested, (2) willing but could do more, (3) "just too difficult," and (4) knowing, interested, and prepared.

Given that the TPB is a theoretical model that defines different constructs, the formulation and estimation of a SEM is appropriate. Unlike the applications of the TPB to disaster planning described above, we propose the use of a dynamic SEM adapted to a multi-period dataset that allows for a check against perceived reality (i.e., comparing past behavior vs reality) and predict future behavior (i.e., intended behavior toward a future disruption), which to the best of our knowledge has not been performed before in the supply chain literature in a disaster context. The assumption behind this model is that businesses actually make disaster preparedness decisions in a dynamic environment, learn from the past, and adapt to changing circumstances—both internal and external. These characteristics are aligned with the definition of a resilient supply chain, in which individual businesses strive to adapt to new states that are in most cases different from the original one (Ponomarov & Holcomb, 2009). Additionally, the approach provides the possibility of adding direct measures of business recovery, as well as controlling for business interruption, property damage, and supply chain issues. This allows for the assessment of post-event actualized behaviors and intended behaviors toward future potential disruptions.

4.2.1 Definition of Components

According to the TPB framework, individual behaviors or decisions to implement mitigation and adaptation actions to cope with disasters are guided by three kinds of considerations or constructs:

1. A favorable or unfavorable evaluation or appraisal of the behavior in question (attitude toward the behavior),
2. Perceived social pressure to perform or not to perform the behavior (subjective norm), and/or
3. Perceived ease or difficulty of performing the behavior (perceived behavioral control), which reflects past experiences as well as anticipated impediments and obstacles (Ajzen, 1991).

These constructs represent latent variables that rely on sets of items geared toward providing direct measures evaluated by means of confirmatory factor analysis (Ajzen, 2020). In combination, attitude toward the behavior (ATT), subjective norm (SN), and perception of behavioral control (PBC) lead to the formation of a behavioral intention. In a context of supply chain resilience planning, the more favorable the attitude and subjective norm with respect to preparing the business for disasters, and the greater the perceived behavioral control, the stronger an individual's intention to perform the behavior in question should be (i.e., more likely to implement mitigation and adaptation actions to disasters). In this regard, the relative importance of ATT, SN, and PBC in the prediction of intention is expected to vary across individual decision-makers.

Additionally, the TPB model usually incorporates composites of behavioral, normative, and control belief. These are not indirect measures of the main constructs (i.e., ATT, SN, and PBC), but formative indicators and determinants (i.e., causal factors) of behavioral intention whose effects are mediated by ATT, SN, and PBC, respectively. Attitude toward implementing mitigation and adaptation actions to cope with disasters is assumed to be determined by beliefs about the consequences of preparing the business for disasters (behavioral beliefs, BB), with each belief weighted by the subjective value of the outcome in question given by the decision-maker. Likewise, subjective norms that exert pressure on the individual to prepare or not to prepare the business for disasters are assumed to be determined by the perceived behavioral expectations of important referent individuals or groups influencing individual's decision such as family, friends, suppliers, customers, and regulators (normative beliefs, NB). These beliefs in combination with the individual's motivation to comply with the different referents or groups determine the prevailing subjective norm regarding disaster preparedness. Finally, perceived behavioral control is assumed to be determined by the perceived presence of factors that can facilitate or impede performance of a behavior (control beliefs, CB). It is assumed that the perceived power of each control factor to impede or facilitate preparing the business for disasters contributes to perceived control of this behavior

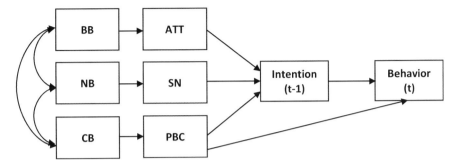

Fig. 5 Path analysis for the theory of planned behavior

in direct proportion to the person's subjective probability that the control factor is present (Ajzen, 2020).

Finally, intention is assumed to be the immediate antecedent of disaster preparedness behavior. However, as it may be the case that difficulties or barriers emerge in implementing mitigation and adaptation actions or in building resilience capacity, which limit the possibility of a business to act. This is also considered an influence of PBC, on to intention. The basic schematic representation of the theory is shown in Fig. 5.

4.2.2 Measures

Given the goal of the proposed model, which is to predict individual intentions and behaviors of implementing mitigation and adaptation actions in a business disaster context, there is no need to incorporate the formative measures (i.e., BB, NB, and CB) as these are more related to the design of behavior change interventions (Ajzen, 2020). The major constructs in the proposed model are obtained through a well-structured survey instrument. Such a questionnaire may follow those developed by Davis et al. (2002) and Fishbein and Ajzen (2010). The major theoretical constructs and direct measures proposed are: *(1) intention to implement mitigation actions within specific domain and more generally, (2) attitude toward mitigation (ATT_m), (3) subjective norms related to mitigation (SN_m), and (4) perceived behavioral control related to mitigation (PBC_m).* These constructs are similar for adaptation-related behavior, including: *(5) intention to implement adaptation actions within specific domain and more generally, (6) attitude toward adaptation (ATT_a), (7) subjective norms related to adaptation (SN_a), and (8) perceived behavioral control related to adaptation (PBC_a).*

In terms of the number of items aimed at measuring each construct, the literature suggests that single item measures are more adequate for simple or concrete constructs that are well understood; however, when internal consistency cannot be achieved with just one item because of the complexity embedded in the construct definition, adequate internal consistency reliability can be obtained with four or five items per scale (Harvey et al., 1985; Hinkin & Schriesheim, 1989). The items will be modeled using a principal axis factor analysis with orthogonal rotation (e.g., Davis et al., 2002). After this procedure, the findings will be considered in combination with business interruption and recovery data also obtained through the survey. Each measure is reviewed below.

Intention (Intent). Five items will assess individual intentions to implement mitigation and five items will assess intentions to implement adaptation. Participants will indicate, on 5-point agree–disagree scales, to what extent they expect to, intend to, will try to, are determined to, and might not (reverse scored) implement mitigation and adaptation actions that are prone to improve the resilience of the business against disasters.

Attitudes Toward Behavior (ATT). Attitudes toward "implementing mitigation and adaptation actions" will be assessed by means of a series of 5-point evaluative semantic differential scales for each construct (i.e., ATT_m and ATT_a). The anchors of these scales will be as follows: negative–positive, useless–useful, harmful–beneficial, difficult-easy, and boring–exciting.

Subjective Norms (SN). For each construct (i.e., SN_m and SN_a), three items will assess subjective norms with respect to "implementing mitigation and adaptation actions." Participants will indicate, on 5-point agree–disagree scales, to what extent he/she, family and friends, and customers think the business should prepare for disasters.

Perceived Behavioral Control (PBC). For each construct (i.e., PBC_m and PBC_a), three items will assess perceived behavioral control over "implementing mitigation and adaptation actions." Participants will indicate, on 5-point agree–disagree scales, to what extent they think that preparing the business for disasters is completely up to them or is beyond their control.

4.2.3 SEM Model

The definition of business resilience and thus, supply chain resilience, assumes a connection between resilience and recovery (see, e.g., Chang & Rose, 2012). It suggests that businesses that implement pre-disaster planning, particularly those predetermined actions geared toward building capacities and developing capabilities that help to maintain operations and keep business continuity can respond more effectively to disruptions and thus, recover more quickly (McManus et al., 2008; Lee et al., 2013). At the same time, it also suggests that firms may use learning from past disruptions and integrate disaster and recovery experiences to their resilience planning mechanisms to cope with future disruptive events (Helgeson et al., 2020). Therefore, the premise of the model is that business disaster experience does not necessarily predict future disaster planning behavior. The intended behaviors toward future potential disruptions will depend instead on the resilience of the business via its recovery function, which compares pre- and post-disaster business conditions and measures the length of time at which the business achieves pre-disaster levels

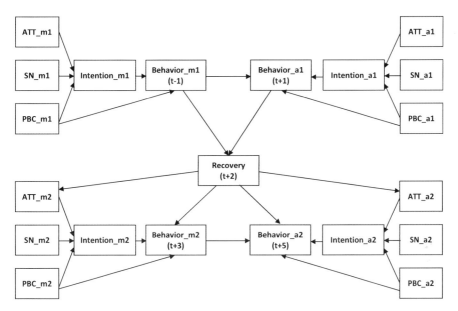

Fig. 6 Path analysis for the model proposed on the relationship between the multi-period mitigation and adaptation behaviors and business resilience and recovery

performance or return to normalcy (see, e.g., Webb et al., 2002; Graveline & Gremont, 2017).

The theoretical model proposed in this chapter is a multi-period or dynamic structural equation model that assumes that the recovery of a business in the aftermath of a disaster relies on individual behavior toward pre-disaster mitigation and post disaster adaptation, after controlling for different factors including business size, age, and sector. The model in Fig. 6 assumes that a business manager is faced with the decision of implementing a mitigation action in $t - 1$, before the occurrence of a potential disruptive event in t. If the disruption would occur in t, the business manager would face another decision in $t + 1$ related to the adaptation action to be implemented in that period. However, the feasibility of implementing an adaptation action in $t + 1$ might be conditioned on the type of decision made in the previous period, $t - 1$. This occurs because the choice option set, which is the potential set of options that may be implemented by the decision-maker, is reduced after mitigation-related decisions in $t - 1$ have been made (Helgeson et al., 2020). The learning mechanism is represented when the individual incorporates the experience of mitigation and adaptation actions through recovery, which influences risk perceptions, attitude, and behaviors in the next round of mitigation and adaptation decisions in $t + 3$ and $t + 5$, respectively.

The schematic representation of this model is presented in Fig. 6.

5 Conclusions and Future Directions

There is a consensus in the literature that supply chain resilience fundamentally involves adaptation and learning. However, most supply chain scholars have focused on the adaptive capabilities at the supply chain and organizational level of analysis and less research has been devoted to explaining the role of post-event actualized behaviors and intended behaviors toward future potential disruptions. One reason for this is that until recently, disaster preparedness decisions and emergency management activities were largely disconnected from supply chain resilience. In more current literature, however, building supply chain resilience capabilities is considered to be a fundamental aspect of disaster preparedness and emergency management. Not only is it important the early detection leading to implementing resilience actions but also in understanding the business recovery function and decision-maker's intention to implement mitigation.

This chapter argues that more research is needed to understand this learning process and provides a theoretical framework based on Ajzen's Theory of Planned Behavior (Ajzen, 1991). It suggests that the capacity of the firm to learn from past disruptions by integrating those experiences in future mitigation decisions improves preparedness for future events, enhance firm's resilience, and hastens recovery.

The framework proposed in this chapter bridges between espoused theory and theory-in-use. Espoused theories are those that are indicated by an actor that they claim to follow, while theories-in-use are those that can be inferred from action taken that can be observed by an external party without self-reporting by an agent. Espoused theory and theory-in-use may or may not be consistent, and the agent may not be aware of any inconsistencies. This is a challenge faced across the board for the use of the TPB.

In future iterations it is important to incorporate other elements that could be influential in this context, like an actor's risk tolerance (which influences risk perceptions and behavior attitudes), susceptibility to social norms, status quo/habit "stickiness" (influences behavior attitudes), previous experience and event or risk salience, and the extent of sufficient money and time to undertake mitigation (influences actual behavior).

In future iterations consideration should be made for how best to control for the extent of a disruptive event that leads to the next iteration of the firm's disaster planning. In addition, there are clear differences in the domains and objective functions considered in resilience planning by the firm, which should be considered in applications of the presented framework. Within each of the potential domains the categories of mitigation and adaptation behaviors that can be taken are further differentiated by whether they are available given characteristics of the agent (i.e., firm), such as geographic location, sector type, position in the supply chain, and capacity constraints. This is left for future work.

References

Adger, W. N. (2000). Social and ecological resilience: Are they related? *Progress in Human Geography, 24*(3), 247–364.
Ajzen, I. (1991). The theory of planned behavior. *Organizational Behavior and Human Decision Processes, 50*(2), 179–211.
Ajzen, I. (2006). Constructing a TPB questionnaire: Conceptual and methodological considerations. September 2002 (revised January 2006).
Ajzen, I. (2020). The theory of planned behavior: Frequently asked questions. *Human Behavior and Emerging Technologies, 2*(4), 314–324.
Ajzen, I., & Fishbein, M. (1977). Attitude-behavior relations: A theoretical analysis and review of empirical research. *Psychological Bulletin, 84*(5), 888.
Ajzen, I., & Fishbein, M. (1980). *Understanding attitudes and predicting social behavior*. Prentice-Hall.
Alesch, D. J., Holly, J. N., Mittler, E., & Nagy, R. (2001). *Organizations at risk: What happens when small businesses and not-for-profits encounter natural disasters*. Public Entity Risk Institute.
Al-Swidi, A., Huque, S. M. R., Hafeez, M. H., & Shariff, M. N. M. (2014). The role of subjective norms in theory of planned behavior in the context of organic food consumption. *British Food Journal, 116*(10), 1561–1580.
Ameli, P., & Kayes, D. C. (2011). Triple-loop learning in a cross-sector partnership: The DC Central Kitchen partnership. *The Learning Organization, 18*(3), 175–188.
Argyris, C. (1982). *Reasoning, learning, and action: Individual and organisational*. Jossey Bass.
Argyris, C., & Schön, D. (1978). *Organizational learning: A theory of action perspective*. Addison-Wesley.
Augier, M., & Teece, D. J. (2009). Dynamic capabilities and the role of managers in business strategy and economic performance. *Organization Science, 20*(2), 410–421.
Blanchard, T. C., Tolbert, C., & Mencken, C. (2012). The health and wealth of US counties: How the small business environment impacts alternative measures of development. *Cambridge Journal of Regions, Economy & Society, 5*(1).
Cao, M., & Zhang, Q. (2011). Supply chain collaboration: Impact on collaborative advantage and firm performance. *Journal of Operations Management, 29*(3), 163–180.
Chang, S. E., & Rose, A. Z. (2012). Towards a theory of economic recovery from disasters. *International Journal of Mass Emergencies & Disasters, 30*(2).
Choi, T. Y., Dooley, K. J., & Rungtusanatham, M. (2001). Supply networks and complex adaptive systems: Control versus emergence. *Journal of Operations Management, 19*(3), 351–366.
Chopra, S., & Sodhi, M. S. (2004). Managing risk to avoid supply chain breakdown. *MIT Sloan Management Review, 46*(1), 53–61.
Christopher, M. (1998). *Logistics and supply chain management: Strategies for reducing cost and improving service* (2nd ed.). Pitman Publishing.
Christopher, M. (2005). Managing risk in the supply chain. In M. Christopher (Ed.), *Logistics & supply chain management* (3rd ed., pp. 231–258). Prentice-Hall.
Christopher, M., & Lee, H. (2004). Mitigating supply chain risk through improved confidence. *International Journal of Physical Distribution & Logistics Management, 34*(5), 388–396.
Christopher, M., & Peck, H. (2004). Building the resilient supply chain. *International Journal of Logistics Management, 15*(2), 1–13.
Clarke, D. J., Mahul, O., Poulter, R., & Teh, T. L. (2018). Evaluating sovereign disaster risk finance strategies: A framework. *The Geneva Papers on Risk and Insurance-Issues and Practice, 42*(4), 565–584.
Craik, K. J. W. (1943). *The nature of explanation*. Cambridge University Press.
Currah, A., & Wrigley, N. (2004). Networks of organizational learning and adaptation in retail TNCs. *Global Networks, 4*(1), 1–23.

Cutter, S. L. (2018). Compound, cascading, or complex disasters: what's in a name? *Environment: Science and Policy for Sustainable Development, 60*(6), 16–25.

Daellenbach, K., Parkinson, J., & Krisjanous, J. (2018). Just how prepared are you? An application of marketing segmentation and theory of planned behavior for disaster preparation. *Journal of Nonprofit & Public Sector Marketing, 30*(4), 413–443.

Dahlhamer, J. M., & D'Souza, M. J. (1997). Determinants of business disaster preparedness in two U.S. metropolitan areas. *International Journal of Mass Emergencies and Disasters, 15*(2), 265–281.

Davis, L. E., Ajzen, I., Saunders, J., & Williams, T. (2002). The decision of African American students to complete high school: An application of the theory of planned behavior. *Journal of Educational Psychology, 94*(4), 810.

Dormady, N., Roa-Henriquez, A., & Rose, A. (2019). Economic resilience of the firm: A production theory approach. *International Journal of Production Economics, 208*, 446–460.

Eagly, A. H., & Chaiken, S. (1993). *The psychology of attitudes*. Harcourt Brace Jovanovich.

Fahimnia, B., Pournader, M., Siemsen, E., Bendoly, E., & Wang, C. (2019). Behavioral operations and supply chain management: A review and literature mapping. *Decision Sciences, 50*(6), 1127–1183.

FEMA. (2019, April). *Supply chain resilience guide*. Retrieved July 17, 2022, from https://www.fema.gov/sites/default/files/2020-07/supply-chain-resilience-guide.pdf

Fiksel, J. (2006). Sustainability and resilience: Toward a systems approach. *Sustainability: Science, Practice and Policy, 2*(2), 14–21.

Fiksel, J. (2015). *Resilient by design*. Island Press.

Fishbein, M., & Ajzen, I. (1975). *Belief, attitude, intention, and behavior: An introduction to theory and research*. Addison-Wesley.

Fishbein, M., & Ajzen, I. (2010). *Predicting and changing behavior: The reasoned action approach*. Psychology Press.

Giampietri, E., Verneau, F., Del Giudice, T., Carfora, V., & Finco, A. (2018). A theory of planned behaviour perspective for investigating the role of trust in consumer purchasing decision related to short food supply chains. *Food Quality and Preference, 64*, 160–166.

Graveline, N., & Gremont, M. (2017). Measuring and understanding the microeconomic resilience of businesses to lifeline service interruptions due to natural disasters. *International Journal of Disaster Risk Reduction, 24*, 526–538.

Grimm, D. (2013). Whole community planning: Building resiliency at the local level. *Journal of Business Continuity & Emergency Planning, 7*, 253–259.

Grinyer, P. H., & Spender, J. C. (1979). Recipes, crises, and adaptation in mature businesses. *International Studies of Management & Organization, 9*(3), 113–133.

Hale, J. L., Householder, B. J., & Greene, K. L. (2003). The theory of reasoned action. *The Persuasion Handbook: Developments in Theory and Practice, 14*(2002), 259–286.

Hale, T., & Moberg, C. R. (2005). Improving supply chain disaster preparedness. *International Journal of Physical Distribution & Logistics Management, 35*(3), 195–207.

Harvey, R. J., Billings, R. S., & Nilan, K. J. (1985). Confirmatory factor analysis of the job diagnostic survey: Good news and bad news. *Journal of Applied Psychology, 70*(3), 461.

Helgeson, J., van der Linden, S., & Chabay, I. (2012). The role of knowledge, learning and mental models in perceptions of climate change related risks. In A. Wals & P. B. Corcoran (Eds.), *Learning for sustainability in times of accelerating change*. Wageningen Academic Publishers.

Helgeson, J. F., Fung, J. F., & Roa-Henriquez, A. R. (2020). Rationally bounded in a storm of complex events: Small businesses facing natural hazard resilience during a pandemic. *Journal of Behavioral Economics for Policy, 4*(S3), 55–65.

Hendricks, K. B., & Singhal, V. R. (2005). An empirical analysis of the effect of supply chain disruptions on long-run stock price performance and equity risk of the firm. *Production and Operations Management, 14*(1), 35–52.

Hinkin, T. R., & Schriesheim, C. A. (1989). Development and application of new scales to measure the French and Raven (1959) bases of social power. *Journal of Applied Psychology, 74*(4), 561–567.

Howe, P. D. (2011). Hurricane preparedness as anticipatory adaptation: A case study of community businesses. *Global Environmental Change, 21*(2), 711–720.

Ilyashov, A. (2020, June 12). *Barely recovered from Sandy, NYC restaurants reflect on lessons learned to cope with covid-19*. Eater NY. Retrieved July 17, 2022, from https://ny.eater.com/2020/6/12/21286755/nyc-coronavirus-restaurants-hurricane-sandy-lessons

Jahre, M., Pazirandeh, A., & Van Wassenhove, L. (2016). Defining logistics preparedness: A framework and research agenda. *Journal of Humanitarian Logistics and Supply Chain Management, 6*(3), 372–398.

Johnson, M. E. (2001). Learning from toys: Lessons in managing supply chain risk from the toy industry. *California Management Review, 43*(3), 106–112.

Jones, N. A., Ross, H., Lynam, T., Perez, P., & Leitch, A. (2011). Mental models: An interdisciplinary synthesis of theory and methods. *Ecology and Society, 16*(1). https://doi.org/10.5751/ES-03802-160146

Josephson, A., Schrank, H., & Marshall, M. (2017). Assessing preparedness of small businesses for hurricane disasters: Analysis of pre-disaster owner, business and location characteristics. *International Journal of Disaster Risk Reduction, 23*, 25–35.

Jüttner, U. (2005). Supply chain risk management: Understanding the business requirements from a practitioner perspective. *The International Journal of Logistics Management, 16*(1), 120–141.

Jüttner, U., & Maklan, S. (2011). Supply chain resilience in the global financial crisis: An empirical study. *Supply Chain Management: An International Journal, 16*(4), 246–259.

Jüttner, U., Peck, H., & Christopher, M. (2003). Supply chain risk management: Outlining an agenda for future research. *International Journal of Logistics: Research and Applications, 6*(4), 197–210.

Kahneman, D. (2011). *Thinking, fast and slow*. Farrar, Straus, and Giroux.

Kovács, G., & Spens, K. M. (2007). Humanitarian logistics in disaster relief operations. *International Journal of Physical Distribution & Logistics Management, 37*(2), 99–114.

Kunreuther, H., Meyer, R., & Michel-Kerjan, E. (2013). Overcoming decision biases to reduce losses from natural catastrophes. In *The behavioral foundations of public policy* (pp. 398–413). Princeton University Press.

Lee, H. L., Padmanabhan, V., & Whang, S. (1997). Information distortion in a supply chain: The bullwhip effect. *Management Science, 43*(4), 546–558.

Lee, A. V., Vargo, J., & Seville, E. (2013). Developing a tool to measure and compare organizations' resilience. *Natural Hazards Review, 14*(1), 29–41.

Lowson, R. H. (2002). Strategic operations management—The new competitive advantage? *Journal of General Management, 28*(1), 36–56.

Lyson, T. A., Torres, R. J., & Welsh, R. (2001). Scale of agricultural production, civic engagement, and community welfare. *Social Forces, 80*(1), 311–327.

Manuj, I., & Mentzer, J. T. (2008). Global supply chain risk management strategies. *International Journal of Physical Distribution & Logistics Management, 38*(3), 192–223.

McKnight, B., & Linnenluecke, M. K. (2019). Patterns of firm responses to different types of natural disasters. *Business & Society, 58*(4), 813–840.

McManus, S., Seville, E., Vargo, J., & Brunsdon, D. (2008). Facilitated process for improving organizational resilience. *Natural Hazards Review, 9*(2), 81–90.

Morrison, A., Breen, J., & Ali, S. (2003). Small business growth: Intention, ability, and opportunity. *Journal of Small Business Management, 41*(4), 417–425.

Najafi, M., Ardalan, A., Akbarisari, A., Noorbala, A. A., & Elmi, H. (2017). The theory of planned behavior and disaster preparedness. *PLoS Currents, 9*.

National Research Council. (2006). *Facing hazards and disasters: Understanding human dimensions*. The National Academies Press.

Oliver, R. K., & Webber, M. D. (1992). Supply chain management: Logistics catches up with strategy. In M. Christopher (Ed.), *Logistics, the strategic issues*. Chapman & Hall.

Passafaro, P., Livi, S., & Kosic, A. (2019). Local norms and the theory of planned behavior: Understanding the effects of spatial proximity on recycling intentions and self-reported behavior. *Frontiers in Psychology, 10*, 744.

Pearson, C. M., & Clair, J. A. (1998). Reframing crisis management. *Academy of Management Review, 23*(1), 59–76.

Peck, H. (2006). Reconciling supply chain vulnerability, risk and supply chain management. *International Journal of Logistics: Research and Applications, 9*(2), 127–142.

Phillips, C. A., Caldas, A., Cleetus, R., Dahl, K. A., Declet-Barreto, J., Licker, R., Merner, L. D., Ortiz-Partida, J. P., Phelan, A. L., Spanger-Siegfried, E., Talati, S., Trisos, C. H., & Carlson, C. J. (2020). Compound climate risks in the COVID-19 pandemic. *Nature Climate Change, 10*(7), 586–588. https://doi.org/10.1038/s41558-020-0804-2

Pierel, E. (2020). Relating resilience, adaptation, and learning. *Figshare*. https://doi.org/10.6084/m9.figshare.12783473.v1

Ponomarov, S. Y., & Holcomb, M. C. (2009). Understanding the concept of supply chain resilience. *The International Journal of Logistics Management, 20*(1), 124–143.

Porter, M. E. (1985). *Competitive advantage*. The Free Press.

Ramish, A., & Aslam, H. (2016). Measuring supply chain knowledge management (SCKM) performance based on double/triple loop learning principle. *International Journal of Productivity and Performance Management, 65*(5), 704–722.

Rose, A. (2004). Defining and measuring economic resilience to disasters. *Disaster Prevention and Management, 13*(4), 307–314.

Rose, A. (2007). Economic resilience to natural and man-made disasters: Multidisciplinary origins and contextual dimensions. *Environmental Hazards, 7*(4), 383–398.

Runyan, R. C. (2006). Small business in the face of crisis: Identifying barriers to recovery from a natural disaster 1. *Journal of Contingencies and Crisis Management, 14*(1), 12–26.

Schaer, C., Bee, S., & Kuruppu, N. (2018). *Developing the business case for adaptation in agriculture: Case studies from the adaptation mitigation readiness project. Private-sector action in adaptation: Perspectives on the role of micro, small and medium size enterprises*, 51.

Schindehutte, M., & Morris, M. H. (2001). Understanding strategic adaptation in small firms. *International Journal of Entrepreneurial Behavior & Research, 7*(3), 84–107.

Scholten, K., & Schilder, S. (2015). The role of collaboration in supply chain resilience. *Supply Chain Management: An International Journal, 20*(4), 471–484.

Scholten, K., Scott, P. S., & Fynes, B. (2014). Mitigation processes – Antecedents for building supply chain resilience. *Supply Chain Management, 19*(2), 211–228.

Schorsch, T., Wallenburg, C. M., & Wieland, A. (2017). The human factor in SCM: Introducing a meta-theory of behavioral supply chain management. *International Journal of Physical Distribution & Logistics Management, 47*(4), 238–262.

Sheffi, Y. (2001). Supply chain management under the threat of international terrorism. *The International Journal of Logistics Management, 12*(2), 1–11.

Sheffi, Y. (2005). *The resilient Enterprise: Overcoming vulnerability for competitive advantage*. MIT Press.

Sheffi, Y. (2020, March 19). *Commentary: It's not premature to set supply chains for a coronavirus recovery*. Retrieved June 25, 2020, from https://www.wsj.com/articles/commentary-its-not-premature-to-set-supply-chains-for-a-coronavirus-recovery-11584619200

Sheffi, Y., & Rice, J. B., Jr. (2005). A supply chain view of the resilient enterprise. *MIT Sloan Management Review, 47*(1), 41.

Skipper, J. B., & Hanna, J. B. (2009). Minimizing supply chain disruption risk through enhanced flexibility. *International Journal of Physical Distribution and Logistics Management, 39*(5), 404–427.

Sodhi, M. S., & Tang, C. S. (2021). *Preparing for future pandemics with a reserve of inventory, capacity, and capability*. Retrieved from https://ssrn.com/abstract=3816606.

Spillan, J., & Hough, M. (2003). Crisis planning in small businesses: Importance, impetus and indifference. *European Management Journal, 21*(3), 398–407.

Svensson, G. (2000). A conceptual framework for the analysis of vulnerability in supply chains. *International Journal of Physical Distribution and Logistics Management, 30*, 731.

Tang, C. S. (2006). Robust strategies for mitigating supply chain disruptions. *International Journal of Logistics: Research and Applications, 9*(1), 33–45.

Tolbert, C. M., Lyson, T. A., & Irwin, M. D. (1998). Local capitalism, civic engagement, and socioeconomic well-being. *Social Forces, 77*(2), 401–427.

Tomlin, B. (2006). On the value of mitigation and contingency strategies for managing supply chain disruption risks. *Management Science, 52*(5), 639–657.

Tosey, P., Visser, M., & Saunders, M. (2011). The origins and conceptualizations of 'triple-loop' learning: A critical review. *Management Learning, 43*(3), 291–307.

Triandis, H. C. (1977). *Interpersonal behaviour.* Brooks/Cole Publishing Company.

Tuler, S. P., Dow, K., Webler, T., & Whitehead, J. (2017). Learning through participatory modeling: Reflections on what it means and how it is measured. In *Environmental modeling with stakeholders* (pp. 25–45). Springer.

Van der Vegt, G., Essens, P., Wahlström, M., & George, G. (2015). Managing risk and resilience. *Academy of Management Journal, 58*(4), 971–980.

Van Wassenhove, L. N. (2006). Blackett memorial lecture: Humanitarian aid logistics: Supply chain management in high gear. *Journal Operational Research Society, 57*(5), 475–489.

Wang, C., Walker, E. A., & Redmond, J. L. (2007). Explaining the lack of strategic planning in SMEs: The importance of owner motivation. *International Journal of Organisational Behaviour, 12*(1), 1–16.

Wang, J., & Ritchie, B. W. (2012). Understanding accommodation managers' crisis planning intention: An application of the theory of planned behaviour. *Tourism Management, 33*(5), 1057–1067.

Webb, G. R., Tierney, K. J., & Dahlhamer, J. M. (2002). Predicting long-term business recovery from disaster: A comparison of the Loma Prieta earthquake and Hurricane Andrew. *Global Environmental Change Part B: Environmental Hazards, 4*(2), 45–58.

Webler, T., Kastenholz, H., & Renn, O. (1995). Public participation in impact assessment: A social learning perspective. *Environmental Impact Assessment Review, 15*(5), 443–463.

World Economic Forum. (2019). *Global risks report 2019* (14th ed.). Retrieved from http://www3.weforum.org/docs/WEF_Global_Risks_Report_2019.pdf

Yoshida, K., & Deyle, R. E. (2005). Determinants of small business hazard mitigation. *Natural Hazards Review, 6*(1), 1–12.

Zhang, Y., Lindell, M. K., & Prater, C. S. (2009). Vulnerability of community businesses to environmental disasters. *Disasters, 33*(1), 38–57.

Zhang, L., Ruiz-Menjivar, J., Luo, B., Liang, Z., & Swisher, M. E. (2020). Predicting climate change mitigation and adaptation behaviors in agricultural production: A comparison of the theory of planned behavior and the value-belief-norm theory. *Journal of Environmental Psychology, 68*, 101408.

Toward a Resilient Supply Chain

Sandeep Ramachandran and Ganesh Balasubramanian

Abstract Supply chains are becoming increasingly global due to outsourcing and global procurements. The deep reach of global supply chains induces supply chain vulnerabilities and increases the supply chain network's disruption risk exposure. Firms need to build resilient supply chains to overcome natural calamities and human-made disruptions. This chapter aims to elucidate the practices involved in a firm's journey toward building a resilient supply chain. Specifically, we focus on three critical aspects of a firm: people, process, and technology. We provide industrial applications and best practices for building resilient supply chains.

1 Introduction

Over the last two decades, firms are exploring ways to make their supply chains more cost-effective and profitable. Consequently, supply chains are becoming increasingly global. For instance, California-based Apple Inc's contract manufacturers have 41 locations in China as of 2019. Also, among the rest of Apple's suppliers, 47.6% are located in China (Nellis, 2019). The deep reach of global supply chains induces supply chain vulnerabilities and increases the supply chain network's disruption risk exposure. A disaster (natural or human-made), in a remote part of the world, may impact a firm's supply chain and may impede its ability to conduct business as usual. For example, Ericsson reported a loss of $400 million due to a fire accident at one of its supplier facilities that disrupted the material supply (Norrman & Jansson, 2004). Land Rover laid off 1400 workers after supplier insolvency in 2001 (Gow, 2002).

S. Ramachandran
Flexport, San Francisco, CA, USA

G. Balasubramanian (✉)
Indian Institute of Management Ahmedabad, Ahmedabad, Gujarat, India
e-mail: phd17ganeshb@iima.ac.in

Recently, the COVID-19 outbreak has caused a significant disruption in many global supply chains. According to the report by Dun and Bradstreet (2020), 938 of Fortune 1000 companies have their tier 1 or tier 2 suppliers based out of the Wuhan region, which is considered to be the origin place of the COVID-19 pandemic. Linton and Vakil (2020) showed that more than 12,000 supply chain facilities are present in COVID-19 quarantined regions. The impact of such unprecedented incidents on a firm's profitability is well documented in the existing literature. For example, Hendricks and Singhal (2005) conducted an empirical study on the long-term stock price effect and equity risk effect of supply chain disruptions based on a sample of 827 disruption announcements made during 1989–2000. Their analysis revealed two significant findings. First, the firms that faced uncertain disruptions underperformed by 33–40 percentage in stock performance compared to the industry benchmark. Second, such firms did not recover quickly from the adverse effects of disruptions.

To overcome such unprecedented incidents and sustain profitability, firms need to build resilient[1] supply chain networks (Sheffi, 2007). The recent COVID-19 outbreak has reiterated the importance of having a resilient supply chain. According to Swiss VR Monitor (2020), a survey of 457 members of the board of directors across Switzerland, though a majority of the board members agreed that they had a crisis management organization before the COVID pandemic, only 32% of them had carried out exercises with the crisis management team or had planned for a pandemic. A subsequent study by Deloitte Switzerland (2020) on the Swiss manufacturing industry revealed that the pandemic has emphasized the importance of analyzing supply chain dependencies frequently. The study summarizes five key considerations toward building a resilient supply chain: supply chain configuration and control, visibility, digitalization, collaboration, and flexibility. Sharma et al. (2020) analyzed Twitter data from NASDAQ 100 firms to highlight the practical challenges faced by firms during the pandemic. They revealed three key challenges: demand–supply mismatch, technology, and the development of a resilient supply chain.

The journey toward supply chain resilience is entwined within the overall organizational capabilities. For example, organizational innovation is crucial for firms to adapt to business landscape changes and to minimize organizational risk. Firms have to develop dynamic capabilities such as adaptation, flexibility, and agility to be resilient to disruptions in the global business environment (Lee & Rha, 2016). Thus, a firm's journey toward supply chain resiliency is not independent of its journey toward organizational resilience.

In the context of the above discussion, we elucidate some key processes involved in a firm's journey toward building a resilient supply chain. Specifically, we focus on three critical aspects of a firm: people, process, and technology. We adopt the 5-stage supply chain maturity model proposed by the Supply Chain Risk Leadership Council, 2013. Figure 1 illustrates the five stages of supply chain maturity.

---

[1] We adopt the definition provided by Christopher and Peck (2004) for resilient supply chain.

Fig. 1 Five stages of supply chain maturity

Firms in the reactive stage are nascent to handling supply chain risk. They might not have any specific process or resource allocated for risk monitoring and assessment. However, as the firms mature in their capability to handle supply chain risks, they move to the aware and proactive stages. By this stage, they would have started mapping their supply chain, identified risky suppliers, and developed their risk profile. However, in these stages, they might still lack an integrated and scalable approach toward monitoring, assessing, and managing supply chain risk. Subsequently, firms move to the integrated and resilient stages. Figure 2 illustrates the maturity model with people, process, and technology characteristics for various stages.

2 People

In this section, we explain people's role in a firm's journey toward building a resilient supply chain. At various stages of maturity, firms display different levels of involvement of individuals/teams that handle supply chain risks. While a firm at the reactive stage would not have any dedicated resources for supply chain risk management, a firm at the aware stage would have at least some part-time resources allocated for managing supply chain risk. Subsequently, as the firm matures to the proactive stage, it will have full-time resources dedicated to managing supply chain risk. Further, in the integrated stage, firms typically would have established a supply chain stakeholder council with all internal business units for supply chain risk management.

Generally, firms in the first two stages of maturity (reactive and aware) would have a piecemeal fashion of addressing supply chain risk. For such firms, the first step in the journey toward resiliency is identifying the people responsible for managing supply chain risk. It is crucial to structure a core team that would assess and address the supply chain risk. We explain the risk assessment processes in Sect. 3. This core team ensures a holistic approach toward managing risk with stakeholders from various functions of the organization. Figure 3 illustrates a suggestive (not exhaustive) list of stakeholders with whom the core team can coordinate for assessing and managing risk.

Continuous risk assessment and risk-sharing with top management support are critical to building supply chain resilience (Ponomarov & Holcomb, 2009). Hence, the core team that runs and manages the risk program should be under the direct line of sight of a C-level (top management) executive. This executive sponsor from the top management should be made accountable for the risk program. The core team

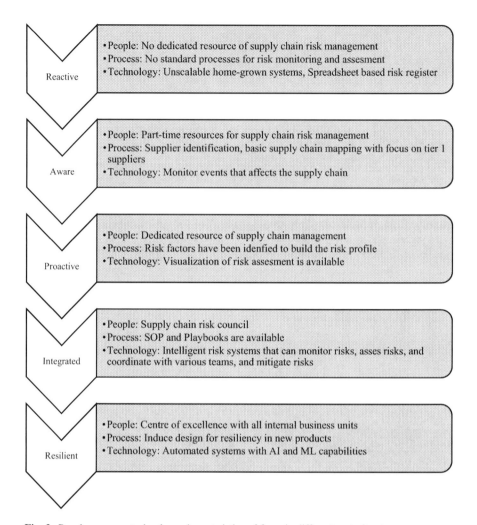

Fig. 2 People, process, technology characteristics of firms in different maturity stages

shall be responsible for the risk program of the organization. A suggestive list of the core team's roles and responsibilities is given below.

1. Define the scope of the risk program
2. Continuous monitoring and triaging of events that could have an impact on the supply chain
3. Setting up of standard operating procedures and playbooks
4. Conducting periodical risk evaluations and ensuring that suppliers have a robust business continuity plan
5. Coordinating with all key stakeholders to mitigate any identified risk
6. Conducting tabletop exercises internally (with various functions) and externally (with suppliers)

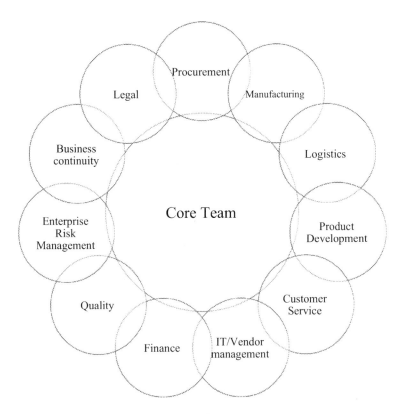

Fig. 3 Functions with which the core team coordinates

7. Ensuring program compliance (internal and external) and periodic status reporting to the risk council
8. Highlighting the importance of risk and influencing new product introductions and supplier selections

Firms at the forefront of supply chain resilience incorporate the participation from the leadership team of their suppliers in their risk programs as a best practice. The risk executives should interact with the suppliers' leadership team regarding the risk program regularly. As the firm moves toward a resilient phase, the risk program's scope should expand to include other business units and entities, and the overall program should be governed and directed by a firm-wide center of excellence (COE). To ensure the risk program's internal adoption, the procurement managers, category managers, and purchasing managers should have risk management as a part of their key responsibilities. In addition to cost-saving targets, strategic sourcing and managing supply chain risk should become a part of a procurement manager's responsibilities.

3 Process

In this section, we focus on the processes involved in assessing and managing risk. Firms at different stages of the maturity level would have different processes established as a part of their risk program. Our objective is to provide a general framework of the key processes involved in assessing and managing risk. Thus, we attempt to provide answers to some of the common questions faced by firms while designing their risk programs.

1. How does a firm select the suppliers for the supply chain risk program?
2. What are the processes involved in supply chain risk mapping?
3. How to ensure continuous monitoring of supply chain risk?
4. How should firms analyze and mitigate risk?

Extant literature on supply chain risk management provides various steps involved in building an effective risk management program. For example, Hallikas et al. (2004) identified a four-step process: risk identification, risk assessment, risk management, and risk monitoring. Note that risk classification is a prerequisite for the effective identification of supply chain risk. At a basic level, one can classify supply chain risk as internal risks (risks arising due to internal processes) and external risks (risks arising from external factors). Figure 4 illustrates a comprehensive classification of supply chain risk provided by Christopher and Peck (2004). Additionally, there are other contemporary risks such as cybersecurity risk, compliance risk, financial market risk, recovery capability, and brand goodwill risk.

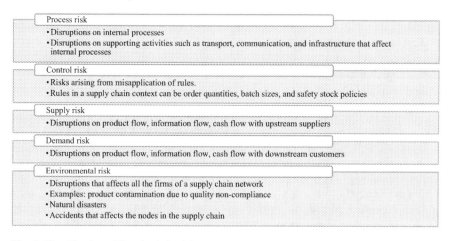

Fig. 4 Classification of Supply chain risk

3.1 Risk Score

Supplier selection for the risk program starts by assigning a risk score for each supplier associated with a firm. Each supplier's risk score shall include risk factors such as sourcing risk, quality risk, location risk, financial risk, technology risk, cybersecurity risk, sustainability practices risk, and compliance risk. Firms often quantify a composite risk score for every supplier (product/components/site), incorporating specific weights to different risk factors mentioned above. For example, a supplier's financial health may contribute say 20% to its overall risk score. At the start of the risk program, firms should select a tractable number of risk factors to compute suppliers' risk scores. Once every supplier has been assigned a risk score, firms can set thresholds to classify their suppliers into low-risk, medium-risk, and high-risk categories. Risk factors and thresholds should be iteratively adjusted so that the number of suppliers in the high-risk category at the start of the program is tractable.

> *Cisco's supplier resiliency index:* Cisco created a supplier resiliency index for measuring the risk associated with its suppliers. The key metrics used were supplier financial health, supplier business continuity plan compliance, preferred supplier status, and new suppliers. Cisco's supplier resiliency score varies from one to ten with one indicating not resilient and ten indicating very resilient.
> *(Source: Miklovic and Witty 2010)*

3.2 Risk Mapping

Risk mapping and supply chain transparency are vital elements for building a resilient supply chain. However, according to Deloitte Global Chief Procurement Officer Survey 2018, only 6% of the CPOs surveyed reported full transparency of the entire supply chain. The survey revealed that 54% have only limited visibility below tier 1 suppliers and 11% have no visibility below their tier 1 suppliers. But, many issues may originate at the sub-tier level (suppliers of tier 1 suppliers). Hence, firms must map their supply chain to the lowest sub-tier level possible. In the process of mapping, it is common to find that the bill-of-materials increases exponentially as a firm starts mapping its products and services from its sub-tier suppliers. Hence, it is pragmatic to conduct this exercise as an iterative process. To start with the supply chain risk program, firms may start mapping the products and services sourced from their tier 1 suppliers. We list some of the industry best practices and common challenges faced by firms during the supplier selection and the risk mapping process below:

- *Unique supplier identification*: Quite often, a single supplier may be listed using different names (using acronyms, subsidiary names, and the region of operation) in the firm's database. Hence, it is crucial to conduct a data cleaning process where every supplier gets a unique identifier/reference.
- *Data cleaning:* Firms often do not update and clean the data related to their bill-of-materials. The presence of obsolete and invalid parts in the Enterprise Resource Planning (ERP) system makes the mapping process difficult for the firm and its suppliers. In such situations, it is a common practice that firms identify and focus initially on the most used parts in the last 6 months and the parts that will be used in the next 2 years. Henceforth, we refer to this list as the clean list of parts.
- *Data collection:* Once a firm identifies the clean list of parts, it should collect as much data as possible regarding these parts from its direct suppliers. Some of the commonly collected data are supplier part number, manufacturing site information, activity conducted at various sites (such as manufacturing, warehousing, assembly, and testing), recovery capability at each of these sites, emergency contact details for each of the sites, alternate sites, contact details, and alternate site's recovery capabilities. The firm should ideally capture the business continuity plan of the supplier's sites in the data along with cybersecurity capability, IT backup, power backup, communication backup, fire and safety plan, and pandemic plan.
- *Sub-contractors:* The firm should collect the details of the sub-contractors, their locations, contact information, and alternate options. It should analyze the dependencies of its suppliers on their respective sub-contractors.
- *Information sharing:* The process of building supply chain transparency includes information collection from suppliers through assessments. A suggestive list of various domains for information collection is as follows: cybersecurity, financial security, policy compliance, quality compliance, and law compliance. While collecting the suppliers' information, firms should be cognizant of the fact that the suppliers may offer their products to multiple firms, and providing compliance reports in different formats to various firms can be exhausting and may lead to supplier fatigue. Many suppliers choose to be members of alliances such as the Responsible Business Alliances (RBA). Membership in such alliances requires the suppliers to be compliant across various categories and hence signals the suppliers' trustworthiness to firms. Such memberships can benefit the suppliers and their clients.
- *Timeline and milestones:* Collection of supplier information and the risk mapping activity can take several months to be completed. Hence, we recommend firms start the risk monitoring process (which we explain in Sect. 3.3) as soon as they finalize the initial list of suppliers for the risk program. Firms should set specific timelines and milestones for risk mapping. For example, collect information from say 80% of tier-I suppliers within the first 90 days.
- *Reluctance toward data sharing:* While many suppliers might be comfortable sharing the data related to their products and their manufacturing capabilities, they might be reluctant in sharing their procurement sources and supplier data.

This reluctance is reasonable since sharing sub-tier supplier data provides opportunities for the firm to bypass their direct suppliers and directly approach the downstream sub-tier suppliers for lower costs. In such situations, firms should put effort into developing trust with the suppliers. For example, initially, firms may request mapping information only for critical parts (based on the revenue impact/product criticality) and then iteratively improve and acquire more data as the suppliers gain confidence in the risk program. Firms should communicate and convince the suppliers that the risk program is a collaborative activity to improve the entire supply chain network's resiliency.

- *Skin in the game for the supplier:* A firm's risk program should not be independent of its tier 1 and sub-tier suppliers. Thus, it is crucial to incentivize suppliers to participate actively and engage in the risk program. For example, firms should include the supplier's supply chain risk program, mapping capabilities, supplier's response to disaster events, and collaboration in the overall supplier scorecard. Suppliers can use their involvement in the risk program as a marketing tool to attract new customers and to get a preferred supplier status for further business development.
- *Refreshing the data:* The risk program's effectiveness depends primarily on refreshing the data and on iterative improvement. Hence, it is crucial to conduct data refresh exercises with iterative improvements. For instance, any change in supplier details such as warehouse location, risk score, and contact information should be reflected in the refreshed data.
- *Revenue-impact criteria:* The fraction of total spend on a part is a traditional criterion used by firms to prioritize the crucial parts for risk mapping. However, firms should also consider the impact of a particular part on the revenue of the firm. For example, a small part (from a sub-tier supplier) may not qualify in the spend criterion, but it might substantially impact the revenue.
- *Audits and Tabletop exercises:* Resilient firms often conduct frequent audits, tabletop exercises, and drills with their major suppliers.

3.3 Risk Monitoring

Risk monitoring is a continuous process that requires a system to monitor, assess, and initiate actions on various risks. Technology plays a crucial role in establishing an effective risk monitoring system. Artificial intelligence and machine learning systems can be used to correlate world events with your supply chain network. These intelligent systems use natural language processing techniques to aggregate events from millions of sources (news and social media) and categorize them based on their potential impact on the supply chain network. Firms can input relevant keywords into the system to track and receive information on various events such as hurricanes, earthquakes, forest-fire, tsunami, floods, flash floods, volcanic eruptions, terrorist attacks, pandemics, coups, labor strikes, political instability, and material shortages. Firms with a global supply chain footprint should deploy monitoring systems that are intelligent to compile different languages and filter relevant information from news, article aggregators, and social media.

Firms should design a crisis response process, in which thresholds are established by prioritizing various risk events. Some of the metrics that can be used to develop the thresholds include the region of impact, the number of sites affected, the number of parts (components) involved, the number of single-sourced items affected, recovery time, the effect on shipments, and importantly, the impact on revenue. The escalation protocols in the system should be defined based on these thresholds. For example, a high-priority event can trigger emails and alarms to the leadership team and all key stakeholders, including the suppliers. The relevant stakeholders can then meet in a virtual war-room, where the mapping information is readily available. The stakeholders conduct a risk assessment using what-if analysis and estimate the possible impact. Finally, the firm initiates the risk mitigation actions and debriefs to learn and improve.

3.4 Risk Prioritization and Mitigation

In Sect. 3.1, we discussed categorizing a firm's suppliers as high-risk, medium-risk, and low-risk suppliers based on each supplier's risk score. Proactive identification of various risk factors associated with suppliers helps firms to assign the risk score. Once the prioritization is complete, the firm assigns high-risk suppliers for the mitigation process.

A typical risk mitigation process consists of the below four steps:

1. Decide the set of high-risk suppliers/materials based on the risk score.
2. Identify the people who will be responsible for various risk mitigation processes and assign individual responsibilities.
3. Periodical reporting of the mitigation progress to the risk program owner.
4. Develop, test, and iteratively improve playbooks and standard operating procedures.

Firms with mature risk management practices establish a technology-enabled system capable of analyzing and prioritizing risk based on pre-defined criteria. Such a system allows the risk team to focus its effort on the mitigation processes. They succeed in establishing a process to manage the mitigation effort that provides visibility on the progress, playbooks launched, and the actions implemented. Once the mitigation is over, they re-evaluate the high-risk supplier's risk score to ensure that the supplier's risk has been reduced to moderate/low risk.

> **Tackling Natural Disasters Through Risk Monitoring and Mitigation**
> Hurricanes Harvey (category 4) and Irma (category 5) disrupted many supply chains when they made their landfall in Texas and Florida in August–October 2017. The white paper released by Resilinc discusses how Biogen, a pharmaceutical company, was able to take proactive actions using the services provided by Resilinc. As per the white paper, Biogen's security and supply chain team members were proactive in understanding that the hurricane Harvey made little impact on its supply chain footprint. This assessment, enabled by Resilinc's services, helped Biogen's employees to revert to business as usual saving several productive manhours. During the time of the hurricane Irma, Biogen was able to proactively advance the shipments from its Florida supplier's site to Kentucky site to avoid shipment delays.
> Source: *Biogen & Resilinc Case Study: Proactive Risk Mitigation in Hurricane Season* (2018, May 29)

4 Technology

Technology enables firms to monitor, prioritize, and mitigate risk. In Sects. 3.3 and 3.4, we discussed the importance of setting up a continuous risk monitoring system and designing a crisis response process leveraging the capabilities of contemporary technological systems. In this section, we highlight some of the best practices and recent technology trends.

Virtual War Rooms:
For firms with a global supply chain footprint, it is important to have the entire impact information of a disruptive event available in one central place. A virtual war room is used as a command center where all the details about the disruptive event, such as the detailed commentary of the event (with live updates), products impacted, suppliers impacted, parts impacted, alternate suppliers/sites list, sub-tier suppliers impacted, and the contact information of the supplier sites are available. Such a facility allows the risk team to quickly identify the high-risk materials/suppliers and gaps in the business continuity plan. The virtual war room facilitates the risk team to conduct what-if analysis to review various possible scenarios. For example, if a hurricane is expected, the risk team can estimate the impact of various paths that the hurricane could take. Technology enables the firms to draw different hurricane trajectories on a map that has mapped the firm's suppliers. This ability to simulate various trajectories helps the firm be prepared and take proactive measures to avoid or minimize the impact.

Interactive Dashboards:
Interactive dashboards provide a customizable view of a firm's supply chain footprint on a map with the locations and other details of the suppliers, sub-tier suppliers, and contractors. Such dashboards allow the firms to filter by regions, supplier type,

and risk type. Smart dashboards can connect multiple systems such as order management, planning, or transportation systems through native or custom integrations.

Spreadsheets to Smart Systems:
Firms should move from spreadsheet-based risk registers to smart systems. Smart systems are logic-based digital workflow systems that can initiate a set of actions based on the severity of incidents. They are capable of creating and tracking multiple tasks, setting automated reminders, and triggering automatically assigned tasks based on some pre-defined criteria. For example, when a supplier is impacted due to a disruptive event, the system can assign a task to a team member and activate a playbook. The member will immediately get notified about the action to be taken. Automated systems can notify suppliers about possible disruption and seek confirmation about their impact. Such automation helps to reduce the time required to analyze the impact so that firms can take proactive measures.

Scalability and Flexibility:
The risk systems should handle a large volume of information such as supply chain mapping, business continuity, cybersecurity, finance, and compliance data. These systems can aggregate data based on set criteria to display supplier profiles along with their risk scores. Since risk management is a continuous and iterative process, these systems must be scalable and yet flexible for changing the criteria of various thresholds.

Collaboration:
Contemporary risk management systems support contextual communication where the internal teams and suppliers can interact over virtual platforms. Communication systems keep digital records of the analysis, actions, and mitigation efforts.

AI and Machine Learning:
AI and machine learning systems leverage previous events' digital records to predict the suppliers/parts/sites that might be impacted due to a future event. For example, intelligent systems can correlate the type of previous events with the purchase order data to predict which purchase orders could be affected in the future due to a similar event. AI systems can also correlate various types of events with the market price of commodities to give useful insights to firms.

5 Conclusions

Supply chain risk continues to be a relevant topic of discussion in industrial practice and academia. Many firms, especially those with a global supply chain footprint, are focusing on improving the resiliency of their supply chain network. The COVID-19 pandemic acted as a litmus test for many firms to check their supply chain resilience. The disruption caused by this pandemic has motivated firms to improve the resiliency of their supply chains. In this context, we discuss the processes and best practices that will help firms manage their supply chain risks and move toward supply chain resiliency. Specifically, to design and implement an effective risk program, we focus on three key aspects of a firm: people, process, and technology. First, we discuss the formation of a risk core team that coordinates with various other functions in a firm to manage its supply chain risk. We present a suggestive list of the core team's roles and responsibilities. Next, we discuss various processes such as assigning risk scores, risk mapping, risk monitoring, and risk mitigation. Here, we highlight the importance of continuous monitoring and the crisis response process. Further, we discuss the key practical challenges faced by the firms while implementing risk assessment, risk monitoring, and risk mitigation processes. Finally, we discuss technology's role in monitoring, prioritizing, and mitigating supply chain risks. Here, we emphasize the importance of technology-enabled virtual war rooms and other decision support systems such as interactive dashboards and smart digital workflow systems. It should be noted that the proposed framework and suggestive list of processes are general and not industry-specific. Hence, every firm should customize its risk program to capture the specific nuances of its industry and market dynamics.

References

Biogen & Resilinc Case Study: Proactive Risk Mitigation in Hurricane Season. (2018, May 29). Retrieved from https://www.supplychainbrain.com/articles/28267-biogen-resilinc-case-study-proactive-risk-mitigation-in-hurricane-season

Christopher, M., & Peck, H. (2004). *Building the resilient supply chain.*

Deloitte Switzerland. (2020). *Building supply chain resilience beyond COVID-19.* Retrieved from https://www2.deloitte.com/content/dam/Deloitte/at/Documents/strategy-operations/deloitte-global-cpo-survey-2018.pdf

Dun & Bradstreet. (2020). *Business impact of the coronavirus.* Retrieved from https://www.dnb.com/content/dam/english/economic-and-industryinsight/DNB_Business_Impact_of_the_Coronavirus_US.pdf

Gow, D. (2002, January 15). Land rover dispute threatens 10,000 jobs. *The Guardian.* Retrieved from https://www.theguardian.com/business/2002/jan/15/carindustry.motoring

Hallikas, J., Karvonen, I., Pulkkinen, U., Virolainen, V. M., & Tuominen, M. (2004). Risk management processes in supplier networks. *International Journal of Production Economics, 90*(1), 47–58.

Hendricks, K. B., & Singhal, V. R. (2005). An empirical analysis of the effect of supply chain disruptions on long-run stock price performance and equity risk of the firm. *Production and Operations Management, 14*(1), 35–52.

Lee, S. M., & Rha, J. S. (2016). Ambidextrous supply chain as a dynamic capability: Building a resilient supply chain. *Management Decision.*

Linton, T., & Vakil, B. (2020). Coronavirus is proving we need more resilient supply chains. *Harvard Business Review, 5.*

Miklovic, D., & Witty, R. J. (2010). *Case study: Cisco addresses supply chain risk management.* Gartner Industry Research, G206060.

Nellis, S. (2019, August 28) Apple's data shows a deepening dependence on China as Trump's tariffs loom. *Reuters.* Retrieved from https://www.reuters.com/article/usa-trade-apple/apples-data-shows-a-deepening-dependence-on-china-as-trumps-tariffs-loom-idINKCN1VI2A0?editionredirect=in

Norrman, A., & Jansson, U. (2004). Ericsson's proactive supply chain risk management approach after a serious sub-supplier accident. *International Journal of Physical Distribution & Logistics Management*.

Ponomarov, S. Y., & Holcomb, M. C. (2009). Understanding the concept of supply chain resilience. *The International Journal of Logistics Management*.

Sharma, A., Adhikary, A., & Borah, S. B. (2020). Covid-19's impact on supply chain decisions: Strategic insights from NASDAQ 100 firms using Twitter data. *Journal of Business Research, 117*, 443–449.

Sheffi, Y. (2007). *The resilient enterprise: Overcoming vulnerability for competitive advantage*. Zone Books.

Supply Chain Risk Leadership Council. (2013). *SCRLC emerging risks in the supply chain 2013* [White paper]. Retrieved October 24, 2021, from https://scrlc.com/articles/Emerging_Risks_2013_feb_v10.pdf

SwissVR Monitor. (2020). *The Board of Directors' perspective on COVID-19: Learning the lessons for the next crisis*. Retrieved from https://www2.deloitte.com/content/dam/Deloitte/ch/Documents/audit/deloitte-ch-en-swissVR-ii-2020.pdf.pdf

The Deloitte Global Chief Procurement Officer Survey. (2018). *Leadership: driving innovation and delivering impact*. Retrieved from https://www2.deloitte.com/content/dam/Deloitte/at/Documents/strategy-operations/deloitte-global-cpo-survey-2018.pdf

Integrated Optimization of Resilient Supply Chain Network Design and Operations Under Disruption Risks

Zhimin Guan, Jin Tao, and Minghe Sun

Abstract In recent years, supply chain disruptions caused by unexpected events have occurred more and more frequently, and these disruptions have been proven to have both short- and long-term negative impacts on supply chain operations and on corporate profitability. Thus, it is imperative to first analyze and understand the effects of these risks and then develop solutions to mitigate their impacts. In this study, an optimization approach is developed for integrated design and operations for resilient supply chain networks with disruption risk considerations. A mixed binary integer programming model is formulated for this purpose. Scenarios are used to describe disruption events of the facilities, and disruption events may take place at multiple facilities at the same time in a scenario. Uncertainties in supplies, demands, and prices are also considered. A region-wide dual-sourcing strategy, strategic emergency inventories, and alternative sourcing facilities are used in the supply chain network design stage to increase network resilience. The Sample Average Approximation method is used to solve the proposed model with disruption risk considerations. An illustrative example is used to demonstrate the validity of the model and sensitivity analysis results are reported to examine the effects of important parameters on the performance of the resulting resilient supply chain networks.

Z. Guan
School of Business Administration, Northeastern University, Shenyang, China

J. Tao (✉)
School of Business Administration, Henan University of Economics and Law, Zhengzhou, China
e-mail: taojintaojintaojin@163.com

M. Sun
Carlos Alvarez College of Business, The University of Texas at San Antonio, San Antonio, TX, USA

1 Introduction

With the economic globalization, many supply chain networks (SCNs) span wide geographic areas and involve many business firms. The relationship among the business firms in a SCN has become increasingly complex. At the same time, SCNs have changed the environment and conditions of the global economy. While helping business firms increase their profits, SCNs also face challenges of supply chain disruption risks. Supply chain disruption risks may cause significant casualties, property losses, ecological environment destructions, and/or serious social harms. According to the origins, characteristics, and mechanisms, disruption risks generally can be divided into natural hazards, accidental disasters, public health emergencies, and public safety events. In spite of the low occurrence probabilities of these disruption risks, the subsequent impacts are huge and the consequences are difficult to control once happened. A disruption event may cause irrevocable damages on the whole SCN. All such disruption events are considered to be low probability and high impact causing significant damages to business operations.

In recent years, supply chain disruptions caused by unexpected events have occurred frequently. Events like the September 11 Attacks and Hurricane Katrina brought huge disasters to different SCNs (Tang, 2006a; Snyder, 2003; Sheffi, 2005; Barrionuevo & Deutsch, 2005; Latour, 2001; Mouawad, 2005). These disruption events showed high risks and changed the characteristics of the modern business environment. Although international SCNs are believed to be stronger and more reliable, in fact, many such international SCNs are fragile and easy to fail when unexpected disruption events happen. For example, the disastrous earthquake and the following tsunami struck Japan in March 2011 not only caused heavy casualties and property losses, but also halted the production in a broad spectrum of the industries in the northeast of the country because of plant ruins, transportation blockages, and/or power outages. As a consequence, the global electronics industries underwent a large supply shortage since Japan is the major supplier for electronics components such as semiconductors, LCD panels, flash memory chips, and so on (Clark & Takahashi, 2011). Many such disastrous SCN disruption events have happened in the last two decades (e.g., Miller, 1992; Christopher & Peck, 2004; Yang et al., 2004; Boyle et al., 2008; Tang, 2006b; Rodrigues et al., 2008; Kleindorfer & Saad, 2005; Prater, 2005; Tomlin, 2006). The COVID-19 pandemic is a more recent noteworthy incident in supply chain disruptions. A hard-hit area is the auto parts and supplies industry. More than 80% of the world's auto parts are made in China. According to data reported, the export value of auto parts from China exceeded 60 billion US dollars in 2019, of which 40% was exported by subsidiaries of foreign-funded enterprises in China. According to reports, Hyundai Motor Company closed a major assembly line in Ulsan, South Korea, due to the disruption of parts supply from China caused by the COVID-19 pandemic, and further aggravation of the impact of the pandemic may force it to stop the operation of three factories in South Korea, which account for 40% of its global production[1].

Traditional SCN design theory and methods are facing new challenges because of the huge destructions caused by disruption events. Compared with the traditional risks such as demand and price uncertainties, these disruption risks are accidental but more destructive to SCNs, which makes the study of SCN optimization under disruption risk considerations particularly important (Park et al., 2013; Fujimoto & Park, 2014; Paul et al., 2016; Gao et al., 2019). So far, most of the studies assume that facilities are always available. Carefully constructed plans from traditional SCN design models can be severely ruined if they fail to consider disruption risks in the design phase and therefore do not have countermeasures when disruptions do strike. However, no matter how secure a SCN is, it cannot completely avoid disruption risks such as natural disasters, operational accidents, etc. Since disruption events cannot be avoided completely, the key is to reduce the impacts of disruption events to SCNs. Therefore, considering the disruption risks in the designing process, giving the SCNs the ability to quickly return to normal operational conditions after disruptions, and designing resilient SCNs will undoubtedly have important value and significance.

The design and operations of SCNs in the petroleum industry motivated this study and the oil industry in northeast China is used as an illustrative example. Crude oil plays a vital role in the development of the world economy because it is one of the most important energy sources and the most important raw material of the petrochemical industry. Two oil crises in the 1970s caused tremendous damages and impacts to the economies and social lives of the oil importing and consuming countries. In recent years, there is an upward trend in the occurrences of unexpected oil supply disruption events causing oil supply failures (Manuj & Mentzer, 2008; Tong et al., 2011; Tverberg, 2012; McKillop, 2005; Mitchell & Mitchell, 2014). An oil SCN is a complex system that connects the business firms of different functions including supplies, production, storage, alternative sourcing facilities, transportation, and sales. Currently, oil SCNs have various problems (Varma et al., 2008; Chen, 2008) such as unreasonable locations of refineries, unreasonable locations of crude oil reserve bases and oil product reserve bases, imperfect transportation networks, lack of the ability to cope with emergencies, and so on. Based on the above reasons, oil energy security has become a strategic issue for the sustainable economic development and national security of China in the new century and has now caused wide concerns of relevant government departments and the general public. Therefore, considering disruption events in the SCN design and operations decisions will enable the SCN to recover quickly from disruptions and ensure stable operations and profits of the business firms involved in oil SCNs.

The remainder of this chapter is organized as follows. In Sect. 2, a general review of the relevant literature is provided. Section 3 describes the problem studied. Section 4 presents the mathematical model. Section 5 describes the solution method.

---

[1] https://www.sohu.com/a/376649906_466840

Section 6 shows an illustrative example for verifying the proposed model and presents the sensitivity analysis results. Most of the data used in the illustrative example are presented in the Appendix. Finally, conclusions are drawn and future works are outlined in Sect. 7.

2 Literature Review

There is an emerging literature dealing with disruption risks in SCN design and operations. About SCN resilience, Muckstadt et al. (2001) considered that enhancing cooperation among SCN members and reducing uncertainty of the operating environment were effective methods to make supply chain resilient. Sheffi (2001) proposed the use of double suppliers to improve SCN resilience and provided a qualitative analysis. More and Babu (2008) developed a unified SCN resilient research framework. Soni and Jain (2011) proposed a new framework for supply chain resilience leveraging existing knowledge and offering a better understanding of the available notion in the literature. Boin et al. (2010) outlined a new method of studying resilient supply chains for extreme situations. Hasani and Khosrojerdi (2016) studied SCN design under disruption and uncertainty considering resilience strategies. Fattahi et al. (2017) considered a responsive and resilient SCN design under operational and disruption risks with delivery lead-time sensitive customers. Ghavamifar et al. (2018) explored the practical application of a bi-level model in a resilient competitive SCN. Azad and Hassini (2019) investigated the design of reliable SCNs to make them resilient to unpredictable disruptions.

Many scholars studied SCN resilience and some of them proposed quantitative models to enhance SCN resilience. However, most of the models consider only one resilience strategy, and cannot increase the SCN resilience obviously when multi disruption events strike.

Many scholars reported studies about SCN design (Bidhandi et al., 2009) and some of them studied SCN design and operations under risks using quantitative models. Klibi et al. (2010) presented a critical review of the optimization models proposed in the literature and also discussed the importance of robustness, responsiveness, and resilience of SCNs. Peidro et al. (2009) reviewed the relevant literature in supply chain planning methods under uncertainty. Cui et al. (2010) developed a mixed integer programming (MIP) model and a continuous approximation of the model for the uncapacitated fixed cost facility location problem under facility disruption risks. Peng et al. (2011) proposed a MIP model minimizing the nominal, *i.e.*, without disruptions, cost while reducing the disruption risks using the *p*-robustness criterion, and solved the problem with a heuristic procedure. They demonstrated the tradeoffs between nominal cost and system reliability concluding that substantial improvements in reliability were often possible with minimal increase in cost. Klibi and Martel (2012) provided a risk modeling approach to facilitate the design of SCNs and to evaluate operations under uncertainty. Two cases were studied to illustrate the key aspects of the approach and to show how the approach could be used to obtain resilient SCNs under disruptions. Baghalian et al. (2013) developed a stochastic mathematical programming formulation for designing a multi-product SCN comprising of several capacitated production facilities, distribution centers (DCs) and retailers under disruptions. Sabouhi et al. (2018) proposed a two-stage possibilistic-stochastic programming model for integrated supplier selection and SCN design under disruption and operational risks. Al-Othman et al. (2008) developed a multi-period stochastic planning model of a petroleum business operating in an oil producing country under uncertain market conditions. In the model, the uncertainties are introduced in market demands and prices. Oliveira and Hamacher (2012) presented a multi-product and multi-period supply investment planning problem considering network design and discrete capacity expansion under demand uncertainty. Hamdan and Diabat (2020) proposed a bi-objective two-stage model using robust optimization techniques under the disaster scenarios. Goh et al. (2007) developed a multi-stage stochastic programming model for a global SCN with objective functions of maximizing profits and minimizing risks, and proposed a solution method based on the Moreau-Yosida regularization (Hiriart-Urruty & Lemarchal, 1993). Mitra et al. (2009) used fuzzy mathematical programming methods for multi-site, multi-product, and multi-period SCN design under an uncertain environment. Georgiadis et al. (2011) used a linear MIP model to study the SCN design problem under time varying demand uncertainty. The global optimal solution of the linear MIP model was obtained by using the standard branch and bound technique. Mirzapour Al-e-hashem et al. (2011) proposed a multi-objective nonlinear MIP model for a SCN with multiple suppliers, multiple manufacturers, and multiple customers, addressing a multi-site, multi-period, and multi-product aggregate production planning problem under uncertainty. The objectives are the minimization of the total cost and the minimization of the total customer dissatisfaction represented by the sum of the maximum shortages. Li et al. (2013) explored a generalized supply chain model with supply uncertainty. Although an approach has not been developed for systematic SCN design under disruption risks, the results of these studies provide theoretical basis and technical methods for this study. However, as Paul et al. (2016) pointed out, most of the previous studies considered only one risk factor such as uncertainty or disruption in a single stage and very little has been done in developing quantitative models to manage other risks, such as imperfect production processes, and disruptions in production, supply, and demand, as well as their combinations.

There is also a growing body of literature addressing strategic emergency inventories (Schmitt, 2011; Sheffi, 2005; Lücker et al., 2019), which should be held throughout the supply chain to withstand disruption risks. Yin and Rajaram (2007) considered the joint pricing and inventory control problem using a Markov chain. In addition, many researchers studied the problem of supplier selection. Sawik (2013, 2014a, b, 2015) studied the problem of optimal selection and protection of portion suppliers and determined order quantity allocation in a supply chain with disruption risks. Considering the static and price-sensitive demand environment, Yu et al. (2009) assessed the effect of disruption risks to the single- and dual-sourcing models through the comparison of the expected profit functions, and illustrated how to make purchase decisions through a numerical example.

In this study, an integrated optimization approach is developed for the design and operations decisions of resilient SCNs under facility disruption risks and conventional risks. A binary MIP model based on stochastic scenarios is proposed to solve the problems. This study is different from previous studies mainly in the following three aspects: (1) According to the specific characteristics of oil supply chains, a four-echelon supply chain structure, including suppliers, plants, DCs, and demand areas, is studied. The optimization of design and operations from the perspective of the whole oil SCN in northeast China is studied as an illustrative example. The conclusions drawn from the numerical analysis provide more insights into the oil SCNs. (2) Rather than considering only disruption risks or conventional risks as in previous studies, both disruption events such as natural hazards and uncertainties such as fluctuations in prices and demands are considered in this study. Disruption events may take place at multiple facilities in a SCN at the same time. Stochastic scenarios are used to describe uncertainty risks which make the solutions of the proposed model closer to reality. (3) Three resilient strategies, *i.e.*, region-wide dual-sourcing, strategic emergency inventories, and alternative sourcing, are used in SCN design. (4) An optimization approach is designed for problems with a large number of scenarios. As pointed out by Santoso et al. (2005), most existing approaches for SCN design under uncertainty are suited for small numbers of scenarios. Considering a SCN with just 50 facilities, each facility is vulnerable to two possible disruption conditions, and each disruption event is independent, then there are a total of 4^{50} possible scenarios, that is far more than most of the existing approaches for SCN design can handle. In this study, a recently proposed sampling strategy, the Sample Average Approximation (SAA) method (Kleywegt et al., 2001; Shapiro & Homem-de-Mello, 2000; Mak et al., 1999), is used to solve the integrated optimization problem of SCN design and operations decisions under disruption risks and conventional risks. The SAA method uses discrete scenarios to handle the randomness in a stochastic optimization model by means of Monte Carlo simulation, so as to facilitate the solution process of the model. As the number of scenarios becomes large, the sample average of the objective function values will approach the expected value of the objective function, and the solution obtained will be a good approximation of the optimal solution, of the model. Furthermore, the number of scenarios can be adjusted in the solution process to achieve the desired precision of the approximation. Given the stochastic nature of the disruptive risks considered in the SCN, the SAA method is suitable in solving the resilient SCN design problem under disruption risks. Because of the stochastic nature of the model and the large number of binary variables involved in the model, exact solution methods directly using a commercial software such as IBM® ILOG® CPLEX® (IBM, 2022) may not be able to solve the problem within a reasonable amount of computation time.

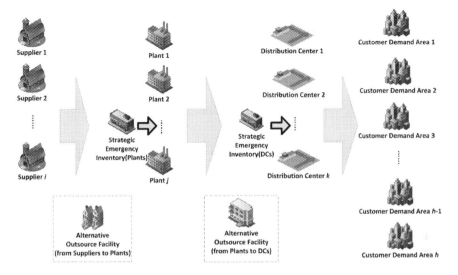

Fig. 1 The SCN structure considered in this study

3 Problem Description

A manufacturer usually purchases its raw materials from a number of suppliers and distributes its final products to many customers through DCs. In the oil industry, a manufacturer needs to choose locations for its refineries and DCs among several candidate sites. With the rapid development of the global economy, the market environment is becoming more dynamic and unstable. Transactions between business firms are becoming complicated. Many disruption risks exist in the business environment. To avoid losses caused by disruptions, a business firm needs to consider the reliability of the whole SCN from a strategic perspective. When disruptions occur, the business firm should have the ability of mitigating risks in the SCN through coordination and cooperation among the members of the SCN.

In this study, a four-echelon SCN optimization problem is considered. The four echelons include suppliers, plants, DCs, and customer demand areas. The suppliers, plants, and DCs are collectively called facilities. In the following, a customer demand area is simply called a customer. The locations of the suppliers and customers are given and candidate or potential sites for plants and DCs are chosen. It is not necessary to use all potential suppliers and it is also not necessary to construct plants or DCs at all potential sites. Therefore, the SCN design problem is to determine which potential suppliers to use, at which potential sites to construct plants and DCs, and the quantities of materials and products to transport among the various facilities and customers. A typical SCN is depicted graphically in Fig. 1. The facilities and customers are called nodes and the roads between facilities and/or customers are called arcs in the SCN.

Two types of risks in a SCN are considered. The first type of risks includes conventional risks coming from uncertainties in supplies, demands, and prices such as production delays, demand variations, and price fluctuations. To deal with the conventional risks, each plant or DC may hold a certain amount of safety stock, a plant may choose more than one supplier to purchase its raw materials, and a DC may choose more than one plant to purchase its products. The second type of risks includes disruption risks stemming mainly from the following three factors: (1) emergencies in daily operations, such as industrial accidents that may cause equipment, vehicle, and/or inventory damages, (2) natural disasters such as earthquakes, hurricanes, storms, debris flows, etc., and (3) terrorism and political instability. Due to disruption risks, the supplies of facilities in a SCN may stagnate.

Scenarios are used to describe the disruptions of the facilities and the fluctuations in supplies, demands, and prices. However, the influence of each scenario on the capacities of the SCN facilities is different. Disruption events may take place at multiple SCN facilities, i.e., at suppliers, plants, and/or DCs, at the same time in one scenario. Three strategies, i.e., dual-sourcing or region-wide dual-sourcing, strategic emergency inventories, and alternative sourcing facilities, are proposed to cope with disruption events. When disruption events happen, strategic emergency inventories can provide different quantities of key raw materials or products to other facilities in the SCN. Therefore, overall collaboration among facilities in the SCN is achieved and the ability of the SCN to cope with failure events is enhanced with low sharing costs. After the disruption events, plants and DCs in the SCN can choose alternative sourcing facilities to provide the unmet demand of raw materials/products to ensure that the facilities in the SCN will continue to operate. A dual-sourcing strategy indicates that a buyer uses two suppliers, one of which may dominate the other in terms of business share, price, reliability, and so on. A region-wide dual-sourcing strategy is on the basis of the dual-sourcing strategy by selecting different upstream facilities from a different region to provide better network coverage and, hence, to enhance the resilience of the network infrastructure in the SCN. Once a supply chain infrastructure is constructed, it will be very difficult and costly to modify. Therefore, it is important to design a SCN with stability and efficiency in the presence of all types of disruption risks.

The following assumptions are made about the operations of the SCN: (1) the occurrence of each scenario is independent; (2) strategic emergency inventories and alternative sourcing facilities will never be disrupted and will be used only when disruption events happen; and (3) a facility loses all its capacity when it is disrupted. The SCN design needs to achieve a reasonable profit level by reducing losses caused by disruptions and by balancing the various SCN parameters.

4 Mathematical Model

The binary MIP formulation is discussed in this section. The notations, including sets and indices, parameters and decision variables, used in the model are introduced first. The model is then formulated with further explanations.

4.1 Sets and Indices

Different sets of facilities and transportation modes are considered in the SCN design. The notations for these sets are presented in Table 1.

4.2 Parameters

Various parameters are used in the binary MIP model. The values of these parameters need to be determined before the model is formulated. These parameters are listed in Table 2.

4.3 Decision Variables

Both continuous and binary decision variables are used in the binary MIP model. These variables are defined in Table 3.

4.4 The Model

The objective function and the constraints of the binary MIP model are presented first. Further explanations are then given.

Table 1 Sets and index

Symbol	Description
I	Index set of the suppliers
J	Index set of the plants
K	Index set of the DCs
H	Index set of the customers
L	Index set of the transportation modes
S	Index set of the scenarios

Table 2 Parameters

Symbol	Description
π_s	Probability of scenario s for $s \in S$
f_i, f_j, f_k	Fixed operating cost of supplier i for $i \in I$, fixed operating cost of plant j for $j \in J$ and fixed operating cost of DC k for $k \in K$, respectively
$a_{si}^S, a_{sj}^M, a_{sk}^D$	Capacities of supplier i for $i \in I$, plant j for $j \in J$ and DC k for $k \in K$, respectively, in scenario s for $s \in S$
P_s	Unit sales price of the final product
c_{si}^R	Unit raw material cost of supplier i for $i \in I$
c_j^P	Unit production cost of plant j for $j \in J$
f^{Ssub}, f^{Msub}	Fixed operating cost of an alternative sourcing facility for a plant and for a DC, respectively
c^{Ssub}, c^{Msub}	Average unit cost (including material and transportation costs) of an alternative sourcing facility for a plant and for a DC, respectively
f^{EXM}, f^{EXD}	Fixed operating costs of using emergency inventories for a plant and for a DC, respectively
c^{EXM}, c^{EXD}	Average unit cost (including material and transportation costs) of strategic emergency inventories for a plant and for a DC, respectively
c^{PEN}	Unit stockout penalty cost for the unmet demand at the DCs
$c_l^{tsm}, c_l^{tmd}, c_l^{tdp}$	Unit transportation cost from a supplier to a plant, from a plant to a DC, and from a DC to a customer, respectively, of transportation mode l for $l \in L$
$a_l^{tsm}, a_l^{tmd}, a_l^{tdp}$	Transportation capacity from a supplier to a plant, from a plant to a DC, and from a DC to a customer, respectively, of transportation mode l for $l \in L$
d_{ij}, d_{jk}, d_{kh}	Distances between supplier i and plant j, between plant j and DC k and between DC k and customer h, respectively
A_{si}, A_{sj}, A_{sk}	Indicators, $A_{si} = 1$ ($A_{sj} = 1, A_{sk} = 1$) if supplier i (plant j, DC k) is disrupted in scenario s and $A_{si} = 0$ ($A_{sj} = 0, A_{sk} = 0$) otherwise for $s \in S$
D_{sh}^{dem}	Demand of customer h for the final product for $h \in H$ in scenario s for $s \in S$
η^M, η^D	Capacities of strategic emergency inventories for the plants and for the DCs, respectively
ℓ^S, ℓ^M	Capacities of alternative sourcing facilities for suppliers and for plants, respectively
A_{kh}^{dis1}	Indicator, $A_{kh}^{dis1} = 1$ if DC k is in the same region as customer h is and $A_{kh}^{dis1} = 0$ otherwise for $k \in K$ and $h \in H$
B	Maximum tolerable stockout quantity of the SCN
ρ	Production transformation coefficient, $\rho > 0$
t_l^{tor}	Tortuosity factor of transportation mode l in the SCN for $l \in L$

Table 3 Decision variables

Symbol	Description
x_i, y_j, z_k	Binary variables, $x_i = 1$ ($y_j = 1$, $z_k = 1$) if supplier i (plant j, DC k) is selected and $x_i = 0$ ($y_j = 0$, $z_k = 0$) otherwise for $i \in I$ ($j \in J$, $k \in K$)
X_{ij}, Y_{jk}, Z_{kh}	Binary variables, $X_{ij} = 1$ ($Y_{jk} = 1$, $Z_{kh} = 1$) if plant j (DC k, customer h) is served by supplier i (plant j, DC k) and $X_{ij} = 0$ ($Y_{jk} = 0$, $Z_{kh} = 0$) otherwise for $i \in I$ and $j \in J$ ($j \in J$ and $k \in K$, $k \in K$ and $h \in H$)
$t_l^{ij}, t_l^{jk}, t_l^{kh}$	Binary variables, $t_l^{ij} = 1$ ($t_l^{jk} = 1$, $t_l^{kh} = 1$) if transportation mode l is chosen to ship from suppliers to plants (from plants to DCs, from DCs to customers) and $t_l^{ij} = 0$ ($t_l^{jk} = 0$, $t_l^{kh} = 0$) otherwise for $l \in L$
s_j, s_k	Binary variables, $s_j = 1$ ($s_k = 1$) if strategic emergency inventory is chosen by plant j (DC k) and $s_j = 0$ ($s_k = 0$) otherwise for $j \in J$ ($k \in K$)
w^M, w^D	Binary variables, $w^M = 1$ ($w^D = 1$) if alternative sourcing facilities are chosen by plants (DCs) and $w^M = 0$ ($w^D = 0$) otherwise
$Q_{slij}, Q_{sljk}, Q_{slkh}$	Quantity shipped in mode l between supplier i and plant j, between plant j and DC k and between DC k and customer h, respectively, in scenario s for $l \in L$ and $s \in S$
w_j^{Rm}, w_k^{Rd}	Quantity provided by strategic emergency inventories to plant j for $j \in J$ or to DC k for $k \in K$, respectively
S_j^{sub}, M_k^{sub}	Quantity provided by alternative sourcing facilities to plant j for $j \in J$ or to DC k for $k \in K$, respectively
b_{sh}	The stockout quantity of customer h for $h \in H$

$$\min \sum_{s \in S} \pi_s \left[P_s \sum_{k \in K} \sum_{h \in H} \sum_{l \in L} Q_{slkh} - \sum_{i \in I} c_{si}^R \sum_{j \in J} \sum_{l \in L} Q_{slij} \right.$$
$$\left. - \sum_{j \in J} c_j^P \sum_{k \in K} \sum_{l \in L} Q_{sljk} - C_s^t - c^{PEN} \sum_{h \in H} b_{sh} \right]. \quad (1)$$
$$- C^e - C^a - F$$

s.t. $C_s^t = \sum_{i \in I} \sum_{j \in J} \sum_{l \in L} Q_{slij} c_l^{tsm} d_{ij} t_l^{tor} + \sum_{j \in J} \sum_{k \in K} \sum_{l \in L} Q_{sljk} c_l^{tmd} d_{jk} t_l^{tor} + \sum_{k \in L} \sum_{h \in H} \sum_{l \in L} Q_{slkh} c_l^{tdp} d_{kh} t_l^{tor},$

$\forall s \in S$

$$\qquad (2)$$

$$C^e = \sum_{j \in J} c^{EXM} w_j^{Rm} + \sum_{k \in K} c^{EXD} w_k^{Rd}. \quad (3)$$

$$C^a = c^{Ssub} S^{subs} + c^{Msub} M^{subm}. \quad (4)$$

$$F = \sum_{i \in I} f_i x_i + \sum_{j \in J} \left(f_j y_j + f^{EXM} s_j \right) + \sum_{k \in K} \left(f_k z_k + f^{EXD} s_k \right) + w^M f^{Ssub}$$
$$+ w^D f^{Msub}. \quad (5)$$

$$\sum_{j \in J}\sum_{l \in L} Q_{slij} - A_{si}a^S_{si}x_i \leq 0 \quad \forall s \in S, \quad i \in I. \tag{6}$$

$$\rho \sum_{i \in I}\sum_{l \in L} Q_{slij} + w^{Rm}_j + S^{sub}_j - \sum_{k \in K}\sum_{l \in L} Q_{sjk} = 0 \quad \forall s \in S, \quad j \in J. \tag{7}$$

$$\sum_{j \in J}\sum_{l \in L} Q_{sljk} + w^{Rd}_k + M^{sub}_k - \sum_{h \in H}\sum_{l \in L} Q_{slkh} = 0 \quad \forall s \in S, \quad k \in K. \tag{8}$$

$$\sum_{k \in K}\sum_{l \in L} Q_{slkh} + b_{sh} = D^{dem}_{sh} \quad \forall s \in S, \quad h \in H. \tag{9}$$

$$\sum_{k \in K}\sum_{l \in L} Q_{sljk} - A_{sj}a^M_{sj}y_j \leq 0 \quad \forall s \in S, \quad j \in J. \tag{10}$$

$$\sum_{h \in H}\sum_{l \in L} Q_{slkh} - A_{sk}a^D_{sk}z_k \leq 0 \quad \forall s \in S, \quad k \in K. \tag{11}$$

$$\sum_{j \in J} S^{sub}_j - \ell^S w^M \leq 0. \tag{12}$$

$$\sum_{k \in K} M^{sub}_k - \ell^M w^D \leq 0. \tag{13}$$

$$\sum_{j \in J} w^{Rm}_j \leq \eta^M. \tag{14}$$

$$\sum_{k \in K} w^{Rd}_k \leq \eta^D. \tag{15}$$

$$w^{Rm}_j \leq \eta^M s_j \quad \forall j \in J. \tag{16}$$

$$w^{Rd}_k \leq \eta^D s_k \quad \forall k \in K. \tag{17}$$

$$\sum_{i \in I}\sum_{j \in J} Q_{slij} \leq a^{tsm}_l t^{ij}_l \quad \forall s \in S, \quad l \in L. \tag{18}$$

$$\sum_{j \in J}\sum_{k \in K} Q_{sljk} \leq a^{tmd}_l t^{jk}_l \quad \forall s \in S, \quad l \in L. \tag{19}$$

$$\sum_{k \in K}\sum_{h \in H} Q_{slkh} \leq a^{tdp}_l t^{kh}_l \quad \forall s \in S, \quad l \in L. \tag{20}$$

$$\sum_{h \in H} b_{sh} \leq B \quad \forall s \in S. \tag{21}$$

$$\sum_{i \in I} X_{ij} = 2 \quad \forall j \in J. \tag{22}$$

$$\sum_{j \in J} Y_{jk} = 2 \quad \forall k \in K. \tag{23}$$

$$\sum_{k \in K} Z_{kh} = 2 \quad \forall h \in H. \tag{24}$$

$$\sum_{k \in K} Z_{kh} A_{kh}^{dis1} = 1 \quad \forall h \in H. \tag{25}$$

$$\sum_{l \in L} Q_{slij} \leq a_l^{tsm} X_{ij} \quad \forall s \in S, \quad i \in I, \quad j \in J. \tag{26}$$

$$\sum_{l \in L} Q_{sljk} \leq a_l^{tmd} Y_{jk} \quad \forall s \in S, \quad j \in J, \quad k \in K. \tag{27}$$

$$\sum_{l \in L} Q_{slkh} \leq a_l^{tdp} Z_{kh} \quad \forall s \in S, \quad k \in K, \quad h \in H. \tag{28}$$

$$X_{ij} \leq x_i, X_{ij} \leq y_j \quad \forall i \in I, \quad j \in J. \tag{29}$$

$$Y_{jk} \leq y_j, Y_{jk} \leq z_k \quad \forall j \in J, \quad k \in K. \tag{30}$$

$$Z_{kh} \leq z_k \quad \forall k \in K, \quad h \in H. \tag{31}$$

The objective function (1) maximizes the total weighted SCN profit. The profit is the difference between sales revenue (the first term in the bracket in (1)) and the total costs. The total costs consist of raw material cost (the second term in the bracket in (1)), production cost (the third term in the bracket in (1)), transportation cost (represented by C_s^t (2)), stockout penalty cost (the last term in the bracket in (1)), strategic emergency inventory holding cost (represented by C^e (3)), alternative sourcing facility cost (represented by C^a (4)), and fixed operating costs of the whole SCN (represented by F (5)). Constraints (2)–(5) are used to compute the different costs. These costs (2)–(5) are substituted into the objective function directly in the solution process.

Constraints (6)–(9) are the conservation of flow constraints for the suppliers, plants, DCs, and customers, respectively, in each scenario. The constraints in (6) also serve as the capacity constraints of the suppliers and the constraints in (9) also serve as the demand constraints of the customers. Constraints (10) and (11) represent the capacities of the plants and the DCs, respectively, in each scenario. Constraints (12) and (13) restrict the total quantities provided by the alternative sourcing facilities to be within their respective capacities for the plants and the DCs, respectively. Constraints (14) and (15) represent the requirements that the strategic emergency inventories used by the plants and DCs, respectively, must be within their respective capacities. Constraints (16) and (17) restrict the strategic emergency inventories to be used only if a facility has chosen to use them for the plants and the DCs, respectively. Constraints (18)–(20) are the capacity constraints of logistics service providers or transportation modes for each scenario. Constraint (21) restricts the total stockout quantity to be within the tolerance for each scenario. Constraints (22)–(24)

represent dual-sourcing strategy restrictions. Each plant is served by exactly 2 suppliers (22), each DC is served by exactly 2 plants (23), and each customer is served by exactly 2 DCs (24). Constraints (24) and (25) are sourcing strategy constraints of the customers. Each customer can only choose 2 DCs (24) one of which is not in the same region (25). Constraints (26)–(28) restrict materials or products to flow on a road only when the corresponding upstream facility is chosen to serve the downstream facility. Constraints (29)–(31) restrict an arc to be chosen only when both the origin and the destination nodes are chosen.

5 A Solution Method

The solution procedure is described in this section. The approach used for scenario generation is discussed first. The SAA method tailored to the binary MIP model in (1)–(31) is then described in detail.

5.1 Scenario Generation

Different types of possible disruption events, such as earthquakes, rainstorms, floods, and other types of environmental issues, may strike the facilities in a SCN. In the scenario generation process, N^d is used to represent the index set of the types of disruption events. Historical data and related statistics are used to determine ϕ_n representing the number of occurrences of disruption event n within time period T. A good estimate of the expected frequency of disruption event n is ϕ_n/T. The occurrences of the disruption events are assumed to follow Poisson distributions, as the Poisson distribution is suitable and is usually used to describe the number of occurrences of a random event per unit of time (Neter et al., 1993). The probability of any disruption event occurring more than once in any one unit of time is set to 0 to more accurately describe the characteristics of disruption events. The probability of disruption event n occurring in a time unit is then given by $p(n) = 1 - e^{-\phi_n/T}$. In the scenario generation process, to determine whether disruption event n occurs in a facility, a random number r_n is generated first, and then the disruption event n occurs if $r_n \leq p(n)$. The facility is disrupted if one or more disruption events occur in the facility.

The supply of the raw material and the demand of the final product are treated as normal random variables. The prices of the raw material and the final product are generated using the Geometric Brown Motion proposed by Awudu and Zhang (2012), a continuous-time stochastic process in which the logarithm of the randomly varying quantity follows a random movement. Let P_t represent the price to be generated at time t and t is a continuous variable, then

---

For $s \in S$

    For $n \in N^d$

        For $i \in I$ ($j \in J$ or $k \in K$)

            If $r_n \leq 1 - e^{-\phi_n/T}$, $A_{si} = 1$ ($A_{sj} = 1$ or $A_{sk} = 1$)

            Otherwise $A_{si} = 0$ ($A_{sj} = 0$ or $A_{sk} = 0$)

        Next i (j or k)

    Next n

    For $i \in I$

$$a_{si}^s = F^{norm}(\mu_i, \sigma_i^2)$$

$$c_{si}^R = c_{si}^{R0}(1 + \mu_r ds + \sigma_r \varepsilon_r \sqrt{ds})$$

    Next i

    For $h \in H$

$$D_{sh}^{dem} = F^{norm}(\mu_h, \sigma_h^2)$$

    Next h

$$P_s = P_0(1 + \mu_p ds + \sigma_p \varepsilon_p \sqrt{ds})$$

Next s

Fig. 2 The scenario generation procedure

$$P_t = P_0\left(1 + \mu dt + \varepsilon \sigma \sqrt{dt}\right), \tag{32}$$

where P_0 is the initial price with $P_0 > 0$, μ and σ are the mean and standard deviation of the price of either the raw material or the final product, ε is a standard normal random variable, and dt is a time interval. Given the mean μ and the standard deviation σ, the values of the prices P_t of either the raw material or the final product in the scenarios are randomly generated by using the function above (32). The procedure of scenario generation is shown in Fig. 2. In this scenario generation procedure, $F^{norm}(\mu_i, \sigma_i^2)$ returns a normal random variable with mean μ_i and a variance σ_i^2.

5.2 The Solution Procedure

Because of the large number of binary variables in the model, the time needed to solve the problem in (1)–(31) using a general purpose mathematical programming software through the calculation of mathematical expectations is unimaginable. Therefore, a scenario-based approach is used to obtain an approximate final solution. However, the computational time needed to solve a real-life problem increases rapidly as the number of scenarios increases. The SAA method (Kleywegt et al., 2001; Shapiro & Homem-de-Mello, 2000; Mak et al., 1999) can be applied to the model to reduce the computation time. The SAA method is a solution approach for stochastic optimization problems with large numbers of scenarios. This approach approximates the expected objective function value utilizing Monte Carlo simulation. The idea of this technique is to generate a sample of scenarios to construct an approximation and to solve the approximation instead of the actual problem.

The approximation is a binary MIP model that is set up by using N scenarios. The objective function of the approximation is given in Eq. (33) in the following

$$F^N = \max \frac{1}{N} \sum_{n=1}^{N} \left[P \sum_{k \in K, h \in H} Q_{skh}^{DP} - \sum_{i \in I, j \in J} c_i^r Q_{sij}^{SM} \right. \quad (33)$$

$$\left. - \sum_{j \in J, k \in K} c_j^p Q_{sjk}^{MD} - C_t - C_h - C_e - C_a - P^{pen} b_s \right] - F$$

and the constraints are the union of all the constraints of the N scenarios. The value of the objective function F^N (33) is an estimate of the expectation of the original problem.

The SAA method tailored to the binary MIP model in Eqs. (1)–(31) is described step-by-step in the following.

Step 1: Randomly generate M sets of, each containing N, scenarios. One binary MIP problem is solved for each set. The objective function of the binary MIP problem is in (33) and the constraints are the union of all the constraints of the N scenarios in the set. As a result, M solutions are obtained. Denote the optimal value of the objective function (33) for set m by F_m^N. The SCN structure whose objective function value is the largest among the M sets is the tentative solution.

Step 2: Compute the average of the M optimal objective function values

$$\overline{F^{MN}} = \frac{1}{M} \sum_{m=1}^{M} F_m^N. \quad (34)$$

The expectations of F^N is not less than the optimal value of the original problem, and $\overline{F^{MN}}$ is an unbiased estimate of the expectation of F^N (Mak et al., 1999). Hence, $\overline{F^{MN}}$

can be considered as the upper bound on the objective function of the original problem.

Step 3: Randomly generate a set of N', with $N' >> N$, scenarios. Use the SCN structure of the tentative solution in Step 1 to form a network optimization problem with the objective function in Eq. (33) but with N' replacing N and with the union of the constraints of the N' scenarios as the constraints. Note that the values of the binary variables are known in this network optimization problem. Solve the network optimization problem to obtain the value $F^{N'}$ for the objective function. Obviously $F^{N'}$ is a lower bound on the objective function of the original problem (Mak et al., 1999). The lower bound $F^{N'}$ will be close to the true objective function value when N' is large.

Step 4: Evaluate the quality of the tentative solution by computing the estimated optimality gap $gap_{N, M, N'}$ using the estimated lower and upper bounds from Steps 2 and 3, as follows:

$$gap_{N,M,N'} = \frac{\overline{F^{MN}} - F^{N'}}{\overline{F^{MN}}}. \qquad (35)$$

Step 5: If the value of $gap_{N, M, N'}$ is unsatisfactory, a larger value of N or M is used and Steps 1–4 are repeated until the value of $gap_{N, M, N'}$ becomes satisfactory (e.g., $gap_{N, M, N'} \leq 0.05$). The tentative solution is then the final solution.

6 An Illustrative Example

In this section, the model and the solution procedure are used to solve an integrated problem of SCN design and operations in an illustrative example. This illustrative example is based on a real-life SCN optimization problem of the oil industry in northeast China. China has several large-scale oilfields with annual output exceeding 100,000 tons (1 ton ≈ 7.40 barrels), such as Daqing, Liaohe, *etc*. Because of the frequent fluctuations in outputs, prices and demands of crude oil and petroleum products as well as all kinds of factors that may lead to disruption risks, it has been a difficult and crucial problem for the oil supply chain to coordinate its production and operations activities to improve efficiency, reliability, and stability. An integrated optimization model will help optimize the facility location, logistics assignment, and inventory strategy to reduce cost and improve efficiency of the whole SCN, so that the SCN participants may then adjust their plans quickly when responding to changes in the market and to disruption events by making the right decisions.

In this illustrative example, the oil SCN optimization problem in Liaoning province is considered. In order to fulfill the needs of provincial gasoline consumption, the oil refineries in Liaoning province purchase crude oil from Liaohe, a local oilfield, and some oilfields in the surrounding provinces, i.e., Daqing (in Daqing,

Heilongjiang), Shengli (in Dongying, Shandong), Dagang (in Tianjin), and Jilin (in Songyuan, Jilin). The gasoline produced by the refineries is distributed by DCs (i.e., gasoline distributors) to the demand areas in the cities of Liaoning province. Due to disruption risks, the emergency facilities including alternative sourcing facilities and strategic emergency inventories are used in this supply chain. Under the above conditions, the following decisions need to be made: (1) locations of the refineries and DCs, (2) the use of strategic emergency inventories and/or alternative sourcing facilities, and (3) the logistics assignments between the SCN facilities as well as the customers. The locations of the 5 major crude oil suppliers (oil fields) and the 14 major gasoline consumption areas (cities), and the candidate locations of the 5 major potential refineries and the 7 major potential gasoline DCs, are shown in Fig. 3.

6.1 Model Inputs

The main input parameters are listed in Tables 4, 5, and 10–18. The values of these parameters are found or estimated from the data on websites or statistical year books of relevant government departments. According to the actual situation of Liaoning province, two transportation modes ($N^L = 2$), i.e., highway and railway, are considered. The distance between any two facilities is estimated using Google Earth. In order to make the distances more realistic, tortuosity factors $c_1^{tor} = 1.1$ and $c_2^{tor} = 1.5$ are used for highway and railway, respectively.

The planning period is one month for this illustrative example. The monthly amortizations of fixed construction costs of refineries and DCs are given by $-f = \left(r/(1 - 1/(1 + r)^t)\right) \cdot f$, where f is the total investment in the facility, $r = 0.0035$ is the interest rate, and $t = 60$ is the asset depreciation period in months. The monthly amortizations of fixed construction costs are reflected in the fixed operating costs shown in Tables 11 and 12.

Table 4 shows the values of the related parameters of facility capacities, productions, and demands. As the crude oil productions of the major oil fields and gasoline demands of different demand areas are available for recent years from the government websites, their means and variances can be estimated. The demand of a demand area is further estimated based on the total demand of the whole province according to the population of each city. The means and variances of the prices of crude oil and gasoline are undifferentiated among all suppliers and demand areas because of market volatility.

Furthermore, the maximum tolerable stockout quantity is $B = 100 \times 10^3$ tons. The capacities of the strategic emergency inventories for refineries and DCs are $\eta^M = 100 \times 10^3$ tons and $\eta^D = 100 \times 10^3$ tons, respectively. The radius of a region is 200 km for the purpose of the region-wide dual-sourcing strategy. The production transformation coefficient is $\rho = 0.7$. Some other input parameters are listed in the Appendix. All quantities and costs are for a month for this illustrative example.

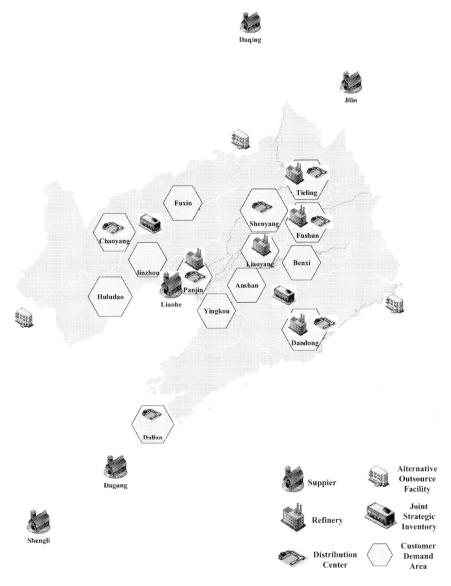

Fig. 3 Locations of the suppliers and customers and candidate locations of the refineries and DCs of the oil SCN in the illustrative example

Scenarios are generated using the method in Sect. 5.1. Disruption events considered in this illustrative example include earthquake (magnitude scale ≥ 5), rainstorm (daily precipitation ≥ 200 mm), flood, and typhoon as shown in Table 5. These are major natural disasters that may take place in Liaoning province. Table 5 also shows the frequency of each disruption risk in each city of Liaoning province, calculated based on the historical data since 1949. The probabilities of the disruption risks can

Table 4 Parameters

Parameters	Mean	Unit	Variance
$[a_{s1}^s, a_{s2}^s, a_{s3}^s, a_{s4}^s, a_{s5}^s]$	[400, 270, 50, 100, 60]	10^3 tons	[20.0, 14.0, 2.5, 5.0, 3.0]
$[a_{s1}^M, a_{s2}^M, a_{s3}^M, a_{s4}^M, a_{s5}^M]$	[1000, 1000, 800, 600, 750]	10^3 tons	
$[a_{s1}^D, a_{s2}^D, a_{s3}^D, a_{s4}^D, a_{s5}^D, a_{s6}^D, a_{s7}^D]$	[500, 300, 300, 250, 250, 200, 150]	10^3 tons	
$[D_{s1}^{dem}, D_{s2}^{dem}, D_{s3}^{dem}, D_{s4}^{dem}, D_{s5}^{dem}, D_{s6}^{dem}, D_{s7}^{dem}, D_{s8}^{dem}, D_{s9}^{dem}, D_{s10}^{dem}, D_{s11}^{dem}, D_{s12}^{dem}, D_{s13}^{dem}, D_{s14}^{dem}]$	[120, 60, 70, 70, 100, 50, 60, 60, 45, 50, 55, 60, 50, 45]	10^3 tons	[6.0, 3.0, 3.5, 3.5, 5.0, 2.5, 3.0, 2.25, 2.5, 2.75, 3.0, 2.5, 2.25]
$[c_1^p, c_2^p, c_3^p, c_4^p, c_5^p]$	[1200, 1200, 1400, 1600, 1500]	¥10^3/10^3 tons	
c_{si}^R	4500	¥10^3/10^3 tons	225
P_s	8200	¥10^3/10^3 tons	410

Table 5 Frequency of disruption risks

Frequency	Earthquake	Flood	Rainstorm	Typhoon
Shenyang	0	0	0.016	0.219
Dalian	0	0	0.016	0.219
Anshan	0.022	0	0.016	0.203
Fushun	0	0	0	0.203
Benxi	0.022	0	0.016	0.219
Dandong	0	0.048	0.047	0.266
Jinzhou	0	0	0	0.234
Yingkou	0	0	0.047	0.234
Fuxin	0	0	0	0.203
Liaoyang	0.067	0	0.016	0.203
Panjin	0	0	0	0.219
Tieling	0	0	0	0.141
Chaoyang	0	0.026	0.031	0.250
Huludao	0	0	0.016	0.188

be estimated by the method described in Sect. 5.1. Because there is no record that any of the crude oil suppliers was disrupted by environmental events, the probabilities of the suppliers being disrupted are set to 0.

6.2 Numerical Results

The binary MIP model is solved with the SAA method on a laptop computer with an Intel i7-5500U processor and 8G RAM. The SAA method is implemented using IBM® ILOG® CPLEX® 12.2.[2] The scale of the model, measured in the number of constraints and the numbers of different types of decision variables, is given in Table 6.

The optimized SCN structure is shown in Fig. 4. Crude oil suppliers chosen in the result are Jilin, Liaohe, Dagang, and Shengli. Because refineries at Liaoyang and Dandong are more exposed to disruption risks, all the downstream DCs served by these refineries have chosen alternative sourcing facilities to secure their supplies. Figure 4 also shows the corresponding assignments of supplies and transportation modes between the different nodes. The quantities shipped between the facilities are not shown because they vary from scenario to scenario. The total profit of the SCN is ¥27222393.73 $\times$ 10^3. Numerical results show that customer demands can still be satisfied within the tolerable stockout quantity when multiple facilities are disrupted. This result shows that the SCN is resilient and the proposed approach is effective.

The computation time used is 115.01 s to solve the illustrative example in CPLEX® by using the SAA method with a gap $= 0.0004$. CPLEX® is also used

---

[2] http://pic.dhe.ibm.com/infocenter/cosinfoc/v12r2/index.jsp.

Table 6 The scale of the model

Constraints	1134301
Continuous variables	530226
Binary variables	193
Non-zero coefficient	4014595

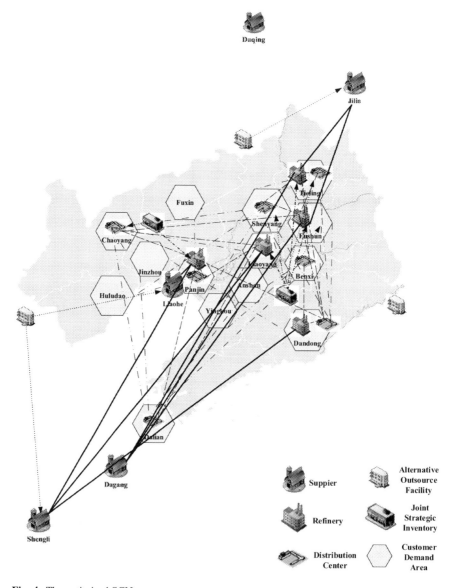

Fig. 4 The optimized SCN

directly to solve the problem without using the SAA method. Unfortunately it is not able to obtain a feasible solution within the running time limit of 6 h.

6.3 Numerical Analysis

Sensitivity analysis results are reported in this section. Specifically, the profit levels with and without disruption risk considerations are compared, customer service levels of using the conventional and resilient strategies are evaluated, the effects of changes in alternative sourcing and strategic inventory costs and of changes in demand on profit levels are examined.

6.3.1 With vs. Without Disruption Risk Considerations

In this subsection, the performances of the SCN with and without disruption risk considerations are compared. With disruption risk considerations, the final solution is obtained with the SAA method as discussed above. Without disruption risk considerations, the model is solved assuming the facilities are never disrupted but the supplies, demands, and prices are stochastic. Disruption events are then introduced into the optimal SCN structure. For comparison purpose, the disruption events in these two cases are the same and $N = 1000$ scenarios are generated by the method described in Sect. 5.1. In order to ensure feasibility of the two cases above, the total stockout quantity in constraint (21) is relaxed by increasing the value of B to $B = 1000 \times 10^3$ tons.

The performances of the final SCNs with and without disruption risk considerations are compared in Fig. 5. The bar labeled Case 1 represents the profit obtained by the SCN without disruption risk considerations when no disruption event strikes. The bars labeled Case 2 and Case 3 represent the profits obtained by the SCNs without and with disruption risk considerations, respectively, when disruption events do happen. As shown in Fig. 5, the SCN without disruption risk considerations performs best when no disruption event occurs but worst when it operates in an environment with disruption risks. On the other hand, the SCN with disruption risk considerations performs better even when disruption events happen. When compared with Case 1, the profit levels of Cases 2 and 3 decrease by 45.84% and 4.04%, respectively. Obviously, the SCN with disruption risk considerations is superior to that without disruption risk considerations when disruption risks exist in the supply chain.

6.3.2 Conventional vs. Resilient Strategies

In this subsection, the advantages of resilient SCNs over the conventional SCNs, *i.e.*, without using the strategic emergency inventories and alternative sourcing, are

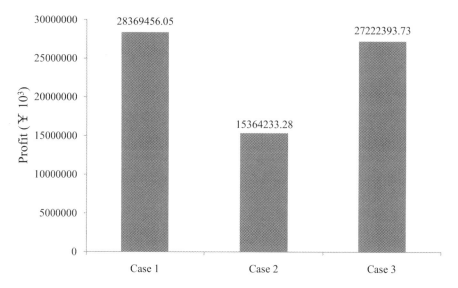

Fig. 5 With vs. without disruption risk considerations

discussed. The profits and service levels of the two cases are compared. The cycle service level is used, which is defined by $1 - \sum_{h \in H} b_{sh} / \sum_{h \in H} D_{sh}^{dem}$ for each scenario s. To ensure feasibility of both cases, the total stockout quantity in constraint (21) is relaxed by increasing the value of B to $B = 1000 \times 10^3$ tons. Constraints associated with resilient SCNs are also relaxed and the values of the parameters related to alternative sourcing facilities (w^m, w^d, S^{subs}, M^{subm}) are all set to zero for the conventional SCNs.

Result shows that there is not much difference in the profit levels between these two SCNs, but there is a big difference in their service levels as shown in Fig. 6. The bars labeled Case 4 represent the minimum and maximum service levels with the conventional strategy. The bars labeled Case 5 represent the minimum and maximum service levels with the resilient strategy. The minimum service level in Case 4 is 21.63%. Although it occurred in only one scenario, it causes a great loss to the entire SCN once it happens. In contrast, the lowest service level in Case 5 with the resilient strategy is 86.32%. Compared with Case 4, Case 5 increases the lowest service level by 299.08%. Apparently, decision makers would prefer the resilient strategy.

6.3.3 Changes in Alternative Sourcing and Strategic Inventory Costs

In this subsection, the effects of changes in variable costs of alternative sourcing and strategic emergency inventories for the DCs are analyzed. The results are graphically shown in Fig. 7. As the supply chain relies on alternative sourcing and strategic emergency inventories to deal with disruption risks, the profit of the SCN linearly

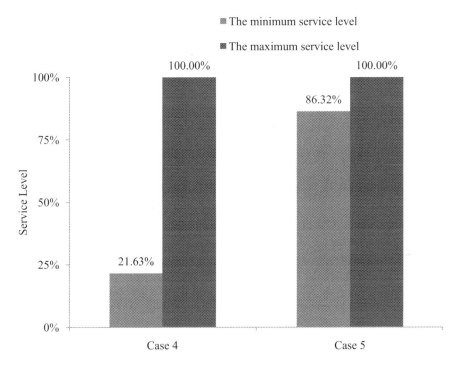

Fig. 6 Conventional vs. resilient strategies

decreases as the average variable costs of these two strategies increase. Furthermore, the effect of the average variable cost of alternative sourcing is much greater than that of strategic emergency inventories. To improve the SCN profit and stability, the decision makers should pay more attention to the average variable cost of alternative sourcing. Finding the appropriate alternative sources and negotiating reasonable prices would help in improving the SCN profit.

6.3.4 Changes in Demands

Changes in demands are always key factors influencing the supply chain performance. Sensitivity analyses are performed when the means and standard deviations change to see their effects on the SCN profit and service levels.

Results are obtained when the means or standard deviations of demands increased and decreased by 20% from their current values. Figure 8 compares the SCN profits using the profit level of the SCN with the current demands as a benchmark. Both of the changes in the means and standard deviations affect the SCN profit. However, the changes in the means of demands have much greater impacts.

Comparisons of the expected service levels are shown in Tables 7 and 8 when the means or standard deviations in demands change. Changes in the means of demands

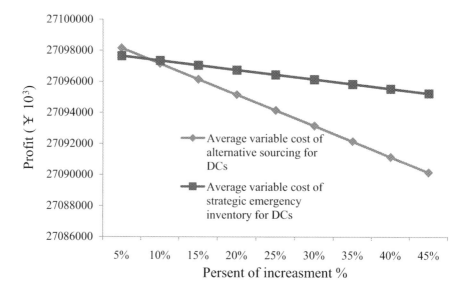

Fig. 7 Effects of average variable costs of alternative sourcing and strategic emergency inventories of DCs on SCN profit

do not affect much in the expected service levels of each demand area. However, when the standard deviations of demands change, the expected service levels fluctuate. Specifically, the expected service levels of some demand areas decrease (increase) when the standard deviations of demands increase (decrease). Table 9 shows the minimum service levels of the demand areas when the means of demands change. Some demand areas suffer much lower (higher) minimum service levels when the means of demands increase (decrease). Hence, the uncertainties in, rather than the averages of, the demands have strong impacts on the SCN performance. These results show the importance of disruption risk considerations in the integrated optimization of design and operations for resilient SCNs. For this illustrative example, the expected service levels are all higher than the service level requirement (85%) under all of the changes.

7 Conclusions

An integrated optimization approach is developed for resilient SCN design and operations with disruption risk considerations and a binary MIP model is formulated for this purpose. Resilient strategies like region-wide dual-sourcing, strategic emergency inventories, and alternative sourcing are used in buffering disruption risks in the SCN designing process to increase the resilience of the resulting SCN. Scenario generation is used to evaluate the effectiveness of the approach. Disruption events are allowed to take place at multiple facilities in a SCN at the same time in one

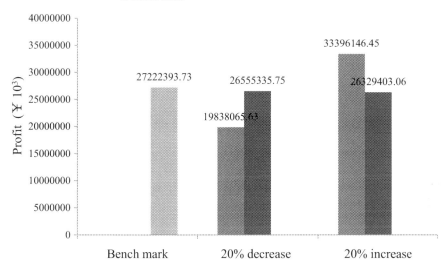

Fig. 8 Effects of changes in demands on profit

Table 7 Effects of changes in the means of demands on expected service levels

Demand areas	Bench mark	20% decreases in the means of demands	20% increases in the means of demands
Shenyang	97.66%	96.71%	98.32%
Dalian	98.98%	100.00%	99.21%
Anshan	85.00%	85.00%	85.00%
Fushun	85.00%	85.00%	85.00%
Benxi	97.11%	98.59%	97.65%
Dandong	85.00%	85.00%	85.00%
Jinzhou	85.00%	85.00%	85.00%
Yingkou	85.00%	85.00%	85.00%
Fuxin	100.00%	100.00%	100.00%
Liaoyang	85.00%	85.00%	85.00%
Panjin	100.00%	100.00%	100.00%
Tieling	100.00%	100.00%	100.00%
Chaoyang	100.00%	96.76%	100.00%
Huludao	85.00%	85.00%	85.00%

scenario. With these scenarios used in the integrated optimization approach, the obtained SCN is much more resilient and much more suitable for the real-life applications.

Table 8 Effects of changes in the standard deviations of demands on expected service levels

Demand areas	Bench mark	20% decreases in the standard deviations of demands	20% increases in the standard deviations of demands
Shenyang	97.66%	98.83%	93.51%
Dalian	99.98%	100.00%	85.42%
Anshan	85.00%	85.00%	85.00%
Fushun	85.00%	85.00%	85.00%
Benxi	97.11%	95.67%	98.36%
Dandong	85.00%	85.00%	85.00%
Jinzhou	85.00%	85.00%	85.00%
Yingkou	85.00%	85.00%	84.15%
Fuxin	100.00%	100.00%	96.32%
Liaoyang	85.00%	85.00%	85.00%
Panjin	100.00%	100.00%	100.00%
Tieling	100.00%	100.00%	100.00%
Chaoyang	100.00%	100.00%	88.56%
Huludao	85.00%	85.00%	85.45%

Table 9 Effects of changes in the means of demands on minimum service levels

Demand areas	Bench mark	20% decreases in the means of demands	20% increases in the means of demands
Shenyang	85.00%	85.00%	85.00%
Dalian	85.00%	85.00%	85.00%
Anshan	85.00%	85.00%	85.00%
Fushun	85.00%	85.00%	85.00%
Benxi	85.00%	85.00%	85.00%
Dandong	85.00%	85.00%	0.00%
Jinzhou	85.00%	90.00%	85.00%
Yingkou	85.00%	85.00%	0.00%
Fuxin	4.46%	15.17%	0.00%
Liaoyang	85.00%	85.00%	85.00%
Panjin	85.00%	85.00%	0.00%
Tieling	85.00%	85.00%	85.00%
Chaoyang	76.53%	85.00%	43.26%
Huludao	85.00%	85.00%	59.61%

The SAA method is used to solve the proposed model under disruption risk considerations. An illustrative example based on a real-life problem is studied to demonstrate the validity and effectiveness of the integrated optimization approach. The results show that a good and resilient solution can be obtained with the SAA method. Profit levels and customer service levels are used to measure the SCN performance. The numerical analysis results show that the SCN obtained with disruption risk considerations performs much better than the conventional SCN. Compared with that of strategic emergency inventories, the average variable cost of alternative sourcing is a more important factor affecting the profit level of the SCN. The uncertainties or fluctuations in demands have a strong impact on the SCN profit and service levels. Therefore, the consideration of demand uncertainties is important in the integrated optimization of SCN design and operations.

There are still some limitations in this study. First, all the disruption events, such as earthquakes, rainstorms, floods, and Typhoons, are assumed to have the same disruption severity. In reality, the impacts of these disruption events may not be the same. Another related limitation is that a facility is assumed to lose all its capacity when disrupted. In reality, the disrupted facility may still be partially operational. For future works, more complex stochastic processes may be considered to overcome these limitations by generating scenarios that are closer to reality when more sophisticated disruption risks are considered. As the scale of the binary MIP model can be extremely large and the SCN integrated optimization problem can be hard to solve when more complex scenarios are considered, more efficient solution methods should be designed to solve the problem. Analytical, simulation, and other hybrid methods may also be helpful in solving these SCN integrated optimization problems.

Acknowledgments This work was partially supported by the Chinese National Natural Science Foundation (No. 70972100).

Appendix

Table 10 Supplier related parameters

	Fixed operating cost (¥10^3)	Unit inventory holding cost (¥10^3/10^3 tons)
Daqing	191851.69	200.00
Shengli	130459.15	
Dagang	23874.87	
Liaohe	48176.09	
Jilin	29843.59	

Table 11 Refinery related parameters

	Fixed operating cost (¥10^3)	Unit inventory holding cost (¥10^3/10^3 ton)
Liaoyang	36159.27	300.00
Fushun	39775. 19	
Panjin	21695.56	
Dandong	12655.74	
Tieling	18079.63	

Table 12 DC related parameters

	Fixed operating cost (¥10^3)	Unit inventory holding cost (¥10^3/10^3 ton)
Shenyang	12790.11	250.00
Tieling	10658.42	
Panjin	14921.79	
Fushun	14921.79	
Dalian	85267.41	
Dandong	14069.12	
Chaoyang	10658.42	

Table 13 Transportation mode related parameters

Transportation mode	S-M		M-D		D-C	
	Unit cost (¥10^3/10^3 tons)	Capacity (10^3 ton)	Unit cost (¥10^3/10^3 tons)	Capacity (10^3 ton)	Unit cost (¥10^3/10^3 tons)	Capacity (10^3 ton)
Highway	120.00	100.00	130.00	100.00	150.00	100.00
Railway	60.00	500.00	70.00	300.00	90.00	350.00

Note: S, M, D, and C represent supplier, refinery, DC, and customer demand area, respectively

Table 14 Strategic emergency inventory related parameters

Facility	Fixed operating costs (¥10^3)	Average variable cost (10^3/10^3 ton)	Capacity limit (10^3 ton)
M	60000.00	800	100
D	35000.00	1100	100

Table 15 Alternative sourcing facility related parameters

Facility	Fixed operating cost (¥10^3)	Average variable cost (¥10^3/10^3 ton)	Capacity (10^3 ton)
S	10000.00	700	500
M	8000.00	1500	500

Table 16 Distances between suppliers and refineries (km)

	Liaoyang	Fushun	Panjin	Dandong	Tieling
Daqing	608	531.2	654.1	717.5	486.9
Shengli	589.8	682.2	513.8	594.7	703.5
Dagang	560.9	651	472.6	611.8	668
Liaohe	92.7	178.6	1	222.4	195.1
Jilin	440.6	363	492.5	553.2	285.6

Table 17 Distances between refineries and DCs (km)

	Shenyang	Tieling	Panjin	Fushun	Dalian	Dandong	Chaoyang
Liaoyang	63.3	127.6	94.2	94.0	293.1	162.7	228.6
Fushun	45.0	45.5	178.3	66.9	385.2	198.2	292.0
Panjin	137.2	196.6	1.0	143.6	249.9	225.0	143.8
Dandong	203.6	244.3	225.0	139.9	273.8	1.0	366.0
Tieling	62.2	1.0	196.7	111.2	420.3	245.3	291.2

Table 18 Distances between DCs and customer demand areas (km)

	Shenyang	Dandong	Jinzhou	Fuxin	Dalian	Yingkou	anshan
Shenyang	1.0	203.0	205.8	148.4	356.8	161.1	86.7
Tieling	62.9	203.7	137.0	183.0	418.3	224.4	150.4
Panjin	136.7	223.7	77.8	110.6	200.0	104.8	76.3
Fushun	63.3	140.6	220.6	192.6	321.4	146.9	67.7
Dalian	356.8	273.4	251.1	247.3	1.0	202.3	269.9
Dandong	203.0	1.0	295.9	312.1	200.0	191.0	160.7
Chaoyang	247.8	369.7	76.3	113.3	312.3	178.8	218.5
	Fushun	Benxi	Liaoyang	Panjin	Tieling	Chaoyang	Huludao
Shenyang	38.0	63.3	62.9	136.7	62.9	247.8	249.0
Tieling	48.0	111.0	126.6	197.2	1.0	291.2	306.5
Panjin	173.9	143.6	94.0	1.0	197.2	145.1	112.6
Fushun	64.9	1.0	50.8	144.6	111.0	279.3	255.0
Dalian	378.0	322.8	393.8	249.3	418.3	312.3	210.0
Dandong	200.5	139.7	167.2	223.7	245.5	368.4	309.3
Chaoyang	285.0	279.3	231.0	145.1	291.2	1.0	100.2

Note: The data in Tables 10–18 are from www.stats.gov.cn, www.cnpc.com.cn, http://www.sinopec.com, www.cnooc.com.cn, and Google Earth

References

Al-Othman, W. B., Lababidi, H., Alatiqi, I. M., & Al-Shayji, K. (2008). Supply chain optimization of petroleum organization under uncertainty in market demands and prices. *European Journal of Operational Research, 189*(3), 822–840.

Awudu, I., & Zhang, J. (2012). Stochastic production planning for a biofuel supply chain under demand and price uncertainties. *Applied Energy, 103*(1), 189–196.

Azad, N., & Hassini, E. (2019). A benders decomposition method for designing reliable supply chain networks accounting for multimitigation strategies and demand losses. *Transportation Science, 53*(5), 1287–1312.

Baghalian, A., Rezapour, S., & Farahani, R. Z. (2013). Robust supply chain network design with service level against disruptions and demand uncertainties: A real-life case. *European Journal of Operational Research, 227*(1), 199–215.

Barrionuevo, A., & Deutsch, C. (2005, September 1). A distribution system brought to its knees. *New York Times*.

Bidhandi, H. M., Yusuff, R. M., Ahmad, M. M. H. M., & Bakar, M. R. A. (2009). Development of a new approach for deterministic supply chain network design. *European Journal of Operational Research, 198*(1), 121–128.

Boin, A., Kelle, P., & Whybark, D. C. (2010). Resilient supply chains for extreme situations: Outlining a new field of study. *International Journal of Production Economics, 126*(1), 1–6.

Boyle, E., Humphreys, P., & McIvor, R. (2008). Reducing supply chain environmental uncertainty through e–intermediation: An organisation theory perspective. *International Journal Production Economics, 114*(1), 347–362.

Chen, M. E. (2008). Chinese national oil companies and human rights. *Orbis, 51*(1), 41–54.

Christopher, M., & Peck, H. (2004). Building the resilient supply chain. *The International Journal of Logistics Management, 15*(2), 1–14.

Clark, D. and Takahashi, Y. (2011, March 12). Quake disrupts key supply chains. *The Wall Street Journal Asia.*

Cui, T., Ouyang, Y., & Shen, Z. J. M. (2010). Reliable facility location design under the risk of disruptions. *Operations Research, 58*(4), 998–1011.

Fattahi, M., Govindan, K., & Keyvanshokooh, E. (2017). Responsive and resilient supply chain network design under operational and disruption risks with delivery lead-time sensitive customers. *Transportation Research Part E: Logistics and Transportation Review, 101,* 176–200.

Fujimoto, T., & Park, Y. W. (2014). Balancing supply chain competitiveness and robustness through "virtual dual sourcing": Lessons from the Great East Japan Earthquake. *International Journal of Production Economics, 147,* 429–436.

Gao, S. Y., Simchi-Levi, D., Teo, C.-P., & Yan, Z. (2019). Disruption risk mitigation in supply chains: The risk exposure index revisited. *Operations Research, 67*(3), 831–852.

Georgiadis, M. C., Tsiakis, P., Longinidis, P., & Sofioglou, M. K. (2011). Optimal design of supply chain networks under uncertain transient demand variations. *Omega–International Journal of Management Science, 39*(3), 254–272.

Ghavamifar, A., Makui, A., & Taleizadeh, A. A. (2018). Designing a resilient competitive supply chain network under disruption risks: A real-world application. *Transportation Research Part E: Logistics & Transportation Review, 115,* 87–109.

Goh, M., Lim, J. Y. S., & Meng, F. W. (2007). A stochastic model for risk management in global supply chain networks. *European Journal of Operational Research, 182*(1), 164–173.

Hamdan, B., & Diabat, A. (2020). Robust design of blood supply chains under risk of disruptions using Lagrangian relaxation. *Transportation Research Part E: Logistics & Transportation Review, 134,* 1–18.

Hasani, A., & Khosrojerdi, A. (2016). Robust global supply chain network design under disruption and uncertainty considering resilience strategies: A parallel memetic algorithm for a real-life case study. *Transportation Research Part E: Logistics & Transportation Review, 87,* 20–52.

Hiriart-Urruty, J. B., & Lemarchal, C. (1993). *Convex analysis and minimisation algorithms II.* Springer.

IBM. (2022). *IBM ILOG CPLEX Optimization Studio, build and solve complex optimization models to identify the best possible actions.* Retrieved from https://www.ibm.com/products/ilog-cplex-optimization-studio?utm_content=SRCWW&p1=Search&p4=43700050328194740&p5=e&gclid=CjwKCAiA5t-OBhByEiwAhR-hmxqKlDEpg09sjw6c-vbrdib-JSDW_pwnfac8F1diTNY4csLCY0K7rRoCneIQAvD_BwE&gclsrc=aw.ds

Kleindorfer, P. R., & Saad, G. H. (2005). Managing disruption risks in supply chains. *Production and Operations Management, 14*(1), 53–68.

Kleywegt, A. J., Shapiro, A., & Homem-de-Mello, T. (2001). The sample average approximation method for stochastic discrete Optimization. *SIAM Journal of Optimization, 12*(2), 479–502.

Klibi, W., & Martel, A. (2012). Scenario-based supply chain network risk modeling. *European Journal of Operational Research, 223*(3), 644–658.

Klibi, W., Martel, A., & Guitouni, A. (2010). The design of robust value–creating supply chain networks: A critical review. *European Journal of Operational Research, 203*(2), 283–293.

Latour, I. A. (2001, January 29). Trial by fire: A blaze in Albuquerque sets off major crisis for cellphone giants. *Wall Street Journal*.

Li, X., Li, Y., & Cai, X. (2013). Double marginalization and coordination in the supply chain with uncertain supply. *European Journal of Operational Research, 226*(2), 228–236.

Lücker, F., Seifert, R. W., & Biçer, I. (2019). Roles of inventory and reserve capacity in mitigating supply chain disruption risk. *International Journal of Production Research, 57*(4), 1238–1249.

Mak, W.-K., Morton, D. P., & Wood, R. K. (1999). Monte Carlo bounding techniques for determining solution quality in stochastic programs. *Operations Research Letters, 24*(1–2), 47–56.

Manuj, I., & Mentzer, J. T. (2008). Global supply chain risk management strategies. *International Journal of Physical Distribution & Logistics Management, 38*(3), 192–223.

McKillop, A. (2005). Oil: No supply side answer to the coming energy crisis. *Refocus, 6*(1), 50–53.

Miller, K. D. (1992). A framework for integrated risk management in international business. *Journal of International Business Studies, 23*(2), 311–331.

Mirzapour Al-e-hashem, S. M. J., Malekly, H., & Aryanezhad, M. B. (2011). A multi-objective robust optimization model for multi-product multi-site aggregate production planning in a supply chain under uncertainty. *International Journal of Production Economics, 134*(1), 28–42.

Mitchell, J. V., & Mitchell, B. (2014). Structural crisis in the oil and gas industry. *Energy Policy, 64*(5), 36–42.

Mitra, K., Gudi, R. D., Patwardhan, S. C., & Sardar, G. (2009). Towards resilient supply chains: Uncertainty analysis using fuzzy mathematical programming. *Chemical Engineering Research & Design, 87*(7A), 967–981.

More, D., & Babu, A. S. (2008). Perspectives, practices and future of supply chain flexibility. *International Journal of Business Excellence, 1*(3), 302–336.

Mouawad, J. (2005, September 4). Katrina's shock to the system. *The New York Times*.

Muckstadt, J. A., Murray, D. H., Rappold, J. A., & Collins, D. E. (2001). Guidelines for collaborative supply chain system design and operation. *Information Systems Frontiers, 3*(4), 427–453.

Neter, J., Wasserman, W., & Whitmore, G. A. (1993). *Applied Statistics* (4th ed.). Allyn and Bacon.

Oliveira, F., & Hamacher, S. (2012). Stochastic benders decomposition for the supply chain investment planning problem under demand uncertainty. *Pesquisa Operacional, 32*(3), 663–678.

Park, Y., Hong, P., & Roh, J. J. (2013). Supply chain lessons from the catastrophic natural disaster in Japan. *Business Horizons, 56*(1), 75–85.

Paul, S. K., Sarker, R., & Essam, D. (2016). Managing risk and disruption in production-inventory and supply chain systems: A review. *Journal of Industrial & Management Optimization, 12*(3), 1009–1029.

Peidro, D., Mula, J., Poler, R., & Lario, F. C. (2009). Quantitative models for supply chain planning under uncertainty: A review. *The International Journal of Advanced Manufacturing Technology, 43*(3–4), 400–420.

Peng, P., Snyder, L. V., Lim, A., & Liu, Z. (2011). Reliable logistics networks design with facility disruptions. *Transportation Research Part B: Methodological, 45*(8), 1190–1211.

Prater, E. (2005). A framework for understanding the interaction of uncertainty and information systems on supply chains. *International Journal of Physical Distribution and Logistics Management, 35*(7–8), 524–539.

Rodrigues, V. S., Stantchev, D., Potter, A., Naim, M., & Whiteing, A. (2008). Establishing a transport operation focused uncertainty model for the supply chain. *International Journal of Physical Distribution and Logistics Management, 38*(5), 388–411.

Sabouhi, F., Pishvaee, M. S., & Jabalameli, M. S. (2018). Resilient supply chain design under operational and disruption risks considering quantity discount: A case study of pharmaceutical supply chain. *Computer & Industrial Engineering, 126*, 657–672.

Santoso, T., Ahmed, S., Goetschalckx, M., & Shapiro, A. (2005). A stochastic programming approach for supply chain network design under uncertainty. *European Journal of Operational Research, 167*(1), 96–115.

Sawik, T. (2013). Selection of resilient supply portfolio under disruption risks. *Omega–International Journal of Management Science, 41*(2), 259–269.

Sawik, T. (2014a). Joint supplier selection and scheduling of customer orders under disruption risks: Single vs. dual sourcing. *Omega–International Journal of Management Science, 43*, 83–95.

Sawik, T. (2014b). Optimization of cost and service level in the presence of supply chain disruption risks: Single vs. multiple sourcing. *Computer & Operation Research, 51*(3), 11–20.

Sawik, T. (2015). On the risk-averse optimization of service level in a supply chain under disruption risks. *International Journal of Production Research, 54*(1), 1–16.

Schmitt, A. J. (2011). Strategies for customer service level protection under multi-echelon supply chain disruption risks. *Transportation Research Part B: Methodological, 45*(8), 1266–1283.

Shapiro, A., & Homem-de-Mello, T. (2000). On rate of convergence of Monte Carlo approximations of stochastic programs. *SIAM Journal on Optimization, 11*, 70–86.

Sheffi, Y. (2001). Supply chain management under the threat of international terrorism. *The International Journal of Logistics Management, 12*(2), 1–11.

Sheffi, Y. (2005). *The resilient enterprise: Overcoming vulnerability for competitive advantage*. MIT Press.

Snyder, L. M. (2003). *Supply chain robustness and reliability: Models and algorithms*. Northwest University.

Soni, U., & Jain, V. (2011). Minimizing the vulnerabilities of supply chain: A new framework for enhancing the resilience. In *IEEE International Conference on Industrial Engineering and Engineering Management (IEEM)* (pp. 933–939). Springer.

Tang, C. S. (2006a). Robust strategies for mitigating supply chain disruptions. *International Journal of Logistics Research and Applications, 9*(1), 33–45.

Tang, C. S. (2006b). Perspectives in supply chain risk management. *International Journal of Production Economics, 103*(2), 451–488.

Tomlin, B. (2006). On the value of mitigation and contingency strategies for managing supply chain disruption risks. *Management Science, 52*(5), 639–657.

Tong, K., Feng, Y., & Rong, G. (2011). Planning under demand and yield uncertainties in an oil supply chain. *Industrial & Engineering Chemistry Research, 51*(2), 814–834.

Tverberg, G. E. (2012). Oil supply limits and the continuing financial crisis. *Energy, 37*(1), 27–34.

Varma, S., Wadhwa, S., & Deshmukh, S. G. (2008). Evaluating petroleum supply chain performance: Application of analytical hierarchy process to balanced scorecard. *Asia Pacific Journal of Marketing and Logistics, 20*(3), 343–356.

Yang, B., Burns, N. D., & Backhouse, C. J. (2004). Management of uncertainty through postponement. *International Journal of Production Research, 42*(6), 1049–1064.

Yin, R., & Rajaram, K. (2007). Joint pricing and inventory control with a Markovian demand model. *European Journal of Operational Research, 182*(1), 113–126.

Yu, H. S., Zeng, A. Z., & Zhao, L. D. (2009). Single or dual sourcing: Decision-making in the presence of supply chain disruption risks. *Omega–International Journal of Management Science, 37*(4), 788–800.

Balancing Sustainability Risks and Low Cost in Global Sourcing

Gbemileke A. Ogunranti and Avijit Banerjee

Abstract In this chapter, we attempt to develop an integrated framework for simultaneously addressing the supplier selection and order allocation problems, considering the trade-off between supply chain sustainability (in environmental, social, and economic terms) related risks and procurement cost. First, we identify supply chain sustainability-related risks for evaluating suppliers based on practice and an extensive literature review. Secondly, these criteria are used to assess suppliers' sustainability risk performance, using principal component analysis. Then, the min-max normalization is applied to the principal component analysis scores to generate the suppliers' adjusted sustainability risk performance scores, which are then utilized to select a set of qualified suppliers. Subsequently, the specific supplier selection and order allocation decisions are determined via a bi-objective mixed-integer programming model, which attempts to maximize the supply chain's sustainability performance while minimizing the procurement cost. This proposed framework forms a decision support system for our sustainable supplier selection and order quantity allocation. Finally, we present an illustrative example of outsourcing contract manufacturers in the apparel industry to demonstrate the applicability of the proposed framework in practice.

1 Introduction

Globalization phenomenon has made supply chains increasingly complex, interconnected, and interdependent. Such supply chains are often influenced by factors beyond the power and purview of global companies. Offshore outsourcing of production or global sourcing is a common cost reduction strategy in global supply chains, but managing such supply chains can be challenging. This makes the supply chain vulnerable to external risks that differ from country to country. These

G. A. Ogunranti (✉) · A. Banerjee
Department of Decision Sciences & MIS, Drexel University, Philadelphia, PA, USA
e-mail: gao32@drexel.edu

types of supply chains are often characterized by external risks such as currency exchange rate fluctuations, import tax rate changes, tariffs, export restrictions, social-cultural issues, natural disasters, terrorism, foreign government political environment, legislation and regulation, health and safety of workers including ethical labor practices, and environmental issues. Some examples of environmental related risk issues are the use and/or disposal practices of toxic or hazardous materials in manufacturing processes (e.g., dyes and softeners), greenhouse gas (GHG) emissions, as well as the availability and costs of natural resources (i.e., energy, cotton, rubber, and water). Moreover, ethical labor practices, such as fair wages, complying with health and safety regulations, and maintaining acceptable working conditions at production facilities across the globe, are critical for success in today's supply chain outsourcing.

A recent McKinsey report by Berg et al. (2017) reveals that country selection and compliance and risk are among the four success factors in apparel sourcing. Global firms are continuously faced with strategic production sourcing decisions, in order to effectively balance the issues of quality, cost, speed, and the challenges of managing social and environmental compliance. Thus, it is important to tackle sustainability-related risks at the very outset of a global sourcing process, leading to eventual supplier selection and order allocation. The purpose of this chapter is to develop a decision support system, based on a combination of two techniques (principal component analysis (PCA) and bi-objective optimization), to help decision-makers integrate supply chain sustainability (environmental, social, and economic) related risks in supplier selection and order allocation decisions under multiple sourcing strategy. Also, multiple diverse global supply sources have the strategic advantage of supply chain resiliency when disruptions, such as the current COVID-19 pandemic, occur. It has greatly disrupted all sectors of the global economy and has had significant negative impacts on most global supply chains. Some of these impacts include shortages of finished products, materials, and labor, regulatory uncertainty, and even logistics disruptions, due to restricted movement and shipping capacity reductions across different modes of transportation, among others. When this pandemic hit the USA, the dependency of this country on China for personal protective equipment (PPE) became obvious and there was a dire need to source PPE items from other countries with manufacturing capacity and little prevalence of the outbreak.

Furthermore, a frequently discussed topic in global outsourcing involves unethical practices on the part of some suppliers and their failure to comply with wage rules and labor standards. A study by Maplecroft (2012) reveals that some companies' supply chains have been exposed to the risk of child labor used in some of the fastest growing economies, such as the Philippines, India, China, Vietnam, Indonesia, and Brazil, all of which are classified as "extreme risk" nations. As such, firms that source or outsource manufacturing to these countries may suffer from a negative impact on their reputations. For instance, Nike's use of foreign manufacturers has periodically tarnished its image, and its campaign to eliminate such problems has not been an easy task (Banjo, 2014). Bangladesh, the second-ranked exporter of apparel after China, is associated with unsafe working conditions, low wages, and persistent fire incidents at many of its garment factories, with hundreds of workers killed over the years (Manik & Yardley, 2012). This country's national economy depends largely on the apparel industry as a source of both employment and foreign currency, as it accounts for about four-fifths of the country's manufacturing exports. Several global retailers, such as Walmart and Sears, outsource production to Bangladesh. In 2014, Walt Disney Co. pulled its manufacturing out of Bangladesh in response to a building collapse in the previous year. Nevertheless, retailers including Wal-Mart Stores Inc., Hennes & Mauritz AB, and 170 others decided to stay, signing five-year agreements vowing to fund improvements and develop meaningful safety standards (Banjo, 2014).

Sustainability risk is still an emerging area in supply chain management, and Fahimnia et al. (2015) note that environmental and social aspects of such risks require more research attention, due to increasing environmental concerns and consumers' awareness of social issues. For global supply chains, effectively managing sustainability risks, complexity, and marketplace dynamics has become crucial for firms to remain competitive. Nonetheless, it is not easy to manage multiple priorities such as cost control, environmental pressures, and maintaining acceptable working conditions. Hence, there is a dire need for firms to take a balanced approach to sourcing by taking into consideration environmental, social, quality, and cost impacts. Although there exist several articles studying the (decision) criteria to be used for the supplier selection process, relatively few papers address criteria related to safety and security issues, which have become important, given the present threats to security and the current "climate" around the world (Sonmez, 2006).

Numerous past studies have combined two approaches for supplier selection and order allocation decisions, but none has combined the principal component analysis (PCA) technique with optimization, simultaneously considering sustainability risks. In this chapter, we examine the trade-off between low-cost outsourced contract manufacturing and the associated sustainability risks. We develop a procedure, where for each supplier under consideration, the associated sustainability risk is measured using PCA. The resulting sustainability performance score is then utilized in a bi-objective optimization model for final supplier selection and order allocation. The results from our illustrative example show that this approach can lead to some useful managerial insights. Also, this procedure allows supply chain risk managers to make optimal decisions based on the cost-sustainability trade-off. Note that the terms contract manufacturer (CM) and supplier will be used interchangeably throughout this chapter.

The remainder of this chapter is organized as follows: In Sect. 2, we review four streams of the extant literature that are closely related to our study. Next, we present a framework for our decision support system, for the sustainable suppliers' selection and order allocation processes. An illustrative example is provided to demonstrate the practical applicability of the proposed framework in Sect. 4. The final section concludes with a summary, outlining some managerial implications of this study and potential future research directions.

2 Literature Review

The literature related to this chapter can be grouped into four different streams, namely: supplier selection and order allocation (SSOA), sustainability-related supply chain risks criteria (SSCRC), green supplier selection and order allocation (GSSOA), and sustainable supplier selection and order allocation (SSSOA). There are two types of supplier selection problems found in the literature. The first type deals with a situation where a single supplier can meet all of the buyer's demands often referred to as single sourcing. The second type is the one in which multiple suppliers are utilized in meeting the buyer's demand, referred to as multiple sourcing. In the latter situation, management splits the total order among selected suppliers for various reasons, such as enhancing competitiveness and encouraging sustainability performance improvements (Demirtas & Üstün, 2008). In this chapter, our focus is on the supplier selection/allocation problem under a multiple sourcing scenario.

2.1 Supplier Selection and Order Allocation

The problem scenario described above is a multi-criteria decision-making (MCDM) problem, involving the selection of a set of best suppliers and their respective order allocations. Several researchers have proposed different approaches to address this problem. Some of these authors suggest integrated hybrid techniques, such as analytic hierarchy process (AHP) and linear programming (LP) (Ghodsypour & O'Brien, 1998), AHP and goal programming (GP) (Kull & Talluri, 2008), analytic network process (ANP) and multi-period goal programming (Demirtas & Ustun, 2009), fuzzy AHP, modified fuzzy TOPSIS, and goal programming (Jolai et al., 2011), fuzzy set theory, TOPSIS, and mixed-integer linear programming (MILP) (Singh, 2014), the joint approach of Taguchi, AHP, and fuzzy multi-objective programming (Azizi et al., 2015), and analytic hierarchy process-quality function deployment (AHP-QFD) and chance-constrained optimization (Scott et al., 2015). Other researchers utilize multi-objective mathematical programming for solving this problem, e.g., Ghodsypour and O'Brien (2001), Wadhwa and Ravindran (2007), Demirtas and Üstün (2008), Amid et al. (2011), Jolai et al. (2011), Jadidi et al. (2014), and Sodenkamp et al. (2016). Note that AHP is often used in most of these hybrid approaches for addressing the supplier selection and order allocation problem, as indicated in a review paper by Singh (2014).

Amin et al. (2011) represent the first work to apply quantified SWOT (strengths, weaknesses, opportunities, and threats) in the context of supplier selection and order allocation. The novelty of their approach is the integration of fuzzy logic and triangular fuzzy numbers with SWOT analysis to account for the vagueness of human thought. Then, the amount of quantity to be ordered from each supplier is determined via a fuzzy linear programming model. Singh (2014) proposes another hybrid algorithm, combining fuzzy set theory, TOPSIS, and MILP methodologies (HFTM), to solve this problem under uncertainty. Another study by Scott et al. (2015) provides a comprehensive method for integrating stochastic multi-stakeholder requirements into a multi-criteria problem by combining the AHP-QFD method with a multi-criteria chance-constrained optimization algorithm. This approach is successfully applied to the emerging biomass to energy conversion industry. Recently, Sodenkamp et al. (2016) proposed a novel meta-approach, which combines multi-criteria decision analysis and linear programming (LP) to address the multi-objective SSOA problem, which is implemented in an agricultural commodity trading firm. Other current works in supplier selection and order allocation are focusing on sustainability and we classified them under GSSOA and SSSOA discussed in Sects. 2.3 and 2.4, respectively.

2.2 Sustainability-Related Supply Chain Risk Criteria

Several sustainability criteria have been considered by researchers for supplier selection and evaluation purposes. Ahi and Searcy (2015) provide a comprehensive review of the metrics used for green and sustainable supply chains, based on a structured content analysis of 445 articles published until the end of 2012. Also, a review by Govindan et al. (2015b) summarizes the criteria used for green supplier selection and evaluation. We refer the interested reader to these works. This section focuses mainly on supply chain sustainability-related risks that buyers might encounter when outsourcing production or procuring products or raw materials from other countries. For example, failure to address such issues at the outset of an outsourcing process can jeopardize the reputation of a buying firm. Hofmann et al. (2014) use a trans-disciplinary approach to describe how supply chain sustainability issues translate to risks for the firm. Based on interviews with industry experts, they contend that the risk sources in the upstream supply chain occur within social, ecological, and ethical business conduct issues (see Table 11 in Appendix 1).

Giannakis and Papadopoulos (2016) identify 30 distinct supply chain risks across environmental, economic, and social aspects of sustainability. These sustainability-related risks are listed in Table 12 in Appendix 1. Their study conducts a correlation test for sustainability-related risks, and their results show that a strong relationship exists between environmental and economic related factors, as well as between social and economic related factors at a statistically significant level. For example, child labor is highly correlated with bribery and financial crises. Similarly, Ogunranti (2018) observes correlation among the outputs used in data envelopment analysis (DEA) for the development of a composite sustainability risk index for an outsourcing country in the apparel industry. Out of the nine outputs considered in this study: political stability and absence of violence (PSAV), rule of law, corruption perception score, inflation, ease of doing business, debt to GDP ratio, human development index (HDI) rank, overall logistics performance index score, world risk index, and environment protection index, the PSAV is strongly correlated with rule of law and corruption perception score. Also, the rule of law and the corruption perception score are highly correlated with each other. Data used in this study are obtained from various reputable international organizations including the World Bank, United Nations Development Programme (UNDP), and Transparency International.

In practice, most companies develop their own sustainability metrics for tracking progress and evaluate the performance of their offshore suppliers/contract manufacturers. A typical example is Nike Corporation. They include four criteria (political risk, economic risk, social/compliance, and infrastructure and climate) in their country risk index for sourcing and manufacturing, which is used to rate the performance of their contract manufacturers' factories (Nike Inc., 2018). These four criteria with some sub-criteria are equally weighted as shown in Table 13 in Appendix 2. Park et al. (2018) select four sustainability indices relevant to supply chain design from publicly available World Bank indices to derive a multi-attribute utility for each supplier considered in their study. These regional sustainability indices are ease of doing business index, logistics performance index, global competitiveness index, and global enabling trade index of potential global suppliers. A summary of sustainability-related supply chain risks based on works by Giannakis and Papadopoulos (2016) and Hofmann et al. (2014) is presented by Foroozesh et al. (2018).

2.3 Green Supplier Selection and Order Allocation

The green concept refers to the environmental aspect of the sustainability concept that also includes the economic and social dimensions (Igarashi et al., 2013; Hamdan & Cheaitou, 2017a). All studies reviewed in this section largely consider only environmental aspects, without any substantive focus on the social dimension of sustainability.

In an early work, Shaw et al. (2012) develop an integrated supplier selection and allocation model which comprises fuzzy AHP and fuzzy multi-objective linear programming. In their model formulation, carbon emission is integrated into the objective function of a multi-objective linear programming model, while a carbon emission cap (Ccap) on sourcing is treated as a constraint in supplier selection. Kannan et al. (2013) propose an integrated approach that combines fuzzy multi-attribute utility theory and multi-objective programming, for green suppliers' selection and order allocation purposes. These authors consider a green criterion in the evaluation of suppliers before allocating the order. This criterion is called environmental competency, which assesses suppliers based on pollution production, resource consumption, environmental management system deployed, and eco-design of products.

Govindan and Sivakumar's (2016) framework integrates fuzzy TOPSIS and multi-objective linear programming (MOLP) methodologies in a heterogeneous decision-making environment for green supplier selection and order allocation in the paper industry. They consider recycling capabilities and GHG emission control as green criteria. Hamdan and Cheaitou (2015, 2017a, b) propose a decision-making (DM) tool that integrates the fuzzy TOPSIS technique, analytic hierarchy process (AHP), and multi-objective optimization to address green supplier selection and order allocation decisions. They also utilize multi-period bi-objective as well as multi-objective optimization for the green suppliers' order allocation process (Hamdan & Cheaitou, 2017a). The optimization part of this DM tool is solved using the weighted comprehensive criterion method and the branch-and-cut algorithm. Their results show that the bi-objective optimization model should be adopted in practice since it performs better than the multi-objective technique in terms of computation time. This approach provides the manager with a DM tool to help balance economic and environmental aspects of sustainability for dealing with the supplier selection and order allocation problem. Also, Hamdan and Cheaitou (2017b) consider multi-period green supplier selection and order allocation with all unit quantity discounts, in which the availability of suppliers differs from one period to another.

Recently, Babbar and Amin (2018) provide a novel two-phase model based on QFD and a stochastic multi-objective mathematical model incorporating environmental issues in supplier selection and allocation. Their multi-objective model is solved using three different approaches, namely: weighted-sums, distance, and ϵ-constraint methods. The green criteria considered for selecting suppliers include green cap or design, recycle and reuse, environmental management system used, and carbon emission. This model is applied in the beverage industry. Another work by Banaeian et al. (2018) integrates fuzzy set theory into three widely used multi-criteria supplier selection techniques (TOPSIS, VIKOR, and GRA) for oil suppliers' selection for a food processing company. The results obtained using these methods yielded similar supplier rankings.

2.4 Sustainable Supplier Selection and Order Allocation

Recently, there has been an increase in the number of studies dealing with the sustainable supplier selection and order allocation problem, but there are relatively few studies with particular attention on sustainability-related supply risks in sustainable supplier selection and order allocation. Azadnia et al. (2015) propose an integrated approach using a rule-based weighted fuzzy method, fuzzy AHP, and multi-objective optimization for sustainable supplier selection and order allocation for multi-period, multi-product lot-sizing problems. They demonstrate the applicability of their approach using a case study of packaging films in the food industry. Another study by Govindan et al. (2015a) integrates sustainability into the supply chain network design (SCND) and the order allocation problem (OAP). These authors develop a novel multi-objective hybrid approach for addressing this problem under a stochastic retailer's demand. Aktin and Gergin (2016) introduce the use of a questionnaire to measure the sustainability scores of potential suppliers and then utilize a mixed-integer linear programming model to allocate demand to appropriately sustainable suppliers. This approach is successfully applied to three different companies from different industries (printing, footwear, and apparel).

As the number of works integrating sustainability into the supplier selection and order allocation problem continues to rise, fuzzy AHP and TOPSIS are commonly combined with optimization techniques to address this problem. Mohammed et al. (2018) combine fuzzy analytical hierarchy process (AHP), the fuzzy technique for order of preference by similarity to ideal solution (TOPSIS), and multi-objective programming model (MOPM) to minimize total costs (transportation, purchasing, and administration), environmental impact measured by CO_2 emissions, and the travel time of products while maximizing social impact and total purchasing value. Ghadimi et al. (2018) apply the multi-agent systems (MASs) approach to address the communication and information exchange challenges in supplier–buyer relationships during the sustainable supplier selection and order allocation. Their proposed MAS model comprises two sub-models. First is the supplier evaluation sub-model which utilizes a proposed fuzzy inference system (FIS) model to evaluate the potential suppliers' sustainability performance. Second is the order allocation sub-model which uses a bi-objective optimization model to allocate a total order quantity among the selected suppliers. More recently, Azadnia and Ghadimi (2018) propose an integrated fuzzy AHP combined with quality function deployment (FAHP-QFD) together with a fuzzy assessment method (FAM), in order to assess suppliers' sustainability scores. Then, a fuzzy multi-objective mixed-integer non-linear programming model (MINLP) is solved for order allocation to suppliers based on the manufacturer's preferences on sustainability.

Vahidi et al. (2018) propose a new framework for addressing the SSSOA problem under operational and disruption risks. They provide a hybrid approach that uses the strengths, weaknesses, opportunities, and threats (SWOT) analysis and the quality function deployment (QFD) method for selecting the most useful sustainability criteria to be used in the supplier selection process. Then, a bi-objective two-stage mixed possibilistic-stochastic programming model is utilized for solving the final supplier selection and order allocation. Another study by Gören (2018) presents a novel hybrid approach, which uses the fuzzy Decision-Making Trial and Evaluation Laboratory (DEMATEL) and Taguchi loss functions for evaluating and ranking sustainable suppliers. Then a bi-objective optimization model is used for order allocation to each selected supplier. Their proposed bi-objective optimization model considers the issue of lost sales during the allocation process.

Park et al. (2018) propose a framework for a sustainable supply chain design using multi-attribute utility theory and multi-objective integer linear programming for sustainable supplier selection and optimal order allocation. Another work by Kellner and Utz (2019) utilized a multi-objective supplier selection and order allocation based on H. Markowitz's investment portfolio theory taking into account sustainability. Moheb-Alizadeh and Handfield (2019) developed a multi-objective MILP for sustainable supplier selection and order allocation with multiple periods, multiple products, and multimodal transportation while taking into account shortage and discount conditions. Their solution approach is the combination of ε-constraint method and Benders decomposition algorithm (BDA) with DEA super-efficiency score to select the final solution. Unlike most recent works, Jia et al. (2020) formulated a distributionally robust goal programming model to solve SSSOA problem for a centralized supply chain with a purchasing company and multiple suppliers, which deal with four conflicting goals concerning sustainability in terms of cost, emissions, society, and suppliers' comprehensive CO_2 value under uncertain environment. According to the authors, their study extends the work by Mohammed et al. (2019).

Despite increasing research attention toward developing integrated approaches for sustainable suppliers selection and order allocation decisions, there are still very few practical quantitative models addressing this problem of sustainability and risks in SSSOA. To the best of our knowledge, there seems a dearth of quantitative models that incorporate supply chain sustainability-related risks in this selection/allocation process. This chapter aims to address this gap in the literature.

3 Proposed Modeling Framework

This section presents our proposed framework, which integrates the PCA, min-max normalization, and bi-objective optimization, to address the joint problems of sustainable supplier selection and order allocation while considering sustainability-related supply chain risks. Typically, supplier selection and evaluation is a multi-criteria decision-making problem which involves the use of numerous dimensions for rating suppliers. Although the PCA is a common dimension reduction technique, it may be useful in generating a composite single score from multiple dimensions. First, we use PCA to estimate the score of each supplier and normalize this score using the min-max normalization method in order to derive the sustainability performance scores for selecting a qualified pool of suppliers. These sustainability performance scores calculated from the sustainability-related risk criteria data for the qualified suppliers are incorporated into the objective function of the bi-objective optimization model. This bi-objective model is used to select the final suppliers and determine the optimal quantity allocated to them, under some relevant constraints. A diagrammatic representation of the proposed framework is presented in Fig. 1.

Based on our proposed modeling framework, we need to obtain a single measure to evaluate the sustainability performance of several suppliers under consideration. Analytical approaches such as data envelopment analysis (DEA), cluster analysis, discriminant analysis, and PCA have been utilized in the literature for estimating a single performance measure for supplier selection purposes (Chen, 2011; Kannan et al., 2013). A study by Zhu (1998), comparing the DEA and PCA approaches, concludes that there is consistency among the rankings obtained using these two techniques. Also, Premachandra (2001) shows that the PCA approach recommended by Zhu (1998) can be improved further for it to work well when the majority of the decision-making units (DMUs) in the sample data considered are efficient.

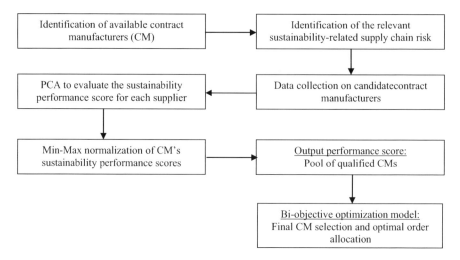

Fig. 1 Framework for the sustainable supplier selection and order allocation model

3.1 Principal Component Analysis (PCA)

PCA is a multivariate method for data dimension reduction. This reduction is achieved by transforming the original data variables into a set of uncorrelated new variables called the principal components. These are a linear combination of the values of the original variable, with the first principal component having the maximum variance. Although there exist relatively few studies that apply PCA in the supplier selection process, Lam et al. (2010) and Petroni and Braglia (2000) point out its usefulness in overcoming the limitations of other popularly used methodologies. The PCA technique eliminates multicollinearity among the criteria (Slottje, 1991), greatly reduces data dimensionality with little or no information loss, and overcomes the subjectivity in assigning weights to the evaluation criteria (Petroni & Braglia, 2000; Lam et al., 2010). For our purposes, a distinct advantage of PCA over other methods is that the various evaluating criteria, or performance measures, do not necessarily have to be in the same units. For simplicity and ease of computation, we choose to use the PCA approach for estimating a single sustainability performance measure for a set of suppliers.

Consider a dataset with dimension $S \times P$, where S is the number of observations in each variable, and P is the number of variables. Let $x = (x_1, x_2, x_3, \ldots, x_P)$ be a S-dimensional vector; then the data matrix can be represented by H, i.e.,

$$H = (x_1, x_2, x_3, \cdots, x_P) = \begin{bmatrix} X_{11} & X_{12} & \cdots & X_{1P} \\ X_{21} & X_{22} & \cdots & X_{2P} \\ . & . & & . \\ . & . & & . \\ . & . & & . \\ X_{S1} & X_{S2} & \cdots & X_{SP} \end{bmatrix}. \quad (1)$$

The principal components $(z_1, z_2, z_3, \ldots, z_p)$, which are the uncorrelated new variables derived from the eigenvectors (ω's), can be estimated using Eq. (2):

$$\begin{bmatrix} z_1 \\ z_2 \\ . \\ . \\ . \\ z_p \end{bmatrix} = \begin{bmatrix} \omega_{11} & \omega_{12} & \cdots & \omega_{1p} \\ \omega_{21} & \omega_{22} & \cdots & \omega_{2p} \\ . & . & & . \\ . & . & & . \\ . & . & & . \\ \omega_{p1} & \omega_{p2} & \cdots & \omega_{pp} \end{bmatrix} X \begin{bmatrix} x_1 \\ x_2 \\ . \\ . \\ . \\ x_p \end{bmatrix}. \quad (2)$$

These principal components are arranged in descending order of importance based on the proportion of variation explained (π_p) by each of them, where $\pi_1 > \pi_2 > \cdots > \pi_P$. A subset of these new variables may be selected by retaining T principal components ($T < P$), which adequately capture the information contained in data matrix H. There are two recommended approaches to select T. One approach is to select the principal components whose eigenvalues are greater than 1. Another approach is to choose the principal components that collectively account for a certain percentage of the total variation explained. Rencher (2003) recommends a total explained variation to be greater or equal to 80%, while Zhu (1998) suggests a value greater than or equal to 90%.

For our purposes, the PCA approach utilized to construct a single measure for evaluating the suppliers' sustainability performance values is summarized in the following steps:

Step 1: Obtain the principal component characteristics vector (ω) and the proportion of variation explained (π_p) by each component.
Step 2: Select the first T principal components (PC) characteristics vector using the total variation explained greater than or equal to 90% approach as per Zhu (1998).
Step 3: Convert the selected principal component characteristics vector to ranks $r(\omega_{pt})$ and normalize the ranks, as proposed by Slottje (1991), using

$$r(\omega_{pt}) = \frac{\text{rank}_p}{\text{max rank}_t} \quad \text{where } p = 1, 2, \ldots, P \text{ and } t = 1, 2, \ldots, T. \quad (3)$$

Step 4: Determine the aggregated weight ($\widehat{\omega}_p$) using the percentage contribution (π_t) of each of the selected principal component characteristics vector and the normalized ranks, $r(\omega_{pt})$, i.e.,

$$\widehat{\omega}_p = \sum_{t=1}^{T} \pi_t r(\omega_{pt}) \quad \text{where} \quad p = 1, 2, \ldots, P. \tag{4}$$

We now define a single measure for evaluating the sustainability performance of the suppliers as follows:

$$\text{PS} = \widehat{\omega}_1 \widehat{x}_1 + \widehat{\omega}_2 \widehat{x}_2 + \cdots + \widehat{\omega}_P \widehat{x}_P = \sum_{p=1}^{P} \widehat{\omega}_p \widehat{x}_p, \tag{5}$$

where $\widehat{x}_p$ is the standardized value of variable p.

3.2 Min-Max Normalization

We utilize the min-max normalization procedure to convert the single measure of sustainability performance described above to a value between 0 and 1 for each supplier. Thus, the possibility of having a negative performance score for any of the suppliers is eliminated. This normalized performance measure score is denoted by η and each supplier's sustainability performance, η_i, can be evaluated using Eq. (6), shown below.

$$\eta_i = \frac{\text{PS}_i - \text{PS}_{\min}}{\text{PS}_{\max} - \text{PS}_{\min}}. \tag{6}$$

This supplier's sustainability performance, η_i, then becomes the coefficient of the maximization part of the bi-objective model outlined in the next section.

3.3 Bi-objective Model

In this section, we present a mathematical optimization model incorporating sustainability-related risks and offshore outsourcing purchasing costs to capture the trade-off between supply chain sustainability risks and low outsourcing costs in contract manufacturer (or supplier) selection and order allocation. The proposed bi-objective mixed-integer programming model attempts to arrive at the supplier selection and order allocation decisions simultaneously, with the dual objectives of maximization of the sustainability performance score and the minimization of the total procurement cost.

3.3.1 Notation

The following notational scheme is used in the optimization model developed below:

Parameters

i	Index for identifying contract manufacturers, or suppliers
η_i	The sustainability performance score of contract manufacturer (supplier) i
c_i	The unit cost of production by contract manufacturer (supplier) i
Q	Total quantity of product outsourced
K_i	Available capacity of contract manufacturer (supplier) i
M	Minimum number of contract manufacturers (suppliers) to use
S	Total number of contract manufacturers (suppliers) available
θ	The minimum proportion of the total quantity to be allocated to any contract manufacturer (supplier)

Decision Variables

y_i - $\begin{cases} 1, & \text{if contract manufacturer } i \text{ is selected} \\ 0, & \text{Otherwise}, i = 1, 2, \ldots, S, \end{cases}$

q_i - Quantity allocated to contract manufacturer i

f_i - Proportion of total order quantity allocated to contract manufacturer i.

$$\text{Maximize} \quad \tau = \frac{\sum_{i=1}^{S} \eta_i y_i}{\sum_{i=1}^{S} y_i}. \tag{7a}$$

$$\text{Minimize} \quad \text{TC} = \sum_{i=1}^{S} c_i q_i. \tag{7b}$$

Subject to:

$$\sum_{i=1}^{S} q_i = Q. \tag{8}$$

$$\sum_{i=1}^{S} y_i \geq M. \tag{9}$$

$$f_i \geq \theta y_i \quad i = 1, 2, \ldots, S. \tag{10}$$

$$\sum_{i=1}^{S} f_i = 1. \tag{11}$$

$$f_i Q \leq q_i \leq K_i y_i \quad i = 1, 2, \ldots, S. \tag{12}$$

$$q_i \geq 0, f_i \geq 0, \quad \text{and} \quad y_i = \{0, 1\} \quad i = 1, 2, \ldots, S. \tag{13}$$

As indicated above, (7a) represents the maximization of the average sustainability performance score (τ) of the selected suppliers and (7b) depicts the minimization of the total procurement cost (TC) for quantities from selected suppliers. Constraint (8) ensures that the total quantity requirement is met. Constraint (9) defines the minimum number of contract manufacturers (suppliers) to select and must always be at least 2 since our focus is on multiple sourcing strategy, while constraint (10) ensures that each selected contract manufacturer is allocated at least a certain proportion of the total quantity required. Constraint (11) ensures that the sum of the contract manufacturers' proportions equals 1. Constraint (12) specifies the minimum quantity allocated to each contract manufacturer (supplier) and the capacity constraints of the contract manufacturers. Constraint (13) defines the binary variables and the non-negativity requirements of the decision variables.

3.3.2 Solution Approach

In our model, we have two conflicting objectives. To solve this bi-objective model, we introduce a new parameter, $\tau \in \{0 - 1\}$, to transform the model into a single objective optimization problem as shown in (14) similar to the epsilon-constraint method. This allows us to easily solve the problem for different values of τ and capture the trade-off between cost and sustainability performance of the selected suppliers.

$$
\begin{aligned}
& \textit{Minimize} \quad \text{TC} = \sum_{i=1}^{S} c_i q_i \\
& \text{Subject to:} \\
& \frac{\sum_{i=1}^{S} \eta_i y_i}{\sum_{i=1}^{S} y_i} \geq \tau \\
& \sum_{i=1}^{S} q_i = Q \\
& \sum_{i=1}^{S} y_i \geq M \\
& f_i \geq \theta y_i \quad i = 1, 2, \ldots, S \\
& \sum_{i=1}^{S} f_i = 1 \\
& f_i Q \leq q_i \leq K_i y_i \quad i = 1, 2, \ldots, S \\
& q_i \geq 0, f_i \geq 0, \quad \text{and} \quad y_i = \{0, 1\} \quad i = 1, 2, \ldots, S.
\end{aligned}
\tag{14}
$$

The next section presents an illustrative example from the global apparel industry, to demonstrate the applicability of our proposed integrated model in practice.

4 Illustrative Example

The apparel industry is unique and one of the most globalized industries in the world. For example, at the end of 2017, Nike Inc.'s value chain had over 500 contract factories in 42 countries with more than 1 million workers, and more than 500,000 different products, each with its own environmental and social footprint (NikeFY1617 Report, 2018). Similarly, VF Corporation, the world's largest publicly owned apparel company, currently manufactures 17% of its products in China, 27% in the Americas, 15% in Vietnam, 3% in India, 6% in Cambodia, 2% in Indonesia, 15% in Bangladesh, and 14% in other countries (VF Corporation, 2018). This firm's global supply chain consists of multiple entities such as a combination of retailers, contractors, subcontractors, merchandisers, buyers, manufacturers, and suppliers. For the entire supply chain, it is, thus, critical to effectively manage the relevant risks in a dynamic and complex environment, in order to remain competitive. This is a supply-driven commodity chain where each player's role is important in the network of supply chains—spanning from fibers to yarn, to fabrics, to accessories, to garments, to trading, and marketing (Ramesh & Bahinipati, 2011).

In our example, we consider a retailer confronted with the problem of selecting at least 5 contract manufacturers from a pool of 10 available suppliers for the supply of 10,000 units of a specified apparel—a cotton for women dress. Contract manufacturers have different production capacities and are located in different countries. The retailer requires that each supplier must meet a minimum supply requirement of at least 10% of total demand.

The first stage in our proposed method is to identify the relevant criteria for rating the sustainability performance of the available contract manufacturers. Then, we collect data on each criterion for the contract manufacturers. Table 1 shows the evaluation criteria considered to be relevant to this study.

4.1 Evaluation Criteria

The criteria selected to evaluate the sustainability performance of a potential contract manufacturer from a particular country are obtained from publicly available indices data provided by various reputable global organizations. These indices utilize both quantitative and qualitative data to examine the economic, environmental, and social performance of different countries. This renders the practical applicability and reliability of our proposed model valid and credible. Data on potential contract manufacturers were collected on these criteria based on the outsourcing country, and the suppliers' normalized data are presented in Table 2.

Table 1 Selected sustainability-related risks criteria

Criteria	Sustainability dimension		
	Economic	Environmental	Social
1. Logistics performance index (LPI) (Arvis et al., 2016)	✓		✓
2. Political stability and absence of violence (PSAV)	✓		✓
3. Rule of law (ROL)	✓		✓
4. Control of corruption (COC)	✓		✓
5. Ease of doing business index (EDB) (World Bank, 2015)	✓		✓
6. Human development index (HDI) (UNDP, 2015)			✓
7. Corruption perception score (CPS) (www.transparency.org)	✓		✓
8. Environmental performance index (EPI) (Hsu et al., 2016)		✓	
9. World risk index (WRI)	✓	✓	✓

Most of the sustainability-related supply chain risks stated in Giannakis and Papadopoulos (2016) are effectively measured by the 9 criteria listed in Table 1. The logistics performance index (LPI) reflects the overall logistics performance of a country in terms of its supply chain reliability and the predictability of service delivery for producers and exporters (Arvis et al., 2016). This is based on five core elements: efficiency of the customs clearance process, quality of trade and transport infrastructure, ease of arranging competitively priced shipments, competence, and quality of logistics services, ability to track and trace consignments, and frequency with which shipments reach the consignee within the scheduled delivery time. Political stability and absence of violence (PSAV), rule of law (ROL), and control of corruption (COC) are part of World Bank's world governance indicators (WGI), which are a good indicative measure of the political, economic, and social status of a country (Kaufmann et al., 2011). Ease of doing business (EDB) and what it measures are provided in Table 14 in Appendix 2. The environmental performance index (EPI) captures all aspects of environmental related issues and is an aggregation of more than 20 indicators from national-level environmental data. Similarly, the world risk index (WRI) is estimated from 28 indicators grouped into four components—exposure (to natural hazards), susceptibility, coping capacities, and adaptive capacities. The WRI is derived from globally available public data. For details on the EPI and WRI indicators, we refer interested readers to Hsu et al. (2016) and WRI (2015), respectively.

4.2 Sustainability Performance

The next stage in our proposed approach is to determine the sustainability performance of each supplier using the steps described in Sect. 3.1. In order to evaluate sustainability performance measures, the first step is to conduct a PCA analysis on

Table 2 Normalized data on sustainability-related criteria

Supplier	Country	CODE	LPI	PSAV	ROL	COC	EDB	HDI	CPS	EPI	WRI
CM1	Bangladesh	BGD	−1.541	−0.827	−1.461	−1.642	−2.121	−1.877	−1.811	−2.209	1.212
CM2	China	CHN	1.246	0.268	0.042	0.658	0.101	0.741	0.675	0.111	−0.566
CM3	India	IND	0.577	−0.577	0.653	0.500	−0.823	−1.136	0.675	−1.034	−0.532
CM4	Indonesia	IDN	−0.649	0.678	−0.662	0.143	−0.438	−0.066	0.142	0.186	−0.031
CM5	Mexico	MEX	−0.287	−0.201	−1.226	−1.483	1.310	1.136	−1.101	0.955	−0.625
CM6	Philippines	PHL	−0.984	−0.860	−0.897	−0.571	−0.219	−0.181	−0.213	0.966	2.259
CM7	South Africa	ZAF	1.580	1.277	1.216	1.570	0.331	−0.445	1.562	0.650	−0.679
CM8	Thailand	THA	0.131	−0.484	0.935	−0.016	1.052	0.774	−0.213	0.553	−0.594
CM9	Turkey	TUR	0.577	−1.144	0.277	0.777	0.799	1.218	0.852	0.368	−0.732
CM10	Vietnam	VNM	−0.649	1.870	1.123	0.064	0.008	−0.165	−0.568	−0.545	0.288

the supplier's normalized data shown in Table 2 for obtaining the principal component characteristics vector and the proportion of variation explained. The results are shown in Table 3.

In step 2, we selected the first four (4) principal components, accounting for 93.8% of the variation in the contract manufacturers' sustainability-related risk criteria, based on the total percentage of variation explained rule. Next, following step 3, the selected princial components characteristics vectors are converted to normalized ranks, $r(\omega_{pt})$, based on the procedure suggested by Slottje (1991), as described earlier in Sect. 3.1. The results are presented in Table 4.

In Step 4, we determine the aggregated weight $(\widehat{\omega}_p)$ from the normalized ranks in Table 4 and the proportion of variation explained by the chosen principal components in Table 3. Using Eq. (4), we obtain the results shown in Table 5.

Now, we can evaluate a single sustainability performance score (PS) for each contract manufacturer from Eq. (5). Using the percentage of contribution (π_p) of the selected principal components and the aggregated weight from step 4, Eq. (5) becomes

$$PS = 0.814\widehat{x}_1 + 0.250\widehat{x}_2 + 0.476\widehat{x}_3 + 0.548\widehat{x}_4 + 0.593\widehat{x}_5 + 0.555\widehat{x}_6 \\ + 0.697\widehat{x}_7 + 0.476\widehat{x}_8 + 0.279\widehat{x}_9. \qquad (15)$$

From Eq. (15) above, the performance score for each contract manufacturer is computed using corresponding normalized variable values from Table 2. Then, the sustainability performance score (η_i) in Table 6 is obtained by applying min-max normalization shown in Eq. (6) to the performance score (PS) derived using the PCA approach.

The sustainability performance scores (η_i) become the coefficients of the first objective function in our proposed integrated model. Note that the contract manufacturer 1 (CM1) is eliminated at this stage because it has the worst sustainability performance.

4.3 Selection and Order Allocation

To apply our integrated model in (14), the contract manufacturers' costs are obtained from the top 20 exporters of women's and girl's cotton dresses (not knitted or crocheted) found in Weil (2006). We assume that the various CMs' capacities range from 2000 to 3000. Both of these input parameters are shown in Table 7.

As mentioned earlier, the following parameter values are selected: the minimum proportion of total quantity to be delivered by each contract manufacturer $(\theta) = 0.10$, the minimum number of contract manufacturers $(M) = 5$, and the total product quantity required $(Q) = 10,000$ units. Using these parameters, the resulting model from (14) is solved using an open-source R-programming package called "GLPK." This application software is known to be appropriate and efficient for solving large-

Table 3 Eigen analysis for the suppliers

Eigenvalue	2.169	1.444	0.963	0.851	0.579	0.369	0.282	0.080	0.056
Proportion (π_p)	0.522	0.232	0.103	0.081	0.037	0.015	0.009	0.001	0.000
Variable	PC1	PC2	PC3	PC4	PC5	PC6	PC7	PC8	PC9
LPI	−0.411	0.143	−0.274	−0.094	−0.280	−0.432	0.673	0.085	0.047
PSAV	−0.159	0.263	0.862	0.013	−0.373	0.054	0.094	−0.078	−0.072
ROL	−0.343	0.312	0.188	−0.061	0.785	−0.126	0.026	0.230	−0.245
COC	−0.388	0.316	−0.112	0.255	0.017	0.403	−0.107	0.075	0.701
EDB	−0.332	−0.449	0.162	−0.091	0.233	−0.254	−0.055	−0.673	0.281
HDI	−0.280	−0.514	0.040	−0.106	0.045	0.657	0.376	0.152	−0.216
CPS	−0.389	0.219	−0.289	0.353	−0.185	0.143	−0.248	−0.404	−0.560
EPI	−0.281	−0.447	0.118	0.484	−0.131	−0.340	−0.296	0.503	−0.011
WRI	0.345	−0.010	0.090	0.738	0.240	0.013	0.483	−0.192	0.023

Table 4 Normalized ranks for the Eigenvectors for the selected principal components

Variable	PC1	PC2	PC3	PC4
LPI	1.000	0.556	0.889	0.889
PSAV	0.222	0.333	0.111	0.556
ROL	0.667	0.222	0.222	0.667
COC	0.778	0.111	0.778	0.444
EDB	0.556	0.889	0.333	0.778
HDI	0.333	1.000	0.667	1.000
CPS	0.889	0.444	1.000	0.333
EPI	0.444	0.778	0.444	0.222
WRI	0.111	0.667	0.556	0.111

Table 5 Aggregated weight for variables

LPI	PSAV	ROL	COC	EDB	HDI	CPS	EPI	WRI
0.814	0.250	0.476	0.548	0.593	0.555	0.697	0.476	0.279

Table 6 CM's sustainability performance

Supplier	Country	Code	PCA score (PS)	SP score (η_i)
CM1	Bangladesh	BGD	−7.334	0.000
CM2	China	CHN	2.299	0.835
CM3	India	IND	−0.378	0.603
CM4	Indonesia	IDN	−0.714	0.574
CM5	Mexico	MEX	−0.760	0.570
CM6	Philippines	PHL	−1.046	0.545
CM7	South Africa	ZAF	4.204	1.000
CM8	Thailand	THA	1.425	0.759
CM9	Turkey	TUR	2.458	0.849
CM10	Vietnam	VNM	−0.154	0.622

scale linear programming (LP) and mixed-integer linear programming (MILP) problems (Theussl et al., 2017).

4.4 Results and Discussion

This section presents the results of the suppliers' sustainability performance measures, the selected suppliers, and optimal order allocations from the bi-objective model and sensitivity analysis on the minimum supply quantity from the supplier (θ). Figure 2 shows the sustainability performance relative to the product's unit cost for each supplier (or contract manufacturer).

Figure 2 suggests that there seems to be no clear relationship between sustainability performance and unit product cost in this case study. Thus, there is a need for a trade-off between low cost and sustainability performance in the contract manufacturer selection and order allocation decisions.

Table 7 Contract manufacturer's cost and capacity

Supplier	Country	CODE	Unit cost ($)	Capacity (K_i)
CM1	Bangladesh	BGD	3.83	2500
CM2	China	CHN	11.25	2300
CM3	India	IND	6.00	2800
CM4	Indonesia	IDN	5.17	2400
CM5	Mexico	MEX	5.58	2100
CM6	Philippines	PHL	5.00	2800
CM7	South Africa	ZAF	4.33	2950
CM8	Thailand	THA	4.42	2000
CM9	Turkey	TUR	6.25	2400
CM10	Vietnam	VNM	3.83	3000

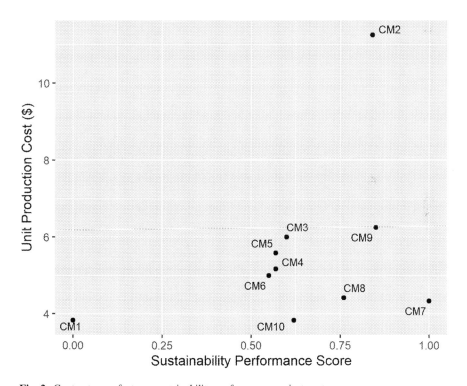

Fig. 2 Contract manufacturer sustainability performance against cost

4.4.1 Optimization Results

To obtain our optimal results, we solved the model for different values of τ, ranging from 0 to 1 in increments of 0.01. The parameter τ can be referred to as the average sustainability performance of the selected contract manufacturers. Table 8 presents the optimal solution to the supplier selection and allocation problem considered. For

Table 8 Selected CMs and order quantity allocated

Supplier	$0 \leq \tau \leq 0.7$			$\tau = 0.71$			$0.72 \leq \tau \leq 0.75$			$\tau = 0.76$			$0.77 \leq \tau \leq 0.81$		
	q_i	y_i	$f_i(\%)$	q_i	y_i	$f_i(\%)$	q_i	y_i	$f_i(\%)$	q_i	y_i	$f_i(\%)$	q_i	y_i	$f_i(\%)$
CM2													1000.0	1.0	10.0%
CM3				1000.0	1.0	10.0%									
CM4	1000.0	1.0	10.0%	1050.0	1.0	10.5%				1050.0	1.0	10.5%			
CM5															
CM6	1050.0	1.0	10.5%				1050.0	1.0	10.5%						
CM7	2950.0	1.0	29.5%	2950.0	1.0	29.5%	2950.0	1.0	29.5%	2950.0	1.0	29.5%	2950.0	1.0	29.5%
CM8	2000.0	1.0	20.0%	2000.0	1.0	20.0%	2000.0	1.0	20.0%	2000.0	1.0	20.0%	2000.0	1.0	20.0%
CM9							1000.0	1.0	10.0%	1000.0	1.0	10.0%	1050.0	1.0	10.5%
CM10	3000.0	1.0	30.0%	3000.0	1.0	30.0%	3000.0	1.0	30.0%	3000.0	1.0	30.0%	3000.0	1.0	30.0%

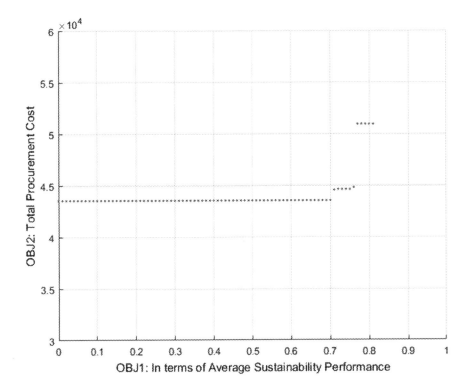

Fig. 3 Pareto frontier for this example

$0 \leq \tau \leq 0.70$, the results obtained remain unchanged, while they change for $0.71 \leq \tau \leq 0.81$. Nevertheless, for values of $\tau > 0.81$, no feasible solutions exist. The resulting Pareto optimality frontier is depicted in Fig. 3.

From Table 8, CM7, CM8, and CM10 are selected in all the optimal solutions in the Pareto set and utilized to full capacity. This is because CM7 and CM10 are non-dominated by others and once their capacity is used up, the next relatively low-cost supplier is selected.

Furthermore, we consider the optimization model without the sustainability performance objective. The minimum total procurement cost is $ 43,523.50 with the same optimal solution as when τ is between 0 and 0.7. This is labeled as the base total procurement cost for the decision-maker to appraise the trade-off between cost and sustainability performance. Our results show that as the average sustainability performance required (τ) increases beyond 0.7, CMs with relatively low sustainability performance are replaced with ones with better sustainability performance, albeit, resulting in increasing total procurement cost as shown in Fig. 3 (e.g., at $\tau = 0.71$, $\tau = 0.72 - 0.75$, 0.76, and $0.77 - 0.81$, the optimal total procurement costs are $44,532.00, $44,603.50, $44,782.00, and $50,916.00, respectively).

4.4.2 Sensitivity Analysis

In this section, we conduct a sensitivity analysis of the minimum number of suppliers' requirements and their capacities to gain some useful managerial insights. Since the sustainability performance measure is a function of the number of suppliers selected, we compare the scenario when a minimum of two suppliers are required as against a minimum of five in the original problem. For both scenarios, we consider the case when suppliers have limited capacity (referred to as the capacitated supplier's case) and the scenario when each supplier has unlimited capacity to supply all the required quantities ordered by the buyer (referred to as the uncapacitated supplier's case). The results for the capacitated and uncapacitated cases with a minimum of two suppliers are presented in Tables 9 and 10, respectively.

Table 9 shows that a costly supplier with high sustainability performance will only be selected when the other suppliers' capacities are used up. Also, it can be deduced that the maximum sustainability performance achievable, when suppliers' capacities are limited, is 0.82 in this example. While in the uncapacitated suppliers' case, a sustainability performance of 0.92 at a lower cost is possible as shown in Table 10. This occurs because more quantity is allocated to a relatively low-priced supplier with high sustainability performance when high sustainability performance is required. In contrast, a relatively large quantity is allocated to a low-priced supplier with moderate sustainability performance, when the required sustainability performance is less than 0.83 in our example problem, as shown in Table 10.

4.4.3 Model Modification

It is important to note that our proposed modeling framework is capable of handling the evaluation of a large number of suppliers using numerous criteria and constraints in the optimization model can easily be modified, if needed. For instance, a buyer may want to tie the suppliers' minimum supply requirement to their sustainability performance, if its evaluation criteria directly measure the suppliers' sustainability performance. To accommodate for such a change, we can define different values of θ for different suppliers (i.e., θ now becomes θ_i, $\forall\ i$).

One of the major characteristics of global supply sourcing is that contract manufacturers often specify a minimum order quantity, which may be due to economies of scale in production or shipment. This means that the buyer must not order less than a certain quantity from each selected supplier or needs to allocate round lots. If we define L_i as the minimum order restriction for supplier i, we can easily replace Eqs. (10) and (12) in the original model with

$$f_i Q \geq L_i \quad i = 1, 2, \ldots, S, \tag{16}$$

$$L_i y_i \leq q_i \leq K_i y_i \quad i = 1, 2, \ldots, S, \tag{17}$$

where $L_i y_i$ and $K_i y_i$ are the lower and the upper bounds for q_i, respectively.

Table 9 Selected CMs and order quantity allocation for the capacitated case when $M \geq 2$

Supplier	$0 \leq \tau \leq 0.73$		$0.74 \leq \tau \leq 0.75$		$\tau = 0.76$		$0.77 \leq \tau \leq 0.80$		$\tau = 0.81$		$\tau = 0.82$	
	q_i	$f_i(\%)$	q_i	$f_i(\%)$	q_i	$f_i(\%)$	q_i	$f_i(\%)$	q_i	$f_i(\%)$	q_i	$f_i(\%)$
CM2									1000.0	10.	1650.0	16.5
CM3												
CM4					1050.0	10.5						
CM5												
CM6	2050.0	20.5	1050.0	10.5								
CM7	2950.0	29.5	2950.0	29.5	2950.0	29.5	2950.0	29.5	2950.0	29.5	2950.0	29.5
CM8	2000.0	20.0	2000.0	20.0	2000.0	20.0	2000.0	20.0	2000.0	20.0		
CM9			1000.0	10.0	1000.0	10.0	2050.0	20.5	1050.0	10.5	2400.0	24.0
CM10	3000.0	30.0	3000.0	30.0	3000.0	30.0	3000.0	30.0	3000.0	30.0	3000.0	30.0
TC	$43,353.50		$44,603.50		$44,782.00		$45,916.00		$50,916.00		$57,826.00	

Table 10 Selected CMs and order quantity allocation for the uncapacitated case when $M \geq 2$

Supplier	$0 \leq \tau \leq 0.81$			$\tau = 0.82$			$0.83 \leq \tau \leq 0.87$			$0.88 \leq \tau \leq 0.92$		
	q_i	y_i	f_i (%)	q_i	y_i	f_i (%)	q_i	y_i	f_i (%)	q_i	y_i	f_i (%)
CM7	1000.0	1.0	10.0%	1000.0	1.0	10.0%	9000.0	1.0	90.0%	9000.0	1.0	90.0%
CM8							1000.0	1.0	10.0%			
CM9				1000.0	1.0	10.0%				1000.0	1.0	10.0%
CM10	9000.0	1.0	90.0%	8000.0	1.0	80.0%						
Total cost	$ 38,800.00			$ 41,220.00			$ 43,390.00			$ 45,220.00		

5 Conclusions

In this chapter, we propose an integrated methodology that combines the PCA technique, min-max normalization, and a bi-objective optimization model for the supplier selection and order allocation problem considering sustainability-related risks in the context of global sourcing. The bi-objective optimization model realistically captures the trade-off between sustainability-related risks and low-procurement cost. The results from our example show that the Pareto frontier can be used as the trade-off curve in global sourcing decision-making since it generates the optimal combination of suppliers. The optimization model allocates an order quantity to selected suppliers and ensures that they have enough capacity to produce and meet minimum supply requirements. This approach can help decision-makers in taking a balanced approach to sustainable supplier selection and order allocation while mitigating the associated supply chain risks related to sustainability.

Additionally, we are in the era of big data and firms have vast amounts of data at their disposal to explore for evaluating supplier performance. Integrating the PCA technique in our proposed procedure can be very useful because of its unique advantages over traditional supplier evaluation methods. Thus, our proposed approach is likely to work well even when evaluating several suppliers (or contract manufacturers) using numerous criteria. Based on our illustrative example, we believe that this approach to evaluating sustainability performance is capable of exposing sustainability-related risks in supplying countries that may adversely affect sourcing firms, as a result of global sourcing. This premise is similar to Park et al.'s (2018) finding that regional characteristics reflect the characteristics of the suppliers in the region.

Finally, this work can be extended in several directions, e.g., the consideration of multi-period, multi-product scenarios. Another possible extension would be to solve the problem with multiple objective functions such as minimizing risk and cost while also maximizing supply chain visibility and transparency. Other possible future research directions lie in the study of uncertainty in suppliers' capacities, due to failure or unforeseen disruptions in operations. We hope that this work will provide the basis for meaningful further explorations in these important areas of study.

Appendices

Appendix 1: Sustainability-Related Supply Chain Risks

Table 11 Example statements on sustainability risk sources (Hofmann et al., 2014)

Social issues

"Child labor, low wages, unbearable working conditions, extortion, you name it" (TelCo, I)

"There is a social dimension if the carrier forces its employees to excess overtime provoking accidents. Thus, our clients may believe that we rely on cheap and forced-labor workers" (LogIntCo, I)

Ecological issues

"If they [suppliers] don't have water treatment for chemical substances, then it is absolutely not acceptable" (ChemCo, I).

"Suppliers may rely on wasteful processes although ecologically friendly alternatives are available" (TelCo, II)

Ethical business conduct issues

"It's also the extreme political opinion of suppliers that may lead to problems for us" (LogNatCo, II)

Table 12 Sustainability-related supply chain risks (Giannakis & Papadopoulos, 2016)

Endogenous	Exogenous
Environmental	
• Environmental accidents (e.g., fires, explosions)	• Natural disasters (e.g., hurricanes, floods, earthquakes)
• Pollution (air, water, soil)	• Water scarcity
• Non-compliance with sustainability laws	• Heatwaves, droughts
• Emission of greenhouse gases, ozone depletion	
• Energy consumption (unproductive use of energy)	
• Excessive or unnecessary packaging	
• Product waste	
Social	
• Excessive working time; work-life imbalance	• Pandemic
• Unfair wages	• Social instability
• Child labor/forced labor	• Demographic challenges/aging population
• Discrimination (race, sex, religion, disability, age, political views)	
• Healthy and safe working environment	
• Exploitative hiring policies (lack of contract, insurance)	
• Unethical treatment of animals	
Financial/economic	
• Bribery	• Boycotts
• False claims/dishonesty	• Litigations
• Price fixing accusations	• Energy prices volatility
• Antitrust claims	• Financial crises
• Patent infringements	
• Tax evasion	

Appendix 2: Indices Criteria and Description

Table 13 Nike's country risk sourcing and manufacturing index criteria (Source: Nike Inc.)

Criteria	Description	Data source
Political risk (25%)		
Political stability and risk of violence (PSRV)	Nike travel risk rating, political violence, and regime stability	World Bank
Rule of law	Effectiveness of legal and regulatory practices	World Bank
Transparency and corruption	Transparency international corruption perception index	https://www.transparency.org/
Economic risk (25%)		
Workforce risk	Access to food, education, and disease risk	
Business and trade environment	Business regulation, trade risk	World Bank
General economic environment	Including risk of economic instability, foreign exchange, inflation, debt, and GDP	
Social/compliance (25%)		
Compliance with labor standards	Risk of wages, hours, FoA, child labor, etc., violations and SMSI performance	
Civil rights	Freedom of the press arbitrary arrests, minority rights	
Worker vulnerability	UNDP Human Development Index	www.undp.org
Infrastructure & Climate (25%)		
Energy and water	Availability and vulnerability of energy and water resources	
Natural disasters and climate change	Vulnerability and exposure to extreme natural disasters	
Logistics infrastructure	Speed, cost, and security of logistics, e.g., roads, ports, communications	World Bank

Table 14 What doing business measures—11 areas of business regulation (World Bank, 2015)

Indicator set	What is measured
Starting a business	Procedures, time, cost, and paid-in minimum capital to start a limited liability company
Dealing with construction permits	Procedures, time, and cost to complete all formalities to build a warehouse and the quality control and safety mechanisms in the construction permitting system
Getting electricity	Procedures, time, and cost to get connected to the electrical grid, the reliability of the electricity supply, and the cost of electricity consumption
Registering property	Procedures, time, and cost to transfer a property and the quality of the land administration system
Getting credit	Movable collateral laws and credit information systems
Protecting minority investors	Minority shareholders' rights in related-party transactions and in corporate governance
Paying taxes	Payments, time, and total tax rate for a firm to comply with all tax regulations
Trading across borders	Time and cost to export the product of comparative advantage and import auto parts
Enforcing contracts	Time and cost to resolve a commercial dispute and the quality of judicial processes
Resolving insolvency	Time, cost, outcome, and recovery rate for a commercial insolvency and the strength of the legal framework for insolvency
Labor market regulation	Flexibility in employment regulation and aspects of job quality

References

Ahi, P., & Searcy, C. (2015). An analysis of metrics used to measure performance in green and sustainable supply chains. *Journal of Cleaner Production, 86*, 360–377.

Aktin, T., & Gergin, Z. (2016). Mathematical modelling of sustainable procurement strategies: Three case studies. *Journal of Cleaner Production, 113*, 767–780.

Amid, A., Ghodsypour, S. H., & O'Brien, C. (2011). A weighted max-min model for fuzzy multi-objective supplier selection in a supply chain. *International Journal of Production Economics, 131*(1), 139–145.

Amin, S. H., Razmi, J., & Zhang, G. (2011). Supplier selection and order allocation based on fuzzy SWOT analysis and fuzzy linear programming. *Expert Systems with Applications, 38*(1), 334–342.

Arvis, J.-F., Saslavsky, D., Ojala, L., Shepherd, B., Busch, C., Raj, A., & Naula, T. (2016). *Connecting to compete 2016: Trade logistics in the global economy: The logistics performance index and its indicators*. World Bank.

Azadnia, A. H., & Ghadimi, P. (2018). An integrated approach of fuzzy quality function deployment and fuzzy multi-objective programming to sustainable supplier selection and order allocation. *Journal of Optimization in Industrial Engineering, 11*(1), 1–22.

Azadnia, A. H., Saman, M. Z. M., & Wong, K. Y. (2015). Sustainable supplier selection and order lot-sizing: An integrated multi-objective decision-making process. *International Journal of Production Research, 53*(2), 383–408.

Azizi, A., Yarmohammadi, Y., & Yasini, A. (2015). Superior supplier selection-a joint approach of Taguchi, AHP, and fuzzy multi-objective programming. *Australian Journal of Basic and Applied Sciences, 9*(2), 163–170.

Babbar, C., & Amin, S. H. (2018). A multi-objective mathematical model integrating environmental concerns for supplier selection and order allocation based on fuzzy QFD in beverages industry. *Expert Systems with Applications, 92*, 27–38.

Banaeian, N., Mobli, H., Fahimnia, B., Nielsen, I. E., & Omid, M. (2018). Green supplier selection using fuzzy group decision making methods: A case study from the agri-food industry. *Computers & Operations Research, 89*, 337–347.

Banjo, S. (2014). Inside Nike's struggle to balance cost and worker safety in Bangladesh. *Wall Street Journal* [online]. Retrieved June 06, 2018, from https://www.wsj.com/articles/inside-nikes-struggle-to-balance-cost-and-worker-safety-in-bangladesh-1398133855

Berg, A., Hedrich, S., Lange, T., Magnus, K.-H., & Matthews, B. (2017). *Digitization: The next stop for the apparel-sourcing caravan.* McKinsey. [online]. Retrieved April 04, 2018, from https://www.mckinsey.com/industries/retail/our-insights/digitization-the-next-stop-for-the-apparel-sourcing-caravan

Chen, Y.-J. (2011). Structured methodology for supplier selection and evaluation in a supply chain. *Information Sciences, 181*(9), 1651–1670.

Demirtas, E. A., & Üstün, Ö. (2008). An integrated multiobjective decision making process for supplier selection and order allocation. *Omega, 36*(1), 76–90.

Demirtas, E. A., & Ustun, O. (2009). Analytic network process and multi-period goal programming integration in purchasing decisions. *Computers & Industrial Engineering, 56*(2), 677–690.

Fahimnia, B., Tang, C. S., Davarzani, H., & Sarkis, J. (2015). Quantitative models for managing supply chain risks: A review. *European Journal of Operational Research, 247*(1), 1–15.

Foroozesh, N., Tavakkoli-Moghaddam, R., & Mousavi, S. M. (2018). Sustainable-supplier selection for manufacturing services: A failure mode and effects analysis model based on interval-valued fuzzy group decision-making. *The International Journal of Advanced Manufacturing Technology, 95*(9–12), 3609–3629.

Ghadimi, P., Toosi, F. G., & Heavey, C. (2018). A multi-agent systems approach for sustainable supplier selection and order allocation in a partnership supply chain. *European Journal of Operational Research, 269*(1), 286–301.

Ghodsypour, S. H., & O'Brien, C. (1998). A decision support system for supplier selection using an integrated analytic hierarchy process and linear programming. *International Journal of Production Economics, 56*, 199–212.

Ghodsypour, S. H., & O'Brien, C. (2001). The total cost of logistics in supplier selection, under conditions of multiple sourcing, multiple criteria and capacity constraint. *International Journal of Production Economics, 73*(1), 15–27.

Giannakis, M., & Papadopoulos, T. (2016). Supply chain sustainability: A risk management approach. *International Journal of Production Economics, 171*, 455–470.

Gören, H. G. (2018). A decision framework for sustainable supplier selection and order allocation with lost sales. *Journal of Cleaner Production, 183*, 1156–1169.

Govindan, K., & Sivakumar, R. (2016). Green supplier selection and order allocation in a low-carbon paper industry: Integrated multi-criteria heterogeneous decision-making and multi-objective linear programming approaches. *Annals of Operations Research, 238*(1–2), 243–276.

Govindan, K., Jafarian, A., & Nourbakhsh, V. (2015a). Bi-objective integrating sustainable order allocation and sustainable supply chain network strategic design with stochastic demand using a novel robust hybrid multi-objective metaheuristic. *Computers & Operations Research, 62*, 112–130.

Govindan, K., Rajendran, S., Sarkis, J., & Murugesan, P. (2015b). Multi criteria decision making approaches for green supplier evaluation and selection: A literature review. *Journal of Cleaner Production, 98*, 66–83.

Hamdan, S., & Cheaitou, A. (2015). Green supplier selection and order allocation using an integrated fuzzy TOPSIS, AHP and IP approach. In *2015 International Conference on Industrial Engineering and Operations Management (IEOM)*, pp. 1–10.

Hamdan, S., & Cheaitou, A. (2017a). Supplier selection and order allocation with green criteria: An MCDM and multi-objective optimization approach. *Computers & Operations Research, 81*, 282–304.

Hamdan, S., & Cheaitou, A. (2017b). Dynamic green supplier selection and order allocation with quantity discounts and varying supplier availability. *Computers & Industrial Engineering, 110*, 573–589.

Hofmann, H., Busse, C., Bode, C., & Henke, M. (2014). Sustainability-related supply chain risks: Conceptualization and management. *Business Strategy and the Environment, 23*(3), 160–172.

Hsu, A., et al. (2016). *Environmental performance index.* Yale University. [online]. Retrieved July 26, 2018, from http://epi2016.yale.edu/downloads

Igarashi, M., de Boer, L., & Fet, A. M. (2013). What is required for greener supplier selection? A literature review and conceptual model development. *Journal of Purchasing and Supply Management, 19*(4), 247–263.

Jadidi, O., Zolfaghari, S., & Cavalieri, S. (2014). A new normalized goal programming model for multi-objective problems: A case of supplier selection and order allocation. *International Journal of Production Economics, 148*, 158–165.

Jia, R., Liu, Y., & Bai, X. (2020). Sustainable supplier selection and order allocation: Distributionally robust goal programming model and tractable approximation. *Computers & Industrial Engineering, 106267.*

Jolai, F., Yazdian, S. A., Shahanaghi, K., & Khojasteh, M. A. (2011). Integrating fuzzy TOPSIS and multi-period goal programming for purchasing multiple products from multiple suppliers. *Journal of Purchasing and Supply Management, 17*(1), 42–53.

Kannan, D., Khodaverdi, R., Olfat, L., Jafarian, A., & Diabat, A. (2013). Integrated fuzzy multi criteria decision making method and multi-objective programming approach for supplier selection and order allocation in a green supply chain. *Journal of Cleaner Production, 47*, 355–367.

Kaufmann, D., Kraay, A., & Mastruzzi, M. (2011). The worldwide governance indicators: Methodology and analytical issues. *Hague Journal on the Rule of Law, 3*(2), 220–246.

Kellner, F., & Utz, S. (2019). Sustainability in supplier selection and order allocation: Combining integer variables with Markowitz portfolio theory. *Journal of Cleaner Production, 214*, 462–474.

Kull, T. J., & Talluri, S. (2008). A supply risk reduction model using integrated multicriteria decision making. *IEEE Transactions on Engineering Management, 55*(3), 409–419.

Lam, K.-C., Tao, R., & Lam, M. C.-K. (2010). A material supplier selection model for property developers using fuzzy principal component analysis. *Automation in Construction, 19*(5), 608–618.

Manik, J. A., & Yardley, J. (2012). Bangladesh finds gross negligence in factory fire. *The New York Times.* [online]. Retrieved June 06, 2018, from https://www.nytimes.com/2012/12/18/world/asia/bangladesh-factory-fire-caused-by-gross-negligence.html

Maplecroft. (2012). *Brazil, China, India, Indonesia and Philippines expose companies to high levels of supply chain risk.* [online]. Retrieved June 06, 2018, from https://www.maplecroft.com/about/news/child labour 2012.html

Mohammed, A., Setchi, R., Filip, M., Harris, I., & Li, X. (2018). An integrated methodology for a sustainable two-stage supplier selection and order allocation problem. *Journal of Cleaner Production, 192*, 99–114.

Mohammed, A., Harris, I., & Govindan, K. (2019). A hybrid MCDM-FMOO approach for sustainable supplier selection and order allocation. *International Journal of Production Economics, 217*, 171–184.

Moheb-Alizadeh, H., & Handfield, R. (2019). Sustainable supplier selection and order allocation: A novel multi-objective programming model with a hybrid solution approach. *Computers & Industrial Engineering, 129*, 192–209.

Nike Inc. (2018). *Learning from our past.* [online]. Retrieved August 13, 2018, from https://sustainability.nike.com/learning-from-our-past

NikeFY1617 Report. (2018). *NIKE-FY1617-Sustainable Business Report.* [online]. Retrieved June 06, 2018, from https://sustainability-nike.s3.amazonaws.com/wp-content/uploads/2018/05/18175102/NIKE-FY1617-Sustainable-Business-Report_FINAL.pdf

Ogunranti, G. A. (2018). Development of a composite sustainability risk index for outsourcing country in apparel. In *2018 Annual Conference on the Proceedings of Decision Sciences Institute (DSI)*, pp. 522–532.

Park, K., Kremer, G. E. O., & Ma, J. (2018). A regional information-based multi-attribute and multi-objective decision-making approach for sustainable supplier selection and order allocation. *Journal of Cleaner Production, 187*, 590–604.

Petroni, A., & Braglia, M. (2000). Vendor selection using principal component analysis. *Journal of Supply Chain Management, 36*(1), 63–69.

Premachandra, I. M. (2001). A note on DEA vs principal component analysis: An improvement to Joe Zhu's approach. *European Journal of Operational Research, 132*(3), 553–560.

Ramesh, A., & Bahinipati, B. (2011). The Indian apparel industry: A critical review of supply chains. In *International conference on operations and quantitative management (ICOQM)*. Nashik, India, pp. 1101–1111.

Rencher, A. C. (2003). *Methods of multivariate analysis.* Wiley.

Scott, J., Ho, W., Dey, P. K., & Talluri, S. (2015). A decision support system for supplier selection and order allocation in stochastic, multi-stakeholder and multi-criteria environments. *International Journal of Production Economics, 166*, 226–237.

Shaw, K., Shankar, R., Yadav, S. S., & Thakur, L. S. (2012). Supplier selection using fuzzy AHP and fuzzy multi-objective linear programming for developing low carbon supply chain. *Expert Systems with Applications, 39*(9), 8182–8192.

Singh, A. (2014). Supplier evaluation and demand allocation among suppliers in a supply chain. *Journal of Purchasing and Supply Management, 20*(3), 167–176.

Slottje, D. J. (1991). Measuring the quality of life across countries. *The Review of Economics and Statistics*, 684–693.

Sodenkamp, M. A., Tavana, M., & Di Caprio, D. (2016). Modeling synergies in multi-criteria supplier selection and order allocation: An application to commodity trading. *European Journal of Operational Research, 254*(3), 859–874.

Sonmez, M. (2006). *Review and critique of supplier selection process and practices.* Loughborough University.

Theussl, S., Hornik, K., Buchta, C., Schwendinger, F., Schuchardt, H., & Theussl, M. S. (2017). *Package 'Rglpk'.*

UNDP. (2015). *Human development report 2016: Work for human development.* [online]. Retrieved from http://hdr.undp.org/sites/default/files/2015_human_development_report_0.pdf

Vahidi, F., Torabi, S. A., & Ramezankhani, M. J. (2018). Sustainable supplier selection and order allocation under operational and disruption risks. *Journal of Cleaner Production, 174*, 1351–1365.

VF Corporation. (2018). *Supply chain.* [online]. Retrieved from https://www.vfc.com/powerful-platforms/supply-chain#global-balance.

Wadhwa, V., & Ravindran, A. R. (2007). Vendor selection in outsourcing. *Computers & Operations Research, 34*(12), 3725–3737.

Weil, P. (2006). *Lean retailing and supply chain restructuring: Implications for private and public governance.* Boston University School of Management.

World Bank. (2015). *Doing business 2016 - Measuring regulatory quality and efficiency.* World Bank Group. [online]. Retrieved from http://www.doingbusiness.org/reports/global-reports/doing-business-2016

WRI. (2015). *The World Risk Report* [online]. Retrieved from https://weltrisikobericht.de/english-2/

Zhu, J. (1998). Data envelopment analysis vs. principal component analysis: An illustrative study of economic performance of Chinese cities. *European Journal of Operational Research, 111*(1), 50–61.

Part IV
Supply Chain Analysis and Risk Management Applications

A Bi-objective, Risk-Aversion Optimization Model and Its Application in a Biofuel Supply Chain

Krystel K. Castillo-Villar and Yajaira Cardona-Valdes

Abstract This chapter discusses approaches to incorporate risk aversion in supply chain network design. The design of supply chain networks involves multiple uncertainty sources. Most of the previous works took a risk-neutral approach by modeling the problem as two-stage stochastic formulation. However, most decision-makers are not risk neutral, and a better understanding of the risk involved is germane. The contributions of this chapter are threefold: *methodological, algorithmic, and application*. From the *methodological perspective*, we propose a novel mathematical formulation of a bi-objective two-stage stochastic programming model that measures the trade-off between the expected cost and the conditional value at risk (CVaR). From the *algorithmic point of view*, the augmented ε-constraint method is used for solving the model and getting the Pareto solutions set. From the *application side*, a real-life data-driven case study at a state level is solved to optimality to obtain pragmatic and managerial insights that enable the investigation of solutions for different levels of risk aversion; this, in turn, helps to increase the production of reliable and cost-effective biofuel. The mathematical model can be transferable to other applications that seek to provide risk-averse solutions to decision-makers.

K. K. Castillo-Villar (✉)
Mechanical Engineering Department and Texas Sustainable Energy Research Institute, University of Texas at San Antonio, One UTSA Circle, San Antonio, TX, USA
e-mail: Krystel.Castillo@utsa.edu

Y. Cardona-Valdes
Autonomous University of Coahuila, Unidad Camporredondo s/n, Edificio S, Saltillo, Coahuila, Mexico

1 Introduction

When designing and planning a supply chain network, the decision-maker needs to consider the multiple sources of uncertainties (i.e., supply availability, price fluctuation, demand variation). Particularly, for location-allocation problems, two-stage stochastic programming has been a powerful technique to address such problems (Birge & Louveaux, 2011). Previous works in supply chain research that have used the two-stage programming approach include Atashbar et al. (2016), Ekşioğlu et al. (2009), Leduc et al. (2008), Leão et al. (2011), and Roni (2013); however, these models are risk neutral since they consider only the expected cost or profit in the objective function. Ahmed (2006) as well as Ruszczyński and Shapiro (2006) studied risk-neutral approaches and pointed out some inefficiencies in those approaches when just expectation is considered in the objective function. To overcome this, the incorporation of risk measures used in financial engineering has demonstrated their applicability in two-stage problems. Specifically, the conditional value at risk (CVaR) is an effective and well-behaved risk measure that provides more robust solutions compared with risk-neutral approaches.

The effort for controlling risk through optimization methods goes back to the studies of Markowitz (1952) that laid out the foundations of modern financial optimization theory. In his work, the trade-off between risk and return was modeled through a quadratic optimization model to determine the optimal investment in a portfolio of financial assets. The risk measure used by Markowitz (1952) was the variance; however, this metric has some limitations when the outcome distribution is asymmetric (Filippi et al., 2020); in this sense, it considers under- and over-performance equally. Due to the limitations presented by the variance, other metrics have been considered in financial applications. The most popular lately has been the shortfall-based or quantile-based, particularly the conditional value at risk (Rockafellar & Uryasev, 2000).

The recent survey of Filippi et al. (2020) presents the general concept of CVaR and its application in optimization modeling including several applications different from financial optimization, such as supply chain management, scheduling, networks, energy, and healthcare.

From the literature review, it was observed that most of the previous works consider the demand and the supply as the sources of uncertainty. Regarding the objective functions, most of the works consider the expected cost or the expected net present value as the economic performance. With respect to the way CVaR is incorporated in the model, most of the problems are modeled as two-stage stochastic problems. Regarding the approaches used to incorporate risk measures, the majority of previous works consider either a single-objective or a bi-objective approach. The single-objective papers (Noyan, 2012; Soleimani et al., 2014; Hemmati et al., 2016; Rahimi et al., 2019) consider a weighted form or a mean-risk form. The bi-objective works (Claro & de Sousa, 2012; Fernandes et al., 2015; Paterakis et al., 2018; da Silva et al., 2020; Delgado & Claro, 2013) use the ε-constraint method or the augmented ε-constraint method. Notably, Carneiro et al. (2010) is the only work reviewed that uses CVaR as constraint.

To position our book chapter in the state of the art, our chapter presents a bi-objective approach to optimize the risk of implementing a supply chain that uniquely considers biomass quality-related properties (as uncertainty sources) as well as logistics decision variables. In this chapter, we implement an augmented ε-constraint method to efficiently solve a realistic case study. Our work fills the gap in the literature by introducing the concept of risk to an emerging field: biofuel supply chains.

The structure of this book chapter follows. In Sect. 2, we review previous works related to risk management in supply chains with an emphasis on energy systems. Section 3 provides background on the use of CVaR in optimization. Section 4 describes approaches for incorporating risk metrics in optimization models. A two-stage stochastic programming model where only expected cost is considered is described. The extension to a risk-averse framework that manages the trade-off between the expected cost and the CVaR is presented. Optimization models for the two approaches, a mean-risk model, which considers a single-objective function, and a bi-objective model, which minimizes simultaneously both objectives, are discussed. Section 5 describes the case study and presents the computational results for a biofuel supply chain network. Finally, Section 6 presents the concluding remarks.

2 Literature Review

Our literature review focuses on risk management in supply chains with a special focus on two-stage stochastic problems in energy domains. Although we can mention some two-stage stochastic optimization papers in the design and planning of biofuel supply chains (Kim et al., 2011; Marufuzzaman & Ekşioğlu, 2014; Castillo-Villar et al., 2017), all of them are risk neutral. For risk-averse approaches, we focus our literature review on approaches considering the CVaR as the risk measure.

As Filippi et al. (2020) highlighted in their survey, the CVaR has been applied in different areas. Particularly, we concentrate our discussion on problems related to facility location and supply chain management. In facility location problems, the decision-makers face strategic and planning decisions. As the authors stated, the most widely used solution approach considers the problem as a scenario-based two-stage stochastic programming, where usually the first-stage variables are related to strategic decisions (e.g., type and location of facilities); these decisions need to be taken before the realization of uncertain events, whereas planning decisions are modeled by second-stage variables (e.g., flow of commodities). Risk factors usually affect directly only the second-stage decisions. To mitigate the risk associated with these decisions incorporating the CVaR measure, an approach to accomplish that is formulating the two-stage stochastic problem as a mean-CVaR model and solving it through optimization commercial solvers.

Next, we review previous studies in the energy domain. Carneiro et al. (2010) addressed the strategic planning of an oil supply chain through the portfolio optimization problem in the integrated oil supply chain to satisfy both fuel specifications and national demand with the maximum profit. Uncertainty is introduced through the crude oil supply, the Brazilian demand for final products, and product and oil prices in the Brazilian and international markets. To optimize this supply chain, the authors proposed a two-stage stochastic model that maximizes the net present value and return distributions, imposing a lower bound on the value of the CVaR.

Gebreslassie et al. (2012) addressed the optimal design of hydrocarbon biorefinery supply chain, under supply and demand uncertainties. The authors presented a bi-objective, multi-period, two-stage stochastic mixed-integer linear programming model. The objectives considered are the minimization of the expected cost, the financial risk measured by the CVaR, and the downside risk.

Fernandes et al. (2015) addressed the design and planning of the petroleum supply chain under demand uncertainty. The authors presented a bi-objective stochastic mixed-integer linear program that maximizes the expected net present value while simultaneously minimizing the risk measure CVaR.

Hemmati et al. (2016) studied optimal decisions on energy storage and thermal units in a transmission-constrained hybrid wind thermal power system. The authors proposed a risk-constrained two-stage stochastic programming model where the uncertainty is related to the wind power output. Risk aversion is explicitly formulated using the CVaR measure.

Paterakis et al. (2018) aimed to determine the optimal energy and reserve volumes while ensuring that the reserves are sufficient to tackle the plausible realizations of the uncertain wind power production. The authors proposed a bi-objective joint energy and reserve day-ahead market based on two-stage stochastic programming from the point of view of a risk-averse decision-maker considering the CVaR metric. The problem is solved by a weighted optimization function and by the ϵ-constraint method.

Remarkably, there is a lack of works addressing bioenergy challenges using a supply chain risk management approach. This book chapter aims to provide an overview of a bi-objective model approach and its application to biofuels.

3 Background on Risk Measures: The Definition of CVaR and Optimization Methods

Let L be a random variable with cumulative distribution function $F_L(z) = P\{L \leq z\}$. Note that L may have the meaning of loss or gain. As in Filippi et al. (2020), we use the random variable L to represent a loss that has a continuous distribution function.

Definition 1 (Value-at-Risk) For a given confidence level $\alpha \in (0, 1)$, the VaR of L is the α-quantile, that is, $\text{VaR}_\alpha(L) = \min \{z | F\_L(z) \geq \alpha\}$.

Definition 2 (CVaR) For a random variable L with continuous distribution function, CVaR equals the conditional expectation of L given that $L \geq \text{VaR}_\alpha(L)$, that is, $\text{CVaR}_\alpha(L) = E(L | L \geq \text{VaR}_\alpha(L))$.

Some basic properties of the VaR and CVaR are listed below:

- Given the same confidence level α, VaR is a lower bound for CVaR.
- The CVaR only penalizes the negative deviations with respect to an efficiency target; CVaR is sensitive to the worst outcomes (i.e., largest losses).
- The CVaR can capture different individual risk attributes by simply changing the quantile parameter α. Particularly, for a high risk-averse decision-maker (i.e., a person who is willing to be protected as much as possible against uncertainties) it can be captured by CVaR when large values of α are considered.
- The CVaR is a coherent risk measure that satisfies monotonicity, subadditivity, positive homogeneity, and translation invariance.
- The CVaR can be embedded in an optimization model adding linear constraints and continuous variables.

4 Approaches for Incorporating Risk Metrics in Optimization

This section dives into the generic two-stage stochastic programing modeling and how the CVaR can be added to the model to make supply chain risk management decisions.

4.1 Two-Stage Stochastic Programming

In a two-stage stochastic optimization model, there are two types of variables: the design variables and the control variables. The design or first-stage variables are decided before the realization of stochastic parameters and cannot be adjusted after the realization; the control or second-stage variables are subject to adjustment once a specific realization of uncertain parameters is known.

The two-stage stochastic programming framework, in its general form, is formulated as follows (Birge & Louveaux, 2011):

$$\min_{x \in R^n} E(f(x, \omega)) = \min_{x \in R^n} c^T x + E(Q(x, \xi(\omega))),$$

where $f(x, \omega)$ is the total cost function and $Q(x, \xi(\omega))$ is the optimal value of the second-stage problem (also known as recourse function):

$$Q(x, \xi(\omega)) = \min_{y \in R^n} q(\xi)^T y.$$

subject to:

$$T(\xi)^T x + W(\xi)^T y = h(\xi)^T; y \geq 0.$$

In the above problem "x" and "y" are the first- and second-stage decisions. At the first stage, we make "here-and-now" decisions before the realization of the uncertain data ξ. At the second stage, we make "wait-and-see" decisions after the realization of ξ is known. Because of the inherent variability, the second stage contains uncertain variables, and therefore, the total cost objective function is also a random variable. The objective is to choose the first-stage variables in a way that the sum of the first-stage cost and the expected value of the second-stage cost is minimized. The expected cost is a risk-neutral approach where the optimal solution performs well on average. As this standard method does not provide a mechanism to deal with unfavorable outcomes given by the variability of the uncertain parameters, in this sense the expected cost cannot be used to control or manage risk explicitly.

This requires extending the stochastic programming model to risk management to provide control over unfavorable outcomes. The fundamental idea of risk management is incorporating the trade-off between expected cost and financial risk, which offers an opportunity to reduce the impact of unfavorable events of the uncertain parameters (Gebreslassie et al., 2012).

One of the best approaches to incorporate risk parameters in two-stage stochastic programming models is by developing a mean-risk model where the mean-risk function is minimized

$$\min_{x \in R^n} \{E(f(x, \omega)) + \lambda \rho (f(x, \omega))\},$$

where λ is a non-negative weighted coefficient of risk part and $\rho : Z \to R$ is a specific risk measure (Z is a linear space of F-measurable function on probability space (Ω, F, P)). Soleimani et al. (2014) proved three risk measures, among them the VaR and CVaR showing the acceptability of CVaR in terms of quality solution. We cannot talk about CVaR without understanding VaR (Sarykalin et al., 2008). VaR is a popular and widely used risk measure, which can be defined as the maximum loss expected to be incurred over a certain time horizon at a given probability as follows:

$$\text{VaR}_\alpha(Z) = \inf \{\eta \in R : F_Z(\eta) \geq \alpha\},$$

where F represents cumulative distribution function of a set of random variables "Z."

Rockafellar and Uryasev (2002) noticed that the VaR does not provide any indication about the severity of the losses beyond its value. The CVaR, a variation of VaR, tries to overcome this drawback as it measures the conditional expectation of losses above η. CVaR is a mathematically well-behaved risk measure introduced by Rockafellar and Uryasev (2000) for a financial application. It is formulated as follows:

$$\text{CVaR}_\alpha = E(Z|Z \geq \text{VaR}_\alpha(Z)),$$

where CVaR_α denotes the conditional value at risk at level α. CVaR can be expressed by the following linear minimization formula, which is computationally more tractable (Schultz & Tiedemann, 2006):

$$\text{CVaR}_\alpha = \min_{\eta \in \mathbb{R}} f(\alpha, \eta, x),$$

where

$$f(\alpha, \eta, x) = \eta + \frac{1}{1-\alpha} E\{ \max \{Z(x, \omega) - \eta, 0\}\}.$$

f is linear, convex, and finite. As CVaR has unique characteristics of linearity, convexity, and continuity, which can lead to a computationally tractable risk measure in comparison with those non-linear and non-convex, it can be successfully used as a suitable risk criterion. Also, it has demonstrated their applicability in two-stage problems, and that provides more robust solutions compared with risk-neutral approaches.

In optimization problems under uncertainty, CVaR can be part of the objective function or the constraints, or both. Few authors considered the optimization of a model incorporating the CVaR from a computational point of view (Filippi et al., 2020).

4.2 The Optimization of CVaR

We elaborate on two approaches to generate the trade-off between the expected cost and the financial risk (CVaR): (*i*) one approach is to reformulate the two-stage stochastic programming problem with the CVaR measure on the total cost as a large-scale linear programming problem, named mean-CVaR model where the mean-risk function is minimized (Schultz & Tiedemann, 2004, Noyan, 2012, Soleimani & Govindan, 2014, and Rahimi et al., 2019), and (*ii*) the second approach is to consider a bi-objective optimization problem in which the expected economic performance and the financial risk metric are the objective functions to be minimized

(Gebreslassie et al., 2012; Claro & de Sousa, 2010, 2012; Fernandes et al., 2015; Paterakis et al., 2018; da Silva et al., 2020).

4.2.1 The Mean-CVaR Optimization Model

In the first approach, CVaR is considered as the risk measure of the proposed two-stage stochastic programming model:

$$\min_{x \in R^n} \{E(f(x, \omega)) + \lambda CVaR_\alpha\}.$$

For the case of a finite probability space, where $\Omega = \{\omega^1, \omega^2, \ldots, \omega^N\}$ with probabilities $p^1, p^2, \ldots, p^N$, the above model can be formulated as the following linear programming problem leading to a mean-CVaR model (Noyan, 2012):

$$\min CVaR_\alpha = (1+\lambda)c^T x + \sum_{s=1}^{N} p^s (q^s)^T y^s + \lambda \left(\eta + \frac{1}{1-\alpha} \sum_{s=1}^{N} (p^s v^s) \right),$$

subject to:

$$W^s y^s = h^s - T^s x, \quad s = 1, \ldots, N, \quad x \in X,$$

$$y^s \geq 0, \quad s = 1, \ldots, N,$$

$$v^s \geq (q^s)^T y^s - \eta, \quad s = 1, \ldots, N,$$

$$\eta \in R, v^s \geq 0, \quad s = 1, \ldots, N.$$

where
λ: a non-negative risk coefficient to reflect the trade-off between the expected cost or benefits and risk.
η: a variable that provides the VaR and CVaR measures at confidence level of α (%).
α: the confidence level to compute the CVaR measure of risk.
N: the number of scenarios.
v^s: the auxiliary variable to compute the CVaR measure. For loss scenarios exceeding VaR, variable v^s assumes a value greater than 0.
Note that we can interpret the variable η as a first-stage variable and the excess variables, v^s, $s = 1, \ldots, N$, as second-stage variables.

4.2.2 The Bi-objective Optimization Model

The second approach considers a bi-objective linear programing formulation where commonly one objective minimizes the expected cost, and the second objective minimizes the CVaR. Among the revised literature the prevailing method used to solve the bi-objective problem is the ϵ-constraint method where an objective function is optimized, while the other objective functions are set as constraints of the model. It is recognized in the literature as the state-of-the-art method to solve this type of problem, the augmented ϵ-constraint algorithm (Mavrotas, 2009; Mavrotas & Florios, 2013), which is presented below.

To get the Pareto optimal solutions, first the ϵ-constraint algorithm calculates through the ideal and Nadir values, to define the extreme points that define the Pareto curve. To do that, consider $f_1(x)$ as the economic objective function and $f_2(x)$ as the risk objective function, with $x \in X$, where x is a decision variable and X is the feasible region. Then, by using lexicographic optimization, we optimize $\min f_1(x)$ and $\min f_2(x)$, separately. The optimal minimum values of such optimization are denoted by $f_1^{\min}(x)$ and $f_2^{\min}(x)$, respectively, which are the ideal values. To get the Nadir values, we optimize $f_1(x)$ setting as constraint $f_2(x) = f_2^{\min}(x)$, and $f_2(x)$ setting as constraint $f_1(x) = f_1^{\min}(x)$, separately. The optimal minimum values of such optimization are denoted by $f_1^{\max}(x)$ and $f_2^{\max}(x)$, respectively, which are the Nadir values. By following the abovementioned steps, both the Pareto curve extreme points and the range variation of each objective function $(f_1^{\min}(x), f_1^{\max}(x))$ and $(f_2^{\min}(x), f_2^{\max}(x))$ are determined.

To find intermediate points in the Pareto curve (i.e., points that offer a trade-off between the two objectives), we prioritize the economic objective as follows. We solve a single-objective function where $f_1(x)$ is minimized and the risk objective $f_2(x)$ is set as constraint.

$$\min f_1(x) + eps \cdot \frac{\sigma}{r}$$

subject to:

$$f_2(x) + \sigma = f_2^{\min}(x) + \epsilon_{\text{step}};$$
$$x \in X;$$

where σ is considered a slack variable; r corresponds to a $f_2(x)$ range; ϵ_{step} is defined as $\frac{r}{n-1}$ with n the number of points to be evaluated and $(n-1)$ the number of intervals; and eps is a small number 10^{-3} or 10^{-6} that does not affect the objective function.

5 A Biofuel Supply Chain Network Design

To exemplify the applicability of the bi-objective model, we employ a case study to test both the model and the solution procedure.

5.1 Problem Description

The problem consists of designing a supply chain to produce biofuel, which incorporates biomass quality uncertainties and risk aversion. The two-echelon supply chain consists of suppliers and biorefineries. The suppliers provide switchgrass as raw material, which is used to produce biofuel at the biorefineries. The switchgrass exhibits key uncertain characteristics (moisture and ash contents) that affect the conversion process from biomass to biofuel. These biomass quality characteristics vary by county and type of biomass. We consider that the moisture affects the quantity of switchgrass provided by suppliers (by using humid instead of dry tons), so this quantity is also uncertain. The uncertainty will be modeled through a set of scenarios. At biorefineries the production capacity is known. The total biofuel demand is also known. A third-party supplier will cover all the demand not satisfied by our supply chain.

A case study of the state of Texas is introduced to show the model's applicability. The state has transportation resources to move considerable amount of biomass (i.e., railroad infrastructure). Texas also presents potential locations to build high-capacity biorefineries, leading to industrial production of biofuels. Switchgrass is an abundant type of biomass in Texas, and it is a potential raw material to produce biofuels due to its chemical and physical characteristics (Roni et al., 2018). We have based all the data collection and parameters estimation methodology in the work of Aboytes-Ojeda et al. (2019). In this case study, the biomass pre-processing is assumed to happen in the biorefineries.

5.2 Two-Stage Stochastic Program

The first-stage decision variables that are independent scenario variables are used to decide which biorefineries to open, their capacity size, and the biomass conversion technology. The second-stage decision variables that are dependent on scenario variables are employed to determine the biomass quantity that will be sent from suppliers to biorefineries (wet biomass), the biomass quantity before pre-processing (dry biomass), and the quantity of biofuel provided by a third party.

The sets, parameters, and mathematical formulation follow.

Sets

I	Set of suppliers
J	Set of potential locations for biorefineries
L	Set of capacity size for biorefineries
K	Set of technologies
S	Set of scenarios for moisture and ash

Parameters:

f_{jlk}	Fixed cost for opening a biorefinery j of size l using biomass conversion technology k; $j \in J, l \in L, k \in K$
v_{jlk}	Production capacity at biorefinery j of size l using biomass conversion technology k; $j \in J, l \in L, k \in K$
g_{jk}	Conversion factor (liters/tons) of biomass supplied by i to biofuel using technology k; $j \in J, k \in K$
ϵ_{is}	Moisture content of biomass provided by supplier i under scenario s; $i \in I, s \in S$
δ_{is}	Ash content of biomass provided by supplier i under scenario s; $i \in I, s \in S$
q_{is}	The amount of biomass available at supplier i under scenario s; $i \in I, s \in S$
τ_{iks}	Conversion factor from wet biomass provided by supplier i to dry biomass with technology k under scenario s; $i \in I, k \in K, s \in S$
c_{iks}	Logistic cost composed of transportation plus pre-processing cost for reaching moisture target (ϵ_k) and ash target (δ_k) from biomass provided by supplier i, using technology k under scenario s; $i \in I, k \in K, s \in S$
c_k	Operative cost for transforming biomass to biofuel using technology k; $k \in K$
p_s	Probability of realization of scenario s; $s \in S$
d	Total demand for biofuel
ρ_s	Penalization for third-party acquisition under scenario s; $s \in S$

The first-stage or design variables are:

Z_{jlk}	Binary variable equal to 1 if biorefinery j of size l is opened using technology k; $j \in J, l \in L, k \in K$
0	Otherwise

The second-stage or control variables include:

X_{ijks}	Quantity of biomass delivered from supplier i to biorefinery j using technology k under scenario s for ash and moisture; $i \in I, j \in J, k \in K, s \in S$
Y_{ijks}	Quantity of preprocessed biomass that was delivered from supplier i to biorefinery j using technology k under scenario s for ash and moisture; $i \in I, j \in J, k \in K, s \in S$
U_s	Quantity of biofuel liters provided by a third-party supplier under scenario s; $s \in S$

Objective function that minimizes the expected total cost:

$$\min f_1(x) = EC^{FS} + \sum_{s \in S} p_s \cdot EC_s^{SS},$$

where EC^{FS} and EC^{SS} are as follows:

$$EC^{FS} = \sum_{j \in J} \sum_{l \in L} \sum_{k \in K} f_{jlk} \cdot Z_{jlk}$$

$$EC_s^{SS} = \sum_{i \in I}\sum_{j \in J}\sum_{k \in K} c_{iks} \cdot X_{ijks} + \sum_{i \in I}\sum_{j \in J}\sum_{k \in K} c_k \cdot Y_{ijks} + \rho_s \cdot U_s \quad \forall s$$

Subject to:

$$\sum_{j \in J}\sum_{k \in K} X_{ijks} \leq q_{is}; \quad \forall i \in I, s \in S. \tag{1}$$

$$\tau_{iks} X_{ijks} = Y_{ijks}; \quad \forall i \in I, j \in J, k \in K, s \in S. \tag{2}$$

$$\sum_{i \in I} g_{ik} Y_{ijks} \leq \sum_{l \in L} v_{jlk} Z_{jlk}; \quad \forall j \in J, k \in K, s \in S. \tag{3}$$

$$\sum_{i \in I}\sum_{j \in J}\sum_{k \in K} g_{ik} Y_{ijks} + U_s = d; \quad \forall s \in S. \tag{4}$$

$$\sum_{l \in L}\sum_{k \in K} Z_{jlk} \leq 1; \quad \forall j \in J. \tag{5}$$

$$X_{ijks}, Y_{ijks} \geq 0; \quad \forall i \in I, \ j \in J, k \in K, s \in S. \tag{6}$$

$$Z_{jlk} \in \{0, 1\}; \quad \forall j \in J, l \in L, k \in K \tag{7}$$

The objective function $f_1(x)$ minimizes the total expected cost, composed of the first-stage cost (EC^{FS}) that considers the investment cost for opening biorefineries, and the second-stage cost (EC^{SS}) that includes the logistic cost (transportation plus preprocessing costs) as well as the cost incurred by the biofuel third-party supplier. Constraints (1) give an upper bound on the amount of biomass available at supplier i under scenario s. Constraints (2) transform wet biomass to dry biomass. Constraints (3) convert biomass into biofuel if biorefinery is open and also limit biofuel production to the maximum biorefinery capacity. Constraints (4) establish that the quantity satisfied by the third party is determined by the total demand and the product sent from biorefineries in each possible scenario. Constraints (5) limit the selection of one technology and capacity size per facility. Finally, constraints (6) and (7) impose the nature of variables.

5.3 Extension to a Risk-Averse Framework

The CVaR measure is introduced to the risk-neutral stochastic model to manage the level of the risk in the problem. The linear programming model that minimizes the CVaR_α is formulated as follows:

$$\min f_2(x) = \text{VaR} + \frac{1}{1-\alpha} \sum_{s \in S} p_s T_s.$$

subject to constraints (1)–(7), and additionally,

$$T_s \geq EC^{FS} + EC_s^{SS} - \text{VaR}; \quad \forall s \in S \tag{8}$$

$$T_s \geq 0; \quad \forall s \in S \tag{9}$$

$$\text{VaR} \geq 0; \tag{10}$$

where VaR is a first-stage decision variable, and T_s is a second decision variable denoting the tail cost for scenario s, defined as the amount by which costs in scenario s exceed VaR. It is the positive deviation between VaR and cost of scenario s. Note that as T_s is constrained to being positive, the model tries to decrease VaR and, hence, positively impact the objective function. However, a large reduction in VaR may result in more scenarios with positive tail costs.

5.4 The Bi-objective Problem

The fundamental idea of risk management is incorporating the trade-off between financial risk and the total expected cost into the decision-making process of the supply chain network design. This leads to a bi-objective optimization problem which simultaneously considers the minimization of both objectives, expressed as:

$$\min f_1(x) = EC^{FS} + \sum_{s \in S} p_s \cdot EC_s^{SS}$$

$$\min f_2(x) = \text{VaR} + \frac{1}{1-\alpha} \sum_{s \in S} p_s T_s$$

subject to constraints (1)–(10).

To solve the proposed bi-objective problem, we propose that both the total cost and risk be minimized simultaneously. Thus, the augmented ε-constraint method is applied to generate a Pareto solution set reflecting the trade-off between the dual objectives.

5.5 Computational Results

We utilized CPLEX 12.8 to solve the instances on a computer with a processor Intel (R) Core (TM) i9-7980XE @2.60GHz and a RAM memory of 32GB. The package JuMP embedded in Julia was utilized to code the algebraic model.

All the problems have 254 counties as suppliers and 167 potential locations to open biorefineries. We consider two levels of risk: 75% (risk-seeking decision-maker) and 95% (risk-averse decision-maker) with twenty equidistant points in each one of the problems. We define four different problems shown in Table 1. All scenarios are equally probable. The results shown in Table 2 and Fig. 1 correspond to the problems with 4 scenarios, that is, problems 2 and 4.

Table 1 Problem definitions

ID[a]	I	J	L	K	S	No. binary variables	No. continuous variables	Total	α	Pareto points
1	254	167	1	1	2	167	84,838	85,005	95%	20
2	254	167	1	1	4	167	339,348	339,515	95%	20
3	254	167	1	1	2	167	84,838	85,005	75%	20
4	254	167	1	1	4	167	339,348	339,515	75%	20

[a]Problem ID

Table 2 Results of problems 2 and 4 by cost breakdown ['000 USD]

	95% risk level (risk-averse policy)			75% risk level (risk-seeking policy)		
Point	Third party	Tail	VaR	Third party	Tail	VaR
1	0	9821.42	3,006,429.35	0	9821.42	3,006,429.35
5	0	6997.41	3,010,194.70	0	8033.24	3,008,813.59
10	600.62	3596.10	3,014,172.05	0	5798.03	3,011,793.88
15	3192.88	2180.52	3,014,172.05	339.02	3857.23	3,014,140.73
20	5702.07	847.99	3,014,172.05	1934.08	3007.25	3,014,140.73

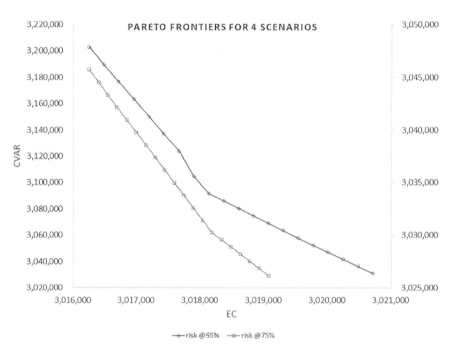

Fig. 1 Pareto Frontiers for different risk levels

Table 2 elaborates on the outputs for third-party cost, tails, and VaR for the two risk levels and for selected points in the Pareto frontiers. We can observe a trade-off between the third party and the tails in Table 2, depending on the risk level. For the

75% risk level, the tails are larger, and the third party needed to satisfy the demand of bioethanol decreases. For the 95% risk level, the tails are smaller and the use of the third party increases. Thus, the decision-maker profile results in different distribution networks and levels of usage of a third-party supplier to meet the demand. If decision-maker is too sensitive to risk (risk averse), he or she will pass on the risk to the third party, but if decision-maker is risk-seeking, he or she will use the third party less and try to meet more demand and absorb the risk of poor biomass supply and its quality.

In Fig. 1, the Pareto frontiers are shown for two risk acceptance levels. In every point of the Pareto frontier, the same number of biorefineries is opened with an investment cost of 1,309,568 ['000 USD]; thus, the basic topology of the supply chain remains unchanged. At 95% risk level, the costs are higher as expected.

6 Concluding Remarks

This chapter aims to introduce the readers to the concept of incorporating risk measures in two-stage stochastic programming. In particular, the approach of modeling the system as a bi-objective optimization problem is addressed. The first objective is the minimization of the expected investment and operational costs, and the second objective is the minimization of the CVaR. A two-stage stochastic mixed-integer linear programming model is formulated to determine the tactical and strategic decisions of a biofuel supply chain network considering the biomass quality variability issues and the risk of investment. The proposed model accounts for biomass quality uncertainties and provides the optimal network configuration that minimizes investment risk and expected system cost. Results reveal the impact of biomass quality variability on the overall network configuration and the trade-off between the third-party contribution and the tails in the CVaR formulation. The managerial insights obtained from this case study aim to help decision-makers to ensure proper resource allocation decisions and optimal production and transportation decisions while designing the biomass supply chain network with the level of investment risk under consideration. The model can be extended to include different transportation modes and more harvesting-related uncertainties.

References

Aboytes-Ojeda, M., Castillo-Villar, K., & Ekşioğlu, S. D. (2019). Modeling and optimization of biomass quality variability for decision support systems in biomass supply chains. *Annals of Operations Research.*

Ahmed, S. (2006). Convexity and decomposition of mean-risk stochastic programs. *Mathematical Programming, 106*(3), 433–446.

Atashbar, N. Z., Labadie, N., & Prins, C. (2016). Modeling and optimization of biomass supply chains: A review and a critical look. *IFAC-PapersOnLine, 49*(12), 604–615.

Birge, F., & Louveaux, J. (2011). *Introduction to stochastic programming* (2nd ed.). Springer Publishing Company, Incorporated.

Carneiro, M. C., Ribas, G. P., & Hamacher, S. (2010). Risk Management in the oil supply chain: A CVaR approach. *Industrial & Engineering Chemistry Research, 49*(7), 3286–3294.

Castillo-Villar, K., Ekşioğlu, S. D., & Taherkhorsandi, M. (2017). Integrating biomass quality variability in stochastic supply chain modeling and optimization for large-scale biofuel production. *Journal of Cleaner Production, 149*, 904–918.

Claro, J., & de Sousa, J. (2010). A multiobjective metaheuristic for a mean-risk multistage capacity investment problem. *Journal of Heuristics, 16*, 85–115.

Claro, J., & de Sousa, J. (2012). A multiobjective metaheuristic for a mean-risk multistage capacity investment problem with process flexibility. *Computers & Operations Research, 39*, 838–849.

da Silva, C., Barbosa-Póvoa, A. P., & Carvalho, A. (2020). Environmental monetization and risk assessment in supply chain design and planning. *Journal of Cleaner Production, 270*.

Delgado, D., & Claro, J. (2013). Transmission network expansion planning under demand uncertainty and risk aversion. *International Journal of Electrical Power & Energy Systems, 44*(1), 696–702.

Ekşioğlu, S. D., Acharya, A., Leightley, L. E., & Arora, S. (2009). Analyzing the design and management of biomass-to-biorefinery supply chain. *Computers & Industrial Engineering, 57*(4), 1342–1352.

Fernandes, L. J., Relva, S., & Barbosa-Póvoa, A. P. (2015). Downstream petroleum supply chain planning under uncertainty. In *12th International Symposium on Process Systems Engineering and 25th European Symposium on Computer Aided Process Engineering* (pp. 1889–1894). Elsevier.

Filippi, C., Guastaroba, G., & Speranza, G. (2020). Conditional value-at-risk beyond finance: A survey. *International Transactions in Operational Research, 27*(3), 1277–1319.

Gebreslassie, B. H., Yao, Y., & You, F. (2012). Design under uncertainty of hydrocarbon biorefinery supply chains: Multiobjective stochastic programming models, decomposition algorithm, and a comparison between CVaR and downside risk. *AICHE Journal, 58*(7), 2155–2179.

Hemmati, R., Saboori, H., & Saboori, S. (2016). Stochastic risk-averse coordinated scheduling of grid integrated energy storage units in transmission constrained wind-thermal systems within a conditional value-at-risk framework. *Energy, 113*(19), 762–775.

Kim, J., Realff, M. J., & Lee, J. H. (2011). Optimal design and global sensitivity analysis of biomass supply chain networks for biofuels under uncertainty. *Computers & Chemical Engineering, 35*(9), 1738–1751.

Leão, R. R. D. C. C., Hamacher, S., & Oliveira, F. (2011). Optimization of biodiesel supply chains based on small farmers: A case study in Brazil. *Bioresource Technology, 102*(19), 8958–8963.

Leduc, S., et al. (2008). Optimal location of wood gasification plants for methanol production with heat recovery. *International Journal of Energy Research, 32*(12), 1080–1091.

Markowitz, H. (1952). Portfolio selection. *The Journal of Finance, 7*(1), 77–91.

Marufuzzaman, M., & Ekşioğlu, S. (2014). Developing a reliable and dynamic intermodal hub and spoke supply chain for biomass. In Y. Guan & H. Liao (Eds.), *IIE Annual Conference. Proceedings* (pp. 2417–2426). Institute of Industrial Engineers-Publisher.

Mavrotas, G. (2009). Effective implementation of the e-constraint method in multi-objective mathematical programming problems. *Applied Mathematics and Computation, 213*(2), 455–465.

Mavrotas, G., & Florios, K. (2013). An improved version of the augmented e-constraint method (AUGMECON2) for finding the exact pareto set in multi-objective integer programming problems. *Applied Mathematics and Computation, 219*(18), 9652–9669.

Noyan, N. (2012). Two-stage stochastic programming involving CVaR with an application to disaster management. *Computers and Operations Research, 39*(3), 541–559.

Paterakis, N. G., Gibescu, M., Bakirtzis, A. G., & Catalão, J. P. S. (2018). A multi-objective optimization approach to risk-constrained energy and reserve procurement using demand response. *IEEE Transactions on Power Systems, 33*(4), 3940–3954.

Rahimi, M., Ghezavati, V., & Asadi, F. (2019). A stochastic risk-averse sustainable supply chain network design problem with quantity discount considering multiple sources of uncertainty. *Computers & Industrial Engineering, 130*, 430–449.

Rockafellar, R., & Uryasev, S. (2000). Optimization of conditional value at risk. *Journal of Risk, 2*(3), 21–42.

Rockafellar, R., & Uryasev, S. (2002). Conditional value-at-risk for general loss distributions. *Journal of Banking & Finance, 26*(7), 1443–1471.

Roni, M. S. (2013). *Analyzing the impact of a hub and spoke supply chain design for long-haul, high-volume transportation of densified biomass*. Mississippi State University.

Roni, M. S., et al. (2018). *Herbaceous feedstock 2018 state of technology report*. Idaho National Laboratory.

Ruszczyński, A., & Shapiro, A. (2006). *Optimization of risk measures. In probabilistic and randomized methods for design under uncertainty* (1st ed.). Springer.

Sarykalin, S., Serraino, G., & Uryasev, S. (2008). Value-at-risk vs. conditional value-at-risk in risk management and optimization. In *Tutorials in operations research* (pp. 270–294). INFORMS.

Schultz, R., & Tiedemann, S. (2004). Conditional value-at-risk in stochastic programs with mixed-integer recourse. Stochastic Programming E-Print Series.

Schultz, R., & Tiedemann, S. (2006). Conditional value-at-risk in stochastic programs with mixed-integer recourse. *Mathematical Programming, 105*(2), 365–386.

Soleimani, H., & Govindan, K. (2014). Reverse logistics network design and planning utilizing conditional value at risk. *European Journal of Operational Research, 237*(2), 487–497.

Soleimani, H., Mirmehdi, S.-E., & Govindan, K. (2014). Incorporating risk measures in closed-loop supply chain. *International Journal of Production Research, 52*(6), 1843–1867.

Conceptualizing and Modeling Supply Chains in the Hazard Context

Douglas S. Thomas and Jennifer F. Helgeson

Abstract This chapter discusses the impact of natural and human-made hazards in the US economy as a whole and the US manufacturing industry in particular. Many studies that examine economic impacts of hazards consider the upstream impact of supply chain disruption. This chapter examines how the economy is affected by a disruption in supplies to determine the magnitude of the downstream effect, which is often referred to as the ripple effect. Additionally, the same analysis is conducted for the manufacturing sector in particular. The goal is to understand whether manufacturers are affected by the ripple effect to a relatively greater extent than is the case for the total US goods economy. This chapter provides evidence that the effect of hazards propagating through the supply chain exceeds that of the localized hazard impacts (i.e., direct impact in the geographic location where the hazard took place). This creates a fundamental incentive misalignment. The establishment within the supply chain that invests in mitigation efforts and experiences the hazard often does not directly experience the majority of the resulting net benefit. Thus, it is necessary that wider systems-level thinking is employed when an establishment along a given supply chain considers vulnerability to disruption and undertakes resilience planning.

1 Introduction

There is a burgeoning literature addressing the design of supply chains that are efficient but also resilient, especially to ever-increasing hazard disruptions and to downstream effects. Industry appears to be increasingly aware of the growing volatility across a range of business parameters from energy cost, to raw material availability, and currency exchange rates (e.g., Neiger et al., 2009; Christopher &

D. S. Thomas (✉) · J. F. Helgeson
Engineering Laboratory, National Institute of Standards and Technology, Gaithersburg, MD, USA
e-mail: douglas.thomas@nist.gov

Holweg, 2011; Vlajic et al., 2013), especially as there is a growing dependency between firms and increasing supply chain complexity (e.g., Kamalahmadi & Parast, 2016).

SwissRe (2018) reports that total economic losses from natural and human-made disasters in 2017 were estimated to be $306 billion, up from $188 billion in 2016. Additionally, even insured losses in 2017 were estimated to be $136 billion, the third highest on record (SwissRe, 2018). In 2020, there were 22 natural disasters with losses greater than $1 billion each in the US alone (NOAA National Centers for Environmental Information (NCEI), 2021). These natural hazards took place in tandem with the global coronavirus disease 2019 (COVID-19) pandemic to create compounded risks and resulting in complex events for many businesses, especially small and medium-sized enterprises (Helgeson et al., 2021). These types of disaster events and the associated losses highlight the shortcomings of using an asset-by-asset approach, which does not consider the larger system, to disaster preparation and resilience planning.

This chapter presents a macro-level analysis of how establishments—at the total economy level and within the manufacturing industry in particular—are affected by a disruption in supplies (i.e., the downstream impact), which is often referred to as the ripple effect. Thus, it is necessary that wider systems-level thinking is employed when an establishment along a given supply chain considers planning for resilience.

The remainder of this chapter is organized as follows: Section 2 provides background and context for the discussion of supply chain resilience. Section 3 provides an econometric model specification by which losses due to supply chain disruption are estimated. There are three estimation models presented and analyzed: (1) the downstream ripple effect at the total economy level when there is a disruption in supplies and (2) the same analysis looking at the manufacturing sector in particular. It is found that the effect on the manufacturing sector is relatively greater than at the total economy level. Section 4 provides a summary and describes areas for further research.

2 Background

The National Centers for Environmental Information (NCEI) within the National Oceanic and Atmospheric Administration (NOAA) estimates that there have been 218 natural weather disaster events, each resulting in at least $1 billion in damage and economic losses in the USA from 1970 to October 2017 (American Society of Civil Engineers ASCE, 2014, 2016). An increase in direct capital losses from such events is documented; however, specific effects on supply chains, especially indirect losses from hazards, are less well developed in the literature.

Over the past decade research has focused increasingly on the potential connections between supply chain risk, supply chain disruption, and supply chain vulnerability; however, the vast majority of contributions are anecdotal or case study-based and frequently do not focus on the potential indirect losses, as noted in Ribeiro and

Barbosa-Povoa (2018). In their study of supply chain characteristics relevant to a firm's exposure to supply chain risk, Wagner and Bode (2006) provide definitions for and a thorough discussion of the concepts of supply chain risk, supply chain disruption, and supply chain vulnerability. In this chapter, we follow the definition of supply chain disruption as "an unintended, untoward situation, which leads to supply chain risk" (Wagner & Bode, 2006.) and acknowledge that supply chain disruptions can materialize from supply-side or demand-side risks and both types of risk can be exacerbated by natural and human-made hazard events.

There have been notable changes over the last few decades to supply chain structures and management through globalization and innovation rates—this also indicates a growing dependency between firms and increasing supply chain complexity (e.g., Kamalahmadi & Parast, 2016). There is evidence that increased extreme weather and associated hazard events will make these vulnerabilities more pronounced in the future (e.g., Karl, 2009; Allison et al., 2009; Bouwer, 2019).

Many of the papers that examine economic impacts of hazards consider the upstream impact of supply chain disruption; they measure the effect of changes in demand for parts, components, and other goods/services. For instance, economic input-output analysis is designed to examine downstream effects (Horrowitz & Planting, 2006) as are computable general equilibrium models, as they use input-output data (Blackburn & Moreno-Cruz, 2020). This chapter examines how manufacturers and establishments in other sectors are impacted, at the total economy level by a disruption in supplies (i.e., the downstream impact), which is often referred to as the ripple effect. Currently, there is limited research on the economic impact of the downstream ripple effect. In this chapter, three models are developed to explore supply chain vulnerability to hazard events across geographic areas of the USA with a focus on the manufacturing sector. The results suggest that manufacturers face greater risk due to supply chain disruption relative to business establishments in other sectors.

2.1 *Manufacturing and Supply Chain*

Manufacturing processes are specific to the type of supply chain analyzed and they depend on the parameters of the study case under consideration. A given facility may consist of multiple plants, where each plant is designated for a specific step of the manufacturing process such as packaging, inspection, and other required steps. The facility can also have a storage space/warehouse to keep the inventory. The number of outputs represents the final product of each facility. The distribution of the final product can be divided across market types, e.g., domestic and global.

Firms sometimes scatter responsibility for assets that make up their supply chain opposed to developing a single, coherent vision of their entire supply chain. Typically, the fundamental element that is overlooked is infrastructure, which supports the movement, activities, and processes that occur throughout the supply chain.

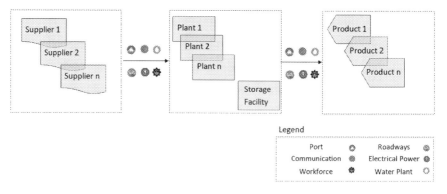

Fig. 1 Infrastructure types that may support a given supply chain

2.2 Infrastructure Components

Multiple components fall under the definition of supply chain infrastructure. The definition contains some physical aspects like the buildings used by manufacturers and distributors, as well as the informational aspects that are essential to run the supply chain (Fallon, 2020). Supply chain infrastructure includes the infrastructure used to move the products across the supply chain components and the other types of infrastructure that are essential to run the facility (see Fig. 1).

Supporting infrastructure is one of the main components of supply chain logistics. That means it is responsible for the movement of materials and information from suppliers to intermediary manufacturers and customers by way of seaports, ports, airports, etc. Infrastructure is also critical to manufacturing operations. The facility requires access to water, electrical power, and communication. Therefore, it is crucial to understand the role of infrastructure in the supply chain and the impact of a potential disruption.

Infrastructure performance under the stress of a disruption will impact the entire supply chain. The performance as a general term considers "the capability of a system to fulfill the need of its functional requirements" (Ayyub, 2014). Infrastructure performance in this formulation is measured through the reliability of the supporting infrastructure.

There are two steps required ahead of supply chain modeling at the institutional level:

1. Construct an inventory of infrastructure assets. There are two constitute steps:

    1.1. Compile information about all facilities, equipment, and information technology applications
    1.2. Characterize each asset in terms of its location and function in the supply chain and take into account its economic life

2. Profile the flow of goods and human services through the supply chain

2.1. Characterize the physical assets in terms of capacity utilization
2.2. Evaluate labor productivity, service levels, and inventory accuracy
2.3. Conduct activity-based costing exercises to understand operating expense

2.3 Failure Modes and Associated Risks

Supply chain resilience is the ability of the supply chain or any of its components to rebound from a setback in the occurrence of disruption (Schmitt & Singh, 2012). Supply chain components must be prepared to face the risk of failure to maintain supply chain performance.

One of the major steps in performing this type of modeling is to develop a strong understanding of how the risk flows across system components (see Fig. 2). In other words, it examines how the failure of one component transfers to other components and eventually affects the output of the supply chain. In this research, the focus is on the risk that threatens the supply chain due to the variation of the reliability of infrastructure and the total failure of the infrastructure responsible for the logistics of materials/goods.

Risk is embedded in the supply chain even under normal conditions due to the variation of the ability of infrastructure to provide the supply chain with services required to maintain its operations. For example, ports and roadways are responsible for the movement of materials; electrical power, water supply, and communication

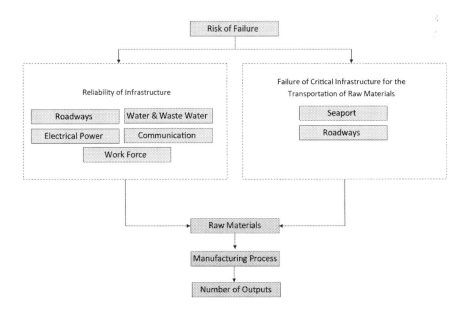

Fig. 2 Risk flow in the supply chain, generalized representation

are susceptible to a disturbance at any time. This variation can affect all the various parts of a system, and it will be reflected in the number of outputs.

3 Estimated Losses Due to Supply Chain Disruption

In a 2019 survey from the Business Continuity Institute, 52% of respondents indicated that they experienced a supply chain disruption within the past 12 months with 49% of them occurring in their tier 1 suppliers (Business Continuity Institute, 2019). Approximately 44% of disruptions were due to unplanned IT or telecommunication outages, making it the largest source; however, weather was the second most common source accounting for 35%. Additionally, the recent coronavirus pandemic has demonstrated how global supply chains can be severely disrupted. It has caused shortages of intermediate goods, delays, and increased transportation costs across many industries (Friesen, 2021).

Businesses struggle to invest in the resilience measures that reduce the risks that disrupt supply chains. There are a number of challenges that contribute to this problem. One is that some of the prevailing approaches for business management involve minimizing inventory. For instance, just-in-time and/or lean production is a dominant theme in manufacturing and both involve reducing inventory. In the short run, these types of approaches are efficient and cost-effective; however, in the long run infrequent events disrupt supply chains, resulting in significant losses to businesses and customers. Businesses are likely to struggle to invest sufficiently in disaster resilience on their own, as they will likely face competition from those that do not invest in resilience. Companies are pressured through competition to either adopt the short-run low-cost approach or possibly be pushed out of the market and out of business. Moreover, in the short run limited investment in resilience is likely to prevail while in the long run business is likely to suffer losses that make investing in resilience cost-effective. Businesses that failed to invest in resilience likely experience greater losses. Thus, underinvestment in resilience may be the result of natural tendencies in the market, resulting in few businesses being sufficiently prepared for natural hazards. There are similar issues in other industries such as banking where there is often the tendency for some to hold as little cash as possible, as this money can be invested to increase profits. Unfortunately, this leaves the bank vulnerable to events that cause people to withdraw their deposits. These types of tendencies make supply chains vulnerable as businesses underinvest in resilience.

Another challenge that contributes to insufficient investment in supply chain resilience is that it is difficult to measure the relative risk. Generally, decision-makers aim to minimize the sum of present value costs and losses from hazards where losses are often calculated as their expected value, which is the loss multiplied by the probability of the loss (Thomas, 2017; Thomas et al., 2017; Thomas & Kandaswamy, 2019; Gilbert et al., 2015). The net present value of a resilience investment, which is used by an estimated 75% of firms for investment analysis, would be calculated as the following:

$$\text{NPV} = \sum_{t=0}^{T} \frac{(I_t - C_t - \text{EL}_t)}{(1+r)^t},$$

where
I_t = Total cash inflow in time period t
C_t = Total cost in time period t
r = Discount rate
t = Time period, which is typically measured in years
EL_t = Expected loss in time period t, which is calculated as the probability P for event e in time t multiplied by the loss L for event e in time t:

$$\text{EL}_t = \sum_{e=1}^{E} P_{t,e} L_{t,e}.$$

The discount rate controls for inflation and the estimated time value of money. Higher net present values tend to be more economical investments. Unfortunately, there is limited data on the probability of hazard events and there is significant uncertainty around these estimates. Resulting losses are not well understood—both ex ante and ex post—and there is a great deal of ambiguity around such estimates, especially as they propagate through a given supply chain. To compound these issues, there is a tendency to underestimate the probability of a disruptive event, where individuals often treat low probability events as though there is no probability. When these losses are occurring in the supply chain and causing disruption, it is difficult to estimate the ramifications, as the resilience of a supplier is often unclear and depends on the resilience of their suppliers. This makes it difficult to identify a robust supply chain. Some investments are also outside of the business's purview, such as the transportation infrastructure and power grid. Thus, there is a misalignment of incentives, as those that make investments in resilient infrastructure (e.g., power companies and governments) do not necessarily experience all the losses when an event occurs.

A critical step in facilitating resilient businesses and infrastructure is understanding the expected loss (i.e., the risk), that is, understanding the combination of probability of an event and the losses that might occur. This section presents an empirical analysis of the impact of natural and human-made hazards on supply chains by modeling the impact of natural and human-made hazards on total GDP, goods GDP, and services GDP. It tests the hypothesis that hazards affect each of these GDP categories and uses a simulation to measure the magnitude of the effect.

3.1 Data

Multiple datasets were used to analyze the impacts of hazards on supply chains. The first is US GDP data on goods, services, and the total economy by county from the Bureau of Economic Analysis (2018). Note that "goods-producing industries consist of agriculture, forestry, fishing, and hunting; mining; construction; and manufacturing." Services-producing industries consist of utilities; wholesale trade; retail trade; transportation and warehousing; information; finance, insurance, real estate, rental, and leasing; professional and business services; educational services, health care, and social assistance; arts, entertainment, recreational, accommodation, and food services; and other services (except public administration)" (Bureau of Economic Analysis, 2006). All dollar figures were adjusted to 2016 values using the consumer price index from the Bureau of Labor Statistics (2018). Data from the Freight Analysis Framework (FAF) accessed through the US Department of Transportation (2018) was also used. The FAF provides shipment data that includes origin and destination data for 122 zones that cover the entire USA. This examination uses 13 categories of commodities classified by the Standard Classification of Transported Goods system. These goods were selected to represent those items that might have low substitutability. Data from the Spatial Hazard Events and Losses Database for the USA or SHELDUS™ (Arizona State University, 2018), which provides information on a range of natural hazards, was used for estimating hazard losses and the number of hazards occurring in each county. This chapter used hazards and perils, as defined in the database. A summary of the data on GDP and hazards is provided in Table 1, and the FAF zones and hazard damage are illustrated in Fig. 3.

3.2 Methods

Three hypotheses regarding supply chain disruption are tested at the total economy level and for manufacturing sectors, specifically.

Table 1 Data summary table

Year	County mean of real goods GDP ($million)	County mean of real services GDP ($million)	County mean of real total GDP ($million)	Mean of property damage by county ($million)	Mean number of hazards by county
2012	997	3631	5272	10.7	5.6
2013	1058	3770	5365	3.1	5.2
2014	1077	3856	5479	1.9	5.0
2015	1097	3954	5637	1.4	4.7

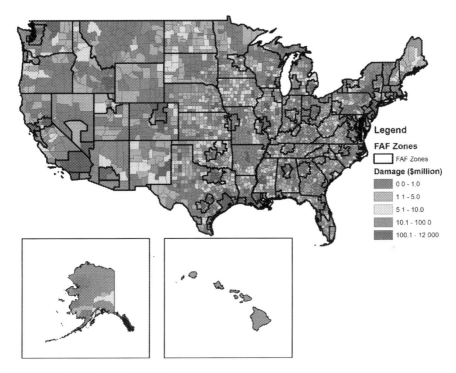

Fig. 3 Hazard damage and FAF zones

1. Hazards in the upstream supply chain have a negative effect on local GDP.
2. Hazards in the upstream supply chain have a negative effect on local goods-producing GDP.
3. Hazards in the upstream supply chain have a negative effect on local service industry GDP.

To address these hypotheses, three models are developed with each one having a different dependent variable: (1) total GDP, (2) goods-producing GDP, and (3) service industry GDP. Note that the local market is not part of the supply chain. The study period is 2013 through 2015. This chapter uses a Cobb–Douglas production function, which has been used by others to examine natural hazard impacts (Mohan et al., 2019). The model includes a 1-year lag of the dependent variable, as the current year is a function of the previous year. Hazards are included both as the number of hazards and the estimated losses. Both of these are included to account for both the positive impacts of a hazard and the negative impacts. Negative impacts are due to damage, while positive ones are due to the increase in economic activity to repair damages and replace total losses (e.g., infrastructure that can no longer function). Hazards include all hazards and perils in SHELDUS, including but not limited to earthquakes, flooding, fog, hail, heat events, hurricanes, tropical storms, landslides, lightning, thunderstorms, tornados, tsunami, volcanos, wildfires, wind events, and winter weather. Two interaction variables are included with one interacting losses with GDP and the other interacting the number of hazards with GDP. These two variables control for the magnitude of potential losses in a particular location.

Two variables account for supply chain effects. The top 24 FAF zones supplying a county's FAF zone were used to represent a county's supply chain, which represents approximately the top 20% of the 122 FAF locations. One variable in the model is an interaction variable between the proportion of the supply chain represented and the damage occurring within that supply chain zone. The total of the interacted variables for the top 24 supply chain zones is, then, summed together. For instance, consider Montgomery County, Maryland. This county falls within the "Washington DC-VA-MD-WV" FAF zone. The supply chain measure for this county is the supply from the largest supplier to that FAF zone, excluding self-supply, divided by the total supplied to the region from all US locations (including FAF region self-supply). The ratio is multiplied by the total hazard damage that occurred in the FAF supplier zone. The top 24 locations are calculated and summed together. This variable, which is an interaction variable, weights the damage occurring at that supply chain zone (i.e., FAF location) by its importance to the destination county. Suppliers are defined as one of the 122 FAF zones. The structural equation for the 3 models is represented as:

$$\ln(GDP_I) = \beta_1 \ln(GDP_{lag,I}) + \beta_2 \ln(SC_{DMG}) + \beta_3 \ln(SC_{CNT}) + \beta_4 \ln(LOC_{DMG}) \\ + \beta_5 \ln(LOC_{CNT}) + \beta_6 IT_{DMG} + \beta_7 IT_{CNT} + \beta_8 \ln(ZERO_{DMG}) \\ + \beta_9 \ln(ZERO_{CNT}) + \beta_{10} + \sum_{s \in S} \beta_s C_s + \varepsilon.$$

GDP_I = County level GDP for industry I where I is services, manufacturing, or total industry

$GDP_{lag,\,I}$ = County level GDP for industry I where I is services, manufacturing, or total industry and where lag indicates a 1 year lag

SC_{DMG} = Estimated damages in SHELDUS for the top 24 supply chain locations

SC_{CNT} = Estimated number of hazards in SHELDUS for the top 24 supply chain locations

LOC_{DMG} = The total damage in the county caused by all hazards and perils listed in the SHELDUS database

LOC_{CNT} = The total number of hazards in the county caused by all hazards and perils listed in the SHELDUS database

IT_{DMG} = An interaction variable between local damage and GDP calculated as

$$\ln(LOC_{DMG}) * \ln(GDP_{lag,I})$$

IT_{CNT} = An interaction variable between the count of hazards and GDP calculated as

$$\ln(\text{LOC}_{\text{CNT}}) * \ln(\text{GDP}_{\text{lag},t})$$

ZERO_{CNT} = Indicator variable for zero hazard incidents locally where ln (ZERO_{CNT}) equals 1 when there are zero hazard incidents and zero otherwise

ZERO_{DMG} = Indicator variable for zero hazard damage locally where ln (ZERO_{DMG}) equals 1 when there are zero hazard incidents and zero otherwise

ε = Error term

β_x = Parameter set to be estimated where x equals 1 through 3016, which includes the $\beta_1, \beta_2, \ldots \beta_{10}$ listed in the equation and the additional 3006 β_s parameters in the summation representing each of the counties in the set of S counties

C_s = An indicator variable for county s, where S is the set of counties

Zero values for LOC_{DMG} and LOC_{CNT} were replaced with 1.0 since the natural log of zero is undefined. To account for this arbitrary change, two indicator variables are included, ZERO_{DMG} and ZERO_{CNT}. This analysis was run in a fixed-effects model.[1]

3.3 Simulation

A simulation was conducted for each of the three models to estimate the impact of damage caused by hazards. A simulation was first run to estimate the impact of hazard damage over the study period. This value is compared to a simulation where no damage occurred in the supply chain. That is, SC_{DMG} is set to zero. This simulates the presence of hazards that cause no damage. The percent change in total GDP is then estimated.

3.4 Results and Discussion

This section examined the impact of supply chain disruption due to hazards on total GDP, goods GDP, and services GDP in the USA using county level data. Table 2 shows results from the regression analysis and shows that the supply chain damage variable (SC_{DMG}) was statistically significant for goods GDP and total GDP at the 0.01 level; however, it was not statistically significant for services GDP. The

---

[1] The Breusch-Pagan and Cook-Weisberg test for heteroskedasticity (Stata, 2013a) was run in three different versions, which indicated that heteroskedasticity was present in the data, meaning that the standard deviations of a predicted variable, monitored over different values of an independent variable or as related to prior time periods, are non-constant. Thus, we fit a fixed-effects model using a "GLS estimator (producing a matrix-weighted average of the between and within results)" to address this issue (Stata, 2013b), which has been shown to provide robust estimates (Hoechle, 2007).

Table 2 Results from regression analysis

Independent variables	Dependent variables		
	GDP_{Goods}	$GDP_{Services}$	GDP_{Total}
$GDP_{lag,x}$	0.304***	0.328***	0.304***
SC_{DMG}	−0.007***	0.00	−0.007***
SC_{CNT}	0.038***	−0.044***	0.038***
LOC_{DMG}	0.005	0.00	0.005
LOC_{CNT}	0.005	−0.003	0.005
IT_{DMG}	0.00	0.00	0.00
IT_{CNT}	−0.001	0.00	−0.001
$ZERO_{DMG}$	0.029**	0.005	0.029**
$ZERO_{CNT}$	−0.016*	−0.003	−0.016*
CON	8.668***	9.077***	8.668***

*Statistically significant at the 0.10 level
**Statistically significant at the 0.05 level
***Statistically significant at the 0.01 level

Table 3 Results from simulation

Dependent variable	Adjusted R^2	Observations	Simulated impact of supply chain damage (2006–2016)		
			Est.	95% Confidence Interval	
GDP_{GOODS}	0.9917	8639	−9.1%	−12.8%	−5.5%
$GDP_{SERVICES}$	0.9976	8639	-	-	-
GDP_{ALL}	0.9982	8982	−2.8%	−4.7%	−0.9%

- Hyphen indicates that the supply chain damage variables were not statistically significant at the 0.1 level

zero-damage variable ($ZERO_{DMG}$) was statistically significant at the 0.05 level for goods GDP and total GDP. Note that the positive value indicates a negative effect from hazards, as this variable indicates zero damage. The zero-count variable ($ZERO_{CNT}$), which was included to control for positive effects from hazards, was statistically significant and negative, which suggests that there are some positive impacts from hazards.

Table 3 shows the results from the simulation along with the adjusted R^2 value and the number of observations. The simulations compare the status quo to a simulated world where no damage occurred from hazards. Hazards in the supply chain had a significant impact on goods GDP with an estimated impact of −9.1% and a 95% confidence interval between −12.8% and −5.5%. The impact of hazards in the supply chain on total GDP was estimated at −2.8% with a 95% confidence interval between −4.7% and −0.9%. Moreover, the impact on goods GDP and total GDP is statistically significant.

The analysis in this section tested three hypotheses:

1. Hazards in the supply chain have a negative effect on local GDP.
2. Hazards in the supply chain have a negative effect on local goods-producing GDP.
3. Hazards in the supply chain have a negative effect on local service industry GDP.

The first one is supported by the statistical significance of the supply chain damage variable (SC_{DMG}) in the model of total GDP. The second one was supported by the statistical significance of the same variable in the goods-producing model. The third hypothesis was not supported, as the supply chain damage variable was not significant in the model of services GDP.

The results suggest that hazards have a greater impact on goods GDP than on services GDP. Recall that the service industry includes utilities, wholesale trade, retail trade, transportation, and warehousing, all of which also depend on or involve physical goods. The lack of statistical significance for the supply chain damage variable (SC_{DMG}) and the zero-damage variable ($ZERO_{DMG}$) suggests that supply chain disruption does not affect these industries as much as goods-producing industries. This makes logical sense, as services can often continue through a disruption. For instance, grocery stores can continue to provide overall service, even if supplies of a few items are interrupted. When stores were unable to keep residential grade toilet paper stocked during the 2020 pandemic, they were still able to maintain most of their operations. Other types of services (e.g., business services, accommodation) have limited reliance on goods. Computer programmers, for instance, do not rely heavily on the physical supplies of goods. Conversely, goods-producing sectors, such as manufacturers, often rely on specific products that have limited substitutability; that is, there are often few substitutes for their current parts and components suppliers.

The simulation results suggest that goods GDP is more readily affected by hazards in the supply chain than is the total economy. Given that the simulation suggests that services GDP is unaffected by hazards in the supply chain, the major contributor to natural hazard effects in the total economy is likely to be goods GDP. The impact on the total economy is approximately 31% of the goods GDP impact, while goods GDP is, similarly, 32% of total GDP. Goods GDP is likely the primary source of impact on total GDP.

The impact of hazards on the supply chain is significant; the simulation results suggest a 9.1% decrease in goods GDP. That is, for goods GDP the expected loss is 9.1% of GDP. As significant as this impact is, hazards are not the only source of supply chain disruption. Recall that IT and telecommunication outages were more frequent; thus, there are additional losses due to supply chain disruption that were not captured in this analysis.

4 Summary and Additional Research

This chapter discusses the impact of natural and human-made hazards in the US economy as a whole and the US manufacturing industry in particular. Many studies that examine economic impacts of hazards consider the upstream impact of supply chain disruption; they measure the effect of changes in demand for parts, components, and other goods/services. This chapter examines how the economy is affected by a disruption in supplies to determine the magnitude of the downstream effect, which is often referred to as the ripple effect. Additionally, the same analysis is conducted looking at the manufacturing sector in particular. A goal is to see whether manufacturers are affected by the ripple effect to a relatively greater extent than is the case for the total US goods economy. Currently, there is limited research on the economic impact of the downstream ripple effect. In Sect. 3 of this chapter, a model is developed to explore supply chain vulnerability to hazard events across geographic areas of the USA. The results suggest that manufacturers face greater risk due to supply chain disruption from hazard relative to business establishments in other sectors.

This chapter provides evidence that the effect of hazards propagating through the supply chain exceeds that of the localized hazard impacts (i.e., direct impact in the geographic location where the hazard took place). This creates a fundamental incentive misalignment. The establishment within the supply chain that invests in mitigation efforts and experiences the hazard often does not directly experience the majority of the resulting net benefit. This, likely, results in a less than optimal level of investment in hazard mitigation.

Thus, it is necessary that wider systems-level thinking is employed when an establishment along a given supply chain considers vulnerability to disruption and undertakes resilience planning.

Business continuity following a natural disaster is an important element of sociotechnical systems that is generally under-researched, especially at the level of individual establishments that constitute part of a larger interactive system of buildings and infrastructure systems. In particular, manufacturing is a key sector in such analysis; according to Thomas (2021), direct and indirect manufacturing accounts for 24.1% of total GDP.. The manufacturing sector directly employs about 15.7 million workers or 9% of total US employment (Thomas, 2021). Additionally, the sector overall supports non-manufacturing jobs up and down the supply chain, from mining to warehousing, as well as engineering, financial, and legal services.

Additionally, this chapter provides evidence that the effect of hazards propagating through the supply chain exceeds that of the localized hazard impacts (i.e., direct impact in the geographic location where the hazard took place). This creates a fundamental incentive misalignment. Some investments are also outside of the business's purview, such as the transportation infrastructure and power grid. Thus, there is a misalignment of incentives, as those that make investments in resilient infrastructure (e.g., power companies and governments) do not experience all the losses when an event occurs.

The findings from this chapter can be used by policy-makers and decision-makers regarding investments in research and in natural hazard mitigation. In terms of research investments, the findings show that the ripple effect from natural hazards has a significant economic impact and that there is a potential market failure regarding investments in risk mitigation due to a misalignment of incentives. In terms of investments in natural hazard mitigation, the findings provide an estimate of the magnitude of impact from natural hazards and the ripple effect. This information can be used to improve investment decisions and demonstrates to individual firm owners and operators the importance of considering resilience along their supply chains, not just at their facility.

References

Allison, E. H., Perry, A. L., Badjeck, M. C., Neil Adger, W., Brown, K., Conway, D., Halls, A. S., Pilling, G. M., Reynolds, J. D., Andrew, N. L., & Dulvey, N. K. (2009). Vulnerability of national economies to the impacts of climate change on fisheries. *Fish and Fisheries, 10*(2), 173–196.

American Society of Civil Engineers (ASCE). (2014). *Seismic evaluation and retrofit of existing buildings*. ASCE.

American Society of Civil Engineers (ASCE). (2016). *Minimum design loads and associated criteria for buildings and other structures*. ASCE.

Arizona State University. (2018). *Spatial hazard events and losses database for the United States*. Retrieved January 28, 2020, from https://cemhs.asu.edu/sheldus

Ayyub, B. (2014). *Risk analysis in engineering and economics*. CRC Press.

Blackburn, C.J. and Moreno-Cruz, J. 2020. *Energy efficiency in general equilibrium with input-output linkages*. BEA Working Paper Series WP2020–1. Retrieved from https://www.bea.gov/system/files/papers/WP2020-1.pdf

Bouwer, L. M. (2019). Observed and projected impacts from extreme weather events: Implications for loss and damage. In R. Mechler, L. Bouwer, T. Schinko, S. Surminski, & J. Linnerooth-Bayer (Eds.), *Loss and damage from climate change. Climate risk management, policy and governance* (pp. 62–83). Springer.

Bureau of Economic Analysis. (2006). *Which industries are included among the goods-producing industries and the services-producing industries?* Retrieved October 10, 2020, from https://www.bea.gov/help/faq/182

Bureau of Economic Analysis. (2018). *Prototype gross domestic product by county, 2012–2015*. Retrieved October 10, 2020, from https://www.bea.gov/news/2018/prototype-gross-domestic-product-county-2012-2015

Bureau of Labor Statistics. (2018). *Consumer Price Index*. Retrieved October 10, 2020, from https://www.bls.gov/cpi/

Business Continuity Institute. (2019). *Supply chain resilience report 2019*. Retrieved from https://www.thebci.org/uploads/assets/e5803f73-e3d5-4d78-9efb2f983f25a64d/BCISupplyChainResilienceReportOctober2019SingleLow1.pdf

Christopher, M., & Holweg, M. (2011). "Supply Chain 2.0": Managing supply chains in the era of turbulence. *International Journal of Physical Distribution & Logistics Management, 41*(1), 63–82. https://doi.org/10.1108/09600031111101439

Department of Transportation. (2018). *Freight analysis framework: Data TABULATION TOOL.* Retrieved June 17, 2020, from https://faf.ornl.gov/fafweb/Extraction0.aspx

Fallon, C. (2020, April 11). *Infrastructure: Supply chain's missing link – Technology – CSCMP's supply chain quarterly.* Retrieved from https://www.supplychainquarterly.com/topics/Technology/20150331-infrastructure-supply-chains-missing-link/

Friesen, G. (2021, September 3). No end in sight for the COVID-led global supply chain disruption. *Forbes.* Retrieved February 22, 2022, from https://www.forbes.com/sites/garthfriesen/2021/09/03/no-end-in-sight-for-the-covid-led-global-supply-chain-disruption/?sh=e6cece53491f

Gilbert, S., Butry, D., Helgeson, J., & Chapman, R. (2015). Community resilience economic decision guide for buildings and infrastructure systems. *Special Publication, 1197.* https://doi.org/10.6028/NIST.SP.1197

Helgeson, J., Fung, J., Henriquez, A. R., Zycherman, A., Nierenberg, C., Butry, D., Ramkissoon, D., & Zhang, Y. (2021, May). Longitudinal study of complex event resilience of small- and medium-sized enterprises: Natural disaster planning and recovery during the COVID-19 pandemic (wave 2). Retrieved from https://www.nist.gov/publications/longitudinal-study-complex-event-resilience-small-and-medium-sized-enterprises-natural

Hoechle, D. (2007). Robust standard errors for panel regressions with cross-sectional dependence. *The Stata Journal, 7*(3), 281–312.

Horrowitz, K. J., & Planting, M. A. (2006). *Concepts and methods of the input-output accounts.* Bureau of Economic Analysis. Retrieved from https://www.bea.gov/sites/default/files/methodologies/IOmanual_092906.pdf

Kamalahmadi, M., & Parast, M. (2016). A review of the literature on the principles of enterprise and supply chain resilience: Major findings and directions for future research. *International Journal of Production Economics, 171,* 116–133.

Karl, T. R. (2009). *Global climate change impacts in the United States.* United States Global Change Research Program, Cambridge University Press.

Mohan, P. S., Spencer, N., & Strobl, E. (2019). Natural hazard-induced disasters and production efficiency: Moving closer to or further from the frontier? *International Journal of Disaster Risk Science, 10,* 166–178. https://doi.org/10.1007/s13753-019-0218-9

Neiger, D., Rotaru, K., & Churilov, L. (2009). Supply chain risk identification with value-focused process engineering. *Journal of Operations Management., 27,* 154–168. https://doi.org/10.1016/j.jom.2007.11.003

NOAA National Centers for Environmental Information (NCEI). (2021). *U.S. billion-dollar weather and climate disasters.* Retrieved from https://www.ncdc.noaa.gov/billions/.

Ribeiro, J. P., & Barbosa-Povoa, A. (2018). Supply chain resilience: Definitions and quantitative modelling approaches – A literature review. *Computers and Industrial Engineering., 115.* https://doi.org/10.1016/j.cie.2017.11.006

Schmitt, A. J., & Singh, M. (2012). A quantitative analysis of disruption risk in a multi-echelon supply chain. *International Journal of Production Economics, 139*(1), 22–32.

Stata. (2013a). *Regress postestimation. Stata user's guide: Release 13.* Retrieved from https://www.stata.com/manuals13/rregresspostestimation.pdf

Stata. (2013b). *Xtreg. Stata user's guide: Release 13.* Retrieved from https://www.stata.com/manuals13/xtxtreg.pdf

SwissRe. (2018). *Preliminary sigma estimates for 2017: Global insured losses of USD 136 billion are third highest on sigma records.* Retrieved from https://www.swissre.com/media/news-releases/2017/nr20171220_sigma_estimates.html.

Thomas, D. (2017). Investment analysis methods: A practitioner's guide to understanding the basic principles for investment decisions in manufacturing. *NIST Advanced Manufacturing Series, 200–5.* https://doi.org/10.6028/NIST.AMS.200-5

Thomas, D. (2021). Annual report on manufacturing industry statistics: 2021. *NIST AMS, 100–42.* https://doi.org/10.6028/NIST.AMS.100-42

Thomas, D., & Kandaswamy, A. (2019). *Economic guide for identifying and evaluating industry research investments: A focus on applied manufacturing research.*

Thomas, D., Butry, D., Gilbert, S., Webb, D., & Fung, J. (2017). The costs and losses of wildfires: A literature survey. *NIST Special Publication, 1215.* https://doi.org/10.6028/NIST.SP.1215

Vlajic, J., van Lokven, S. W. M., Haijema, R., & van der Vorst, J. (2013). Using vulnerability performance indicators to attain food supply chain robustness. *Production Planning & Control, 24*(8/9), 785–799.

Wagner, S. M., & Bode, C. (2006). An empirical investigation into supply chain vulnerability. *Journal of Purchasing & Supply Management, 12*(6), 301–312. https://doi.org/10.6028/NIST.SP.1215

Developing Predictive Risk Analytic Processes in a Rescue Department

Mika Immonen, Jouni Koivuniemi, Heidi Huuskonen, and Jukka Hallikas

Abstract Aging and home care bias in elder care has changed the supply chain environment of rescue departments. The current situation requires data-driven approaches to manage risk and resiliency in such a way that decision-makers can take reactive, proactive, and prescriptive actions. In practice, public service providers should anticipate risks and prevent accidents that arise from the various needs of residents in home environments. The new risk prevention policies should be built on cooperation between the rescue board and the social and healthcare sectors in order to consolidate and process large amounts of data and to develop a foundation for anticipating and managing safety risks in housing. This chapter explores the data structure requirements needed to use the predictive analytics that consolidates information from the logs of the rescue service and social and healthcare agencies as well as the electricity consumption data of residents. In addition, demographic descriptors of the regions should be connected to the process logs. The data sources form a diverse body of data that can be significantly leveraged in three areas of risk management to (1) estimate operation response, (2) create a risk profile of individuals, and (3) understand the chain of events that lead to accidents.

1 Introduction

The chapter aims to assess the role of data sources in managing safety risks in the home environment. From the theoretical perspective, the chapter discusses approaches to manage risk and resiliency in service supply chains and explains the

M. Immonen (✉) · J. Hallikas
School of Business and Management, LUT University, Lappeenranta, Finland
e-mail: Mika.Immonen@lut.fi

J. Koivuniemi
School of Engineering Science, LUT University, Lappeenranta, Finland

H. Huuskonen
South Karelia Rescue Department, Lappeenranta, Finland

© The Author(s), under exclusive license to Springer Nature Switzerland AG 2022
Y. Khojasteh et al. (eds.), *Supply Chain Risk Mitigation*, International Series in Operations Research & Management Science 332,
https://doi.org/10.1007/978-3-031-09183-4_14

vulnerabilities emerging from service chains, networks, and operations. Risk management practices to prevent, detect, respond to, and recover from issues are enhanced by improving risk information sharing in order to increase proactive operations (Christopher & Lee, 2004; Fan et al., 2017). Risk management is based on the identification, assessment, and management of risks. Resilience refers to a system's ability to recognize, anticipate, manage, and recover from disruptions (Francis & Bekera, 2014; Sheffi & Rice, 2005). Predictive resilience is the ability to identify future exposure based on past performance data.

Considering service supply risks, the chapter presents a case study to discuss the data perspective of rescue service management in the future. The results assess the data sources from the perspective of risk prediction and the application of data analysis and model deployment. The rescue service development needs are driven by megatrends, such as the aging population, the increase in the number of dwellers living alone, and the emphasis on decentralized home care plans in the social and healthcare sectors. These environmental changes lead to situations where risks are widely decentralized across households instead of contained in controllable care units. In these circumstances, the risk assessment relies on safety prediction, which emphasizes the importance of information sharing between rescue operations and the social and healthcare sectors to anticipate the chain of events and conditions in diverse home environments that lead to accidents. Furthermore, the safety risk forecasting path provides a framework to connect various types of information from multiple organizations and sources and to consider the timeline of accident prediction.

Using the current case, the chapter describes the process to combine the management of residential safety risks with the identification of the data sources required for predictive analytics. The results increase awareness of the key resources that are needed to predict the chain of events leading to accidents, thereby facilitating the identification of preventive measures for housing risks and improving accuracy of the produced interventions. The appropriate selection of the data interrupts and prevents the risk development by identifying root causes as early as possible. Root causes can be identified by analyzing specific datasets related to hazards, undesirable events, and interactions between the occupant and the operating environment of the households. The data sources within the case study include survey data, event logs from service providers, patient information from electronic health records, and geospatial descriptors. The data sources were reviewed according to their content and availability and in relation to the safety risk forecasting path, the condition of the individual, and the living environment. Generally, the data sources formed a comprehensive and diverse dataset, and the analysis of this data can potentially be used in key areas of risk management to (1) develop the responsiveness of a service network, (2) profile groups and households in terms of risk, and (3) understand the chain of events that may lead to accidents.

2 Predictive Risk Analytics as an Approach

2.1 Predictive Risk Analytics in Supply Chains

Risk management today is often based on the risks and vulnerabilities that emerge from chains and networks as the services of organizations are increasingly viewed as delivery networks or chains. Supply chains refer to the external and internal resources an organization needs for the productive activities of its business. These resources include material and service suppliers, subcontractors, and collaborators (Choi & Wu, 2009). Using the supply chain perspective for the risk management of organizations provides an opportunity to consider chains of sequential operations and processes that often include actors from multiple organizations. Supply chain disruption management is based on understanding the level of vulnerability of the supply chain (Wagner & Neshat, 2012) and the nature of the disruption (Wagner & Bode, 2006).

The ability to respond and recover quickly from disruptions is a key feature of an organization's resiliency (Jüttner & Maklan, 2011). Data analysis can be utilized in operations in descriptive, predictive, and prescriptive ways (Porter & Heppelmann, 2014). Each approach impacts risk management and the resiliency of operations. Based on data analytics, decision-makers can take reactive, proactive, and prescriptive actions to generate resiliency in supply chains. Both proactive and reactive capabilities are important in improving the pre- and post-disaster condition of supply chain resiliency (Chowdhury & Quaddus, 2017). Reactive resilience is the ability of an organization to respond and recover from disruptions (Sheffi & Rice, 2005), but proactive resilience refers to the capabilities that are built into the system, such as flexibility, redundancy, durability, collaboration, and financial stability (Pettit et al., 2013). Consequently, to improve resilience, it should be considered in the design of the supply chain (Christopher & Peck, 2004). Predictive resilience is the ability to predict future exposure based on past performance data, and it allows a user to predict a problem before it occurs and to offer mitigation strategies (Blackhurst et al., 2008). Finally, prescriptive resilience identifies the actions that should be taken to eliminate problems or to adapt to future trends. Figure 1 presents these three concepts of resilience related to risk management and data analytics in the context of the classical life cycle of crisis and recovery (Sheffi & Rice, 2005).

Resilience development requires organizations to orient to disruptions, define resources, and implement risk management infrastructure (Chowdhury & Quaddus, 2017). Resilience can also be enhanced by developing supply chain planning strategies, using collaboration and agility, and creating a risk management culture (Christopher & Peck, 2004). Recent studies have also shown that resiliency is rooted in the structure of the supply chain and the capability to share information throughout the supply chain. Information sharing and integration capabilities should apply to both suppliers and customers, and network structures should be aligned to decrease risks that may potentially harm the overall performance of the supply chain (Ledwoch et al., 2018; Munir et al., 2020). Organizations need processes to support

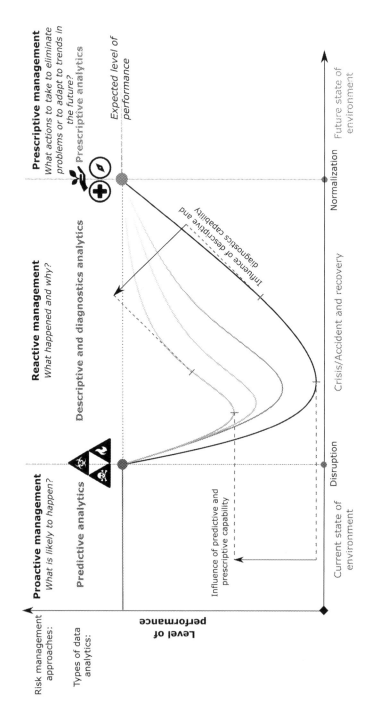

Fig. 1 Risk management and data analytics in relation to the life cycle of a crisis

information management so that big data analytics can be used effectively to anticipate and monitor supply chain disruptions (Gunasekaran et al., 2017). Information capability has a positive effect on resilience, particularly through supply chain visibility (Brandon-Jones et al., 2014). The creation of visibility depends significantly on the development of analytical skills in companies (Srinivasan and Swink, 2018).

2.2 Data Sources for Supply Chain Risk Management

Supply chains produce a large amount of different data from a variety of external and internal sources related to, among other things, transactions, supplier performance, responsibility, risks, and equipment operations. IoT (Internet of Things) and big data platforms generate huge amounts of structured and unstructured data for the stakeholders in the supply chain. These data sources facilitate the use of advanced analytics (Wang et al., 2016), indicating a need to consider approaches to data science in the SCM field. SCM data science is the application of quantitative and qualitative methods that combine various data sources from business environments and utilize the information in decision-making.

Data analytics is needed in supply chains for two reasons. First, there is a need to better describe and predict changes in the supply markets to fulfill the operations requirements. Supply market research refers to the systematic gathering, classification, and analysis of data to consider all relevant factors that influence the procurement of goods and services for the purpose of meeting present and future requirements (Van Weele, 2009). Second, intelligent data analytics is needed to run business operations in the supply chain. Organizations that aim to be successful in a data-driven world must respond quickly to changes in the business environment (Negnevitsky, 2011). The analytics capability helps businesses detect changes in the micro- and macro-environment, thereby leading to the improved fit of competitive strategies (Akter et al., 2016). Competitive strategy fit is the capability to regularly monitor demand trends over time with both numeric and textual data sources (Yang et al., 2019). At the operational level, analytics enables organizations to optimize processes, predict capacity utilization, and experiment with new and competing process constructs (Côrte-Real et al., 2019).

3 Viewpoints into Predicting Safety Risks for Service Supply Development

3.1 Safety Risk Forecasting Path

The production of analytical risk data begins with the definition of a safety risk forecasting path (Fig. 1) as outlined in Heinrich's iceberg model (Heinrich et al., 1980). This model divides the risk accumulation path into three different phases for monitoring and managing situations. During the first phase, latent insecurity occurs among the population due to the functional decline of the aging population. Safety should be maintained during the early steps of the risk path by taking prescriptive actions to recognize and rate the criticality of the prevailing trends of the population. The prescriptive actions may potentially redirect development paths to impede the risk. During the second phase, observable risks and hazards may occur in an individual's daily life that indicate the demand for supportive actions to maintain functional capability. These actions should be proactive, and they should address the sources of the risks and implement measures in the living environment to diminish the effects of the hazards. During the third phase, the realized risks are likely to lead to accidents (e.g., falling), which are signs of imbalance between a person's capabilities and the requirements of the environment. These accidents require reactions to countermeasure the effects of the primary event and minimize the functional decline of the person (Fig. 2).

The safety risk forecasting path forms a time-adjacent framework to depict the different types of data on the timeline in relation to their appearance before the actual accident. The data positioning is based on the contextual understanding of datasets, i.e., their ability to help interpret event information, prevailing operational circumstances, and harmful chains of events. The optimal temporal and content allocation of service development measures significantly impacts cost-effective outcomes in the short term and long term.

The causes and mechanisms of the housing safety risk are often complex, partially due to the multi-level interdependencies of the formation processes of risks. Traditionally, the information needed for accident prediction has been scattered among various actors and data sources, and it has been challenging to create a comprehensive view of the housing safety risks. Accordingly, a multidisciplinary approach to research and development is needed to understand the issues that undermine housing safety and individual well-being and to create additional operating models for increased safety. The safety and well-being of the residents should be at the core of development, and safety needs to be considered using a multidisciplinary and collaborative approach among authorities.

Developing Predictive Risk Analytic Processes in a Rescue Department 317

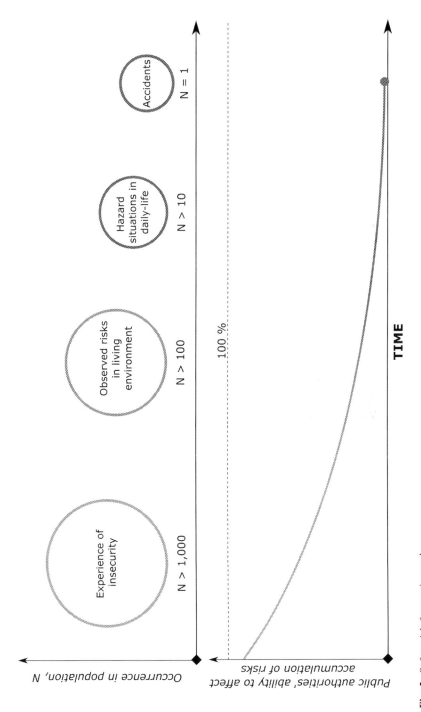

Fig. 2 Safety risk forecasting path

3.2 Trends Influencing Safety Risk Management

The key drivers for the safety risk management in the rescue service supply include demographic information (e.g., age of the residents), residential status (e.g., urbanization, increased number of dwellers living alone), and public service changes that impact the support networks available to the individual. These trends are particularly relevant for the oldest persons living at home. The transformation of the health and social care policies from institutional care to home-centered care for the elderly requires new capabilities of public service providers and individuals living independently at home late in life. As a result of societal changes, the public sector must consider several countermeasures, such as (1) supporting independent living at home with services and technological solutions, (2) ensuring preparedness of homes and districts, particularly in rural areas, and (3) enacting legislation and creating cooperative operating models among public authorities to increase transparency of the various risk profiles of the regions. From the rescue authority's perspective, managing risk is entangled with the challenges of the abilities of the residents to function in their living environments. The reduced ability to live independently at home makes it difficult to manage hazardous situations and to provide services in rural areas because of the distance. Considering the limitations of societal resources, risk management of the housing, as described herein, must occur without duplicate work from various public actors. Therefore, public organizations should review their internal operating models and define a shared knowledge base to enable joint development of responsibilities for risk management.

3.3 Safety Risk Assessment of Individuals

The transition from institutional care to home-centered care changes the safety circumstances for individuals. Further, questions concerning responsibilities for safety must be rebalanced as individual liability is a more significant concern in home environments. Considering regulations related to patient and fire safety, institutionalized care is a more controlled environment, but individual autonomy is a concrete consideration in homes. Several questions are crucial to the analysis of the conditions and motives under which residents are willing to live independently at home. Is it the resident's choice to live in conditions where the level of safety services may be decreased? What are the responsibilities and roles of various public actors and residents themselves for ensuring safety in changed circumstances? How do residents perceive and experience safety and security in their living environments?

The service needs of home care customers are mainly determined by the customer's emotional, cognitive, and physical functioning as impacted by age and diseases. The demand for home care services increases as the customer's functioning decreases. The research explores this concept by analyzing metrics related to instrumental activities of daily living (IADL) and activities of daily living (ADL), both of which are used to measure independent performance in everyday activities. The higher the average level of impairment to ADLs and IADLs, the higher the average coverage and volume of home care services supply. Thus, functional capacities are directly connected to the ability of customers to care for themselves and live independently. Home care customers and resident groups comprise a pivotal risk group for analysis. Home care customers experience an increased risk for home accidents, such as falls.

In the next few decades, the proportion of older people in Finland will increase considerably, and based on the current service structure, the number of home care customers will also increase. An aging population also experiences diseases, such as memory disorders and various musculoskeletal disorders, which further decrease functional ability. This development leads to a growing challenge to ensure fire safety in the homes of elderly people. Age-related changes impact an individual's ability to detect and understand the development of risks and to respond and prepare accordingly. In the future, specific attention should be given to home safety and ways to improve the level of preparedness to respond to reduced functional capacities. The authorities should use an operating model that conveys information about the effects of decreased functioning on safety. The foundation of this model should be a risk assessment based on joint roles. Using this type of model, realistic situational awareness of the home environment can develop.

4 Data Sources for Risk Prediction

4.1 Open Data, Exclusive Data, and Controlled Data

In the analysis of housing safety risks, the aim is to evaluate and describe the usability of various datasets from the perspective of accident prevention. The materials can be roughly divided into three groups (see Fig. 3): (1) exclusively owned (i.e., proprietary), (2) controlled access, and (3) open materials. Table 1 summarizes the datasets and provides the name of the data, a brief content description, the format of data storage, applied primary analysis approaches, and a comment on availability or location. In addition, data sources have been assessed at a more general level in terms of their future usability in Table 2.

The exclusively owned dataset includes survey data collected and analyzed by a research organization. Controlled data sources include event logs from healthcare and rescue services (information about service missions) and healthcare customer registry information. The organizations that provided key information included a national vocational education organization for rescue services, a regional healthcare district, and a local electricity distribution company. Regarding the customer registry data and healthcare event logs, a compulsory research permit was obtained to access the data. The information was anonymized during the research phase by replacing social security numbers of individual citizens with research numbers and replacing

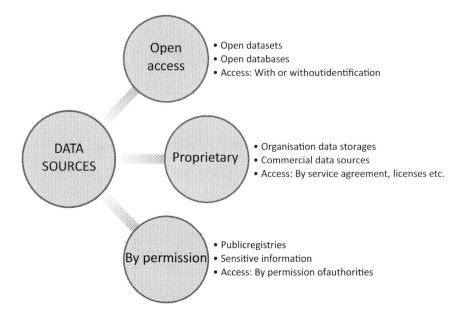

Fig. 3 Types of data sources

specific street addresses of customers' homes with the name of the street or regional identifier. Researchers had no access to original social security numbers in any phase of research. Technically, it was possible to link individual datasets (depending on the dataset) based on an anonymized social security number, spatial data, or time stamps of events.

Open data sources can be divided into two groups: (1) tabular internet distributions and (2) data shared through application programming interfaces (APIs). Tabular data typically includes descriptive regional level information (e.g., population, building stock) represented as summary reports. The APIs provide detailed restrictive information about a specific unit or time of analysis, e.g., weather observations, address coordinates, and traffic information. The information obtained from the APIs over the Internet is in machine-readable form, and its use in the analysis requires further processing. The operation of APIs is based on queries made over the Internet from REST/API servers. A broader utilization of such data is often limited by types of licenses, codes of conduct, or technical capacities.

4.2 Data Sources to Risk Prediction in the Context of Rescue Services

The data can be roughly divided into two parts: (1) data describing the service needs of the population and (2) a spatial dataset describing the area. In this context, the

Table 1 Datasets for analyses

Dataset	Description	Key data content	Analysis approach	Data representative/accessibility
Citizen survey	Structured survey on safety experiences and risk preparedness among 25–75-year-old citizens. $n = 1175$	Perceived quality of life; preparedness for risks; accidents	Quantitative analysis; risk perceptions analysis; risk profiling of households; occurrence of risk profiles in population groups; estimates of risk volumes in population	Research organization (university)/proprietary
PRONTO	Statistics system of Finnish rescue services. $n = 22{,}153$ (number of service events between 2007 and 2016)	Type of mission; time stamp; location	Predictive analytics; modeling and testing emergency service supply scenarios (i.e., service network structure and response times) through neural network analysis	National vocational education organization for rescue services/research permit required
Emergency medical service data	Data on emergency medical service missions (paramedic units). $n = 187{,}755$ (number of emergency missions between 2006 and 2018).	Emergency unit code; type and priority of mission; time stamps, location, social security number[a]	Descriptive analytics for merging emergency medical service mission volumes into regular home care customer risk groups; analysis of mission types	Regional healthcare district/research permit required
Safety helper contacts	Safety helpers are responding to service needs delivered through emergency phone. $n = 69{,}823$ (number of home visits between 2013 and 2017)	Type of mission; time stamps; location; social security number[a]	Descriptive analytics for merging safety helper contact volumes into regular home care customer risk groups	Regional healthcare district/research permit required
HaiPro	Reporting system for safety incidents in healthcare	Written report of incidents or accidents; time stamp	Qualitative text analysis; failure mode and effects analysis (FMEA)	Regional healthcare district/research permit required

(continued)

Table 1 (continued)

Dataset	Description	Key data content	Analysis approach	Data representative/ accessibility
	organizations. n = 1319 (number of reported safety incidents in home environments in 2017)		in terms of reported falls (n = 421) in elderly homes	
Home care contacts	Operations and activity data in regular home care. n = 6,971,634 (number of customer home visits between 2010 and 2017).	Type of service; time stamps; location; social security number[a]	Descriptive analytics to recognize regular home care customer risk groups in terms of contact volumes and aggregate contact time	Regional healthcare district/research permit required
interRAI HC	The interRAI (Resident Assessment Instrument) Home Care Assessment System focuses on the measurement of the person's functioning and quality of life. n = 2300 (number of assessments among regular home care customers within an 8-month period in 2018)	Various measures of person's functioning and quality of life (e.g., IADL); time stamp; location; social security number[a]	Descriptive analytics for merging indicators of functioning (e.g., ADL, IADL) into regular home care customer risk groups	Regional healthcare district/research permit required
Assistive devices	Electricity-dependent assistive devices in use among regular home care customers. n = 3737 (number of assistive devices borrowed by home care customers between 2010 and 2018)	Type of assistive device; time stamps; location; social security number[a]	Descriptive analytics for merging assistive device types and volumes into regular home care customer risk groups	Regional healthcare district/research permit required

(continued)

Table 1 (continued)

Dataset	Description	Key data content	Analysis approach	Data representative/ accessibility
Energy consumption	Measured actual hourly electricity consumption data (kWh) in households. $n = 7$ (number of households among regular home care customers with diagnosed memory disorder)	Electricity consumption (kWh/h); location; social security number[a]	Descriptive analytics for researching changes in circadian rhythms of residents (i.e., analyzing moving average and standard deviation of aggregate hourly energy consumption)	Regional healthcare district; electricity distribution company/ research permit required
Building location	Location data of buildings in Finland. $n = $ app. 60,000 (number of buildings in the case region)	Province; municipality; street address; postal area code; coordinates (WGS84)	Enriching dataset in predictive analytics model for testing emergency service supply scenarios	National population register organization/ open data
PAAVO database	Open data by postal code area	Population structure; buildings and dwellings; workplaces; Main activities of the inhabitants	Enriching dataset in predictive analytics model for testing emergency service supply scenarios	National statistics organization/open data
Grid database	Contains coordinate-based statistical data calculated by map grid.	Includes several data groups, e.g., population structure, size, and stage in life of households	Enriching dataset in predictive analytics model for testing emergency service supply scenarios	National statistics organization/open data (chargeable)
OpenStreetMap (OSM)	OpenStreetMap (OSM) is a collaborative project to create a free editable map of the world	Geo-coding and route machine	Enriching dataset in predictive analytics model for testing emergency service supply scenarios	OpenStreetMap Foundation/ open data

[a]Anonymity of individual citizens was ensured during the research phase by replacing social security numbers with research numbers. Researchers had no access to original social security numbers in any phase of the research

material describing the need for services refers to the safety experience of the population, the number of emergency tasks, and planned customer visits collected from registries. Spatial datasets include open spatial datasets produced by national actors and commercial interfaces from which data regarding the region, population, and urban structure have been compiled. The project's data sources cover the questionnaire data, the authorities' event logs, the healthcare customer registry

Table 2 The role of data sources in relation to the safety risk forecasting path and conditions of residents and living environment

Data source	Individual and environmental conditions		Stage of safety risk forecasting path (see Fig. 1)		
	Condition of living environment	Condition of residents	Experience of insecurity	Perceived risks and hazard situations	Accidents
Citizen survey	Population level estimates for individual or environment risks		Population level estimates for occurrence of risks or insecurity		
PRONTO					Realized risk
Emergency medical service data		Specific mission codes based on individual needs	Psycho-social motives and triggers for emergency contacts (e.g., loneliness)	Risk behavior	Realized risk
Safety helper contacts		Functioning deficiencies; risk behavior	Occurrence of acute need of help and occurrence of hazards		Realized risk
HaiPro	Environmental conditions in homes	Functioning deficiencies; risk behavior		Nearly realized risk	Realized risk
Home care contacts		Breadth and coverage of services	Indicator for actual demand of supportive service and regular monitoring		
interRAI HC	Suitability of living environment	Functioning capacities	Indicator for actual level of housing-related risks		
Assistive devices	Installed base of assistive devices in home environment	Functioning deficiencies; concrete needs for assistance			
Energy consumption	Installed base of electric appliances	Effect of daily activities on energy consumption			
Building location	Structure of the built environment; points of demand				
PAAVO database	Demographics and features of the built environment				
Grid database	Background geometries for spatial models to compute aggregate figures for population				

(continued)

Table 2 (continued)

Data source	Individual and environmental conditions		Stage of safety risk forecasting path (see Fig. 1)		
	Condition of living environment	Condition of residents	Experience of insecurity	Perceived risks and hazard situations	Accidents
OpenStreetMap, address geo-coding	Location data for connecting events to spatial model				
OpenStreetMap, route information	Distance data for service accessibility assessment				

data, and the descriptive information connected to the location data. All registry information was anonymized for security reasons to make it impossible to identify an individual customer. Exact addresses were also removed from the log data, preventing events from being linked to the location with street or neighborhood precision. It was possible to technically link individual datasets (depending on the dataset) based on an anonymized social security number, spatial data, and time stamps. In addition, statistical and regional information was retrieved from open sources across different types of interfaces. The data sources are summarized in Table 1, which provides the name of the data, a brief description, the format of data storage, applied analysis approach, and a comment on availability or location. In addition, a concise assessment of the usability of the data sources is provided in Table 2, which links the datasets with the accident forecasting path and describes which datasets determine the state of the customer and his or her living environment.

The data sources described in Tables 1 and 2 provide rich analytical material that has significant potential to be used for accident prevention and service network development. Based on the analyses that were conducted, the information can be used to develop concurrent service responses among the network of actors, identify risk groups, establish risk profiles, and understand the chain of events behind accidents.

Development of concurrent service response among the network of actors— Service demand forecasting models can be created for rescue services by combining the spatiotemporal variables of PRONTO data, open map and route information, and data on building locations (points of potential demand). A predictive analytics approach using neural network analysis was applied to model and test emergency service scenarios and related response times to anticipate demand locations. In general, the information from forecasting models can be applied to the structural planning of the service network, the decision-making for the location of service sites, and the detailed allocation of resources (types of service units needed).

*Identifying risk groups and establishing risk profiles—*The social security number is the key data field that combines customer-specific information regarding service needs, actual service usage, and functioning abilities. Datasets that were possible to integrate through social security numbers were home care event logs, emergency medical care event logs, safety helper event logs, customers' functional capacities (RAI-HC questions and measures), and information about the usage of assistive devices. In the analysis, data from the home care event log formed the baseline for regular home care customer risk categories using aggregate contact volumes and aggregate contact time. Each risk category was then enhanced by selected metrics related to functioning capacities, emergency medical services data, safety helper data, and usage of assistive devices. The results of the combined multivariate dataset were then used to identify key data-rich risk groups and related anomalies and incidents. Risk group profiling can be used to classify risk areas and define collaborative models of operation and management between rescue services and healthcare organizations (e.g., the evacuation needs of risk groups under storm conditions). Concerning other datasets in the analysis, energy consumption data of households can be used to analyze possible changes in the circadian rhythms of the target residents. Specifically, this information can help identify possible changes about the timing of daily activities, which may signify the need to change services in cases involving memory disorders.

As a supplemental method of risk grouping and profiling, a structured citizen survey was administered to individuals aged 25–75 regarding safety perceptions, risk experiences, and risk preparedness. The statistical analysis divided households into four risk profiles based on their experiences with hazard situations and their need to contact emergency care. Furthermore, the risk occurrence and volume of hazard incidents in the age-based groups were analyzed. The risk profiles helped develop targeted measures for service response.

Understanding the chain of events behind accidents—To implement preventive measures in a cost-effective manner, it is critical to understand the chain of events that lead to accidents (e.g., fires and falls). Accordingly, qualitative textual analysis and failure mode and effects analysis (FMEA) were used to identify the root causes and underlying circumstances related to reported falls in the homes of elderly individuals. The qualitative HaiPro data comprises written reports about risk incidents and accidents in the home environment. Using this analysis, it is possible to determine the factors and circumstances related to the resident and the home environment that potentially contributed to the accidents. Although the occurrence of accidents cannot be completely eliminated, the analysis can be used to find new ways to mitigate the underlying factors, determine the actual root causes, and minimize the negative effects of incidents by facilitating a more effective intervention from the network of the rescue board, social and healthcare authorities, and other relevant actors.

5   Summary and Conclusions

The utilized data sources included the primary data sources from the rescue service board and the social and health operations as well as the material produced in the project. A suitable procedure was developed to analyze the data. Generally, the data sources provided diverse analytical material. Based on the results, this material has the potential to be used significantly in three areas of risk management and service development that include (1) developing a cooperative response with a service network comprised of the rescue board, social and healthcare sectors, and other relevant actors, (2) identifying risk groups and creating risk profiles of occupants, and (3) understanding the chain of events leading to accidents in heterogeneous home environments.

Information sharing and data analytics play a key role in supply chain risk management. This chapter presented a case study of the use of data analytics in risk management by building on the findings of Fan et al. (2017) related to effective risk management practices. These practices include preventing, detecting, responding, and recovering from issues by using capable information systems in the supply chain. As highlighted in the literature, resiliency is an emergent property that relates to the inherent and adaptive capabilities that enable an organization to act proactively to mitigate threats and risks (Burnard & Bhamra, 2011). Regarding our findings, data-driven approaches have potential in building resiliency in supply chains. The contribution of this section to existing literature regarding supply chain risk management is twofold. First, it presents a data-driven risk management case application in the service supply chain. Existing literature has explained the potential use of data analytics in supply chain risk management (e.g., Fan et al., 2017), but there are few actual case studies on the subject. Secondly, the case study highlights the diverse use of different types of structured and unstructured data in proactive risk management. Existing research (e.g., Wang et al., 2016) has explained that the utilization of big data in supply chains is important, but practical examples are still needed for further research.

The results help to identify the chain of events that lead to inadequate safety and that cause accidents to improve understanding of incidents in the home environment and the mutual influence and interdependency of the resident and the operational environment. The information facilitates consideration of risk-based preventive measures to be taken, thereby improving the accuracy of the interventions that are implemented. The goal is to interrupt and prevent risk development at the earliest possible stage by understanding and mitigating the root causes. This requires the concise exchange of information and cooperation between the rescue board, the social and healthcare sectors, and other actors that play a key role in housing safety.

Data management processes and interfaces should be further developed and streamlined to allow a broader and more diverse use of data analysis by individuals in organizations other than the data controllers. Efficient management requires cross-governmental cooperation to create the interface rules and architectures required for entity management. This improves both information security and the usability of the collected data. In the context of risk management and accident prevention in heterogeneous home environments, the data analysis of risks and the root causes provides significant potential to increase risk-specific situational awareness and thus provide more accurate intervention with the cooperation of the rescue board, social and healthcare sectors, and other relevant actors. Consequently, the optimized management of risks in home environments will increase home safety and enhance well-being in general.

References

Akter, S., Wamba, S. F., Gunasekaran, A., Dubey, R., & Childe, S. J. (2016). How to improve firm performance using big data analytics capability and business strategy alignment? *International Journal of Production Economics, 182*, 113–131. https://doi.org/10.1016/j.ijpe.2016.08.018

Blackhurst, J. V., Scheibe, K. P., & Johnson, D. J. (2008). Supplier risk assessment and monitoring for the automotive industry. *International Journal of Physical Distribution & Logistics Management, 38*(2), 143–165.

Brandon-Jones, E., Squire, B., Autry, C. W., & Petersen, K. J. (2014). A contingent resource-based perspective of supply chain resilience and robustness. *Journal of Supply Chain Management, 50*(3), 55–73.

Burnard, K., & Bhamra, R. (2011). Organisational resilience: Development of a conceptual framework for organisational responses. *International Journal of Production Research, 49*(18), 5581.

Choi, T. Y., & Wu, Z. (2009). Taking the leap from dyads to triads: Buyer–supplier relationships in supply networks. *Journal of Purchasing and Supply Management, 15*, 263–266.

Chowdhury, M. M. H., & Quaddus, M. (2017). Supply chain resilience: Conceptualization and scale development using dynamic capability theory. *International Journal of Production Economics, 188*, 185–204.

Christopher, M., & Lee, H. L. (2004). Mitigating supply chain risk through improved confidence. *International Journal of Physical Distribution & Logistics Management, 35*(4), 388–396.

Christopher, M., & Peck, H. (2004). Building the resilient supply chain. *International Journal of Logistics Management, 15*, 1–13.

Côrte-Real, N., Ruivo, P., Oliveira, T., & Popovič, A. (2019). Unlocking the drivers of big data analytics value in firms. *Journal of Business Research, 97*, 160–173. https://doi.org/10.1016/j.jbusres.2018.12.072

Fan, H., Li, G., Sun, H., & Cheng, T. C. E. (2017). An information processing perspective on supply chain risk management: Antecedents, mechanism, and consequences. *International Journal of Production Economics, 185*, 63–75.

Francis, R., & Bekera, B. (2014). A metric and frameworks for resilience analysis of engineered and infrastructure systems. *Reliability Engineering & System Safety, 121*, 90–103.

Gunasekaran, A., Papadopoulos, T., Dubey, R., Fosso Wamba, S., Childe, S. J., Hazen, B., & Akter, S. (2017). Big data and predictive analytics for supply chain and organizational performance. *Journal of Business Research, 70*, 308–317.

Heinrich, H. W., Petersen, D., & Roos, N. (1980). *Industrial accident prevention: A safety management approach* (5th ed.). McGraw-Hill.

Jüttner, U., & Maklan, S. (2011). Supply chain resilience in the global financial crisis: An empirical study. *Supply Chain Management, 16*(4), 246–259.

Ledwoch, A., Yasarcan, H., & Brintrup, A. (2018). The moderating impact of supply network topology on the effectiveness of risk management. *International Journal of Production Economics, 197*, 13–26.

Munir, M., Jajja, M. S. S., Chatha, K. A., & Farooq, S. (2020). Supply chain risk management and operational performance: The enabling role of supply chain integration. *International Journal of Production Economics, 227*, 107667.

Negnevitsky, M. (2011). *Artificial intelligence: A guide to intelligent systems* (3rd ed.). Academic Press.

Pettit, T. J., Croxton, K. L., & Fiksel, J. (2013). Ensuring supply chain resilience: Development and implementation of an assessment tool. *Journal of Business Logistics, 34*, 46–76.

Porter, M. E., & Heppelmann, J. E. (2014). How smart, connected products are transforming competition. *Harvard Business Review, 92*(11), 64–88.

Sheffi, Y., & Rice, J. B. (2005). A supply chain view of the resilient enterprise. *MIT Sloan Management Review, 47*, 41–48.

Srinivasan, R., & Swink, M. (2018). An investigation of visibility and flexibility as complements to supply chain analytics: An organizational information processing theory perspective. *Production and Operations Management, 27*(10), 1849–1867.

Van Weele, A. J. (2009). *Purchasing and supply chain management: Analysis, strategy, planning and practice*. Cengage Learning EMEA.

Wagner, S. M., & Bode, C. (2006). An empirical investigation into supply chain vulnerability. *Journal of Purchasing and Supply Management, 12*, 301–312.

Wagner, S. M., & Neshat, N. (2012). A comparison of supply chain vulnerability indices for different categories of firms. *International Journal of Production Economics, 50*, 2877–2891.

Wang, G., Gunasekaran, A., Ngai, E. W., & Papadopoulos, T. (2016). Big data analytics in logistics and supply chain management: Certain investigations for research and applications. *International Journal of Production Economics, 176*, 98–110.

Yang, Y., See-To, E. W. K., & Papagiannidis, S. (2019). You have not been archiving emails for no reason! Using big data analytics to cluster B2B interest in products and services and link clusters to financial performance. *Industrial Marketing Management, 86*, 1–14. https://doi.org/10.1016/j.indmarman.2019.01.016

A Manufacturing-Supply Chain Risk Under Tariffs Impact in a Local Market

Omar Alhawari and Gürsel Süer

Abstract Nowadays, supply chains experience the risk of the impact of trade tariffs imposed by governments on the imported items. Basically, higher tariffs or import taxes affect global trade through amplifying the risk on supply chain costs which in turn jeopardizes the supply chain networks. This chapter considers a manufacturing company, assumed to be in the USA, with its manufacturing system that is inspired by a jewelry manufacturing industry. The manufacturer produces finished products and procures raw materials required from local and global suppliers that are assumed to be in the USA and China, respectively. In a supply chain risk, considering the global economy competition, a local government needs to improve a local industry; therefore, it imposes tariffs on the imported raw materials. This policy helps local suppliers compete with global suppliers who are negatively impacted by higher imposed tariffs. Additionally, a manufacturer as a buyer is impacted by imposed tariffs. The main objective of this chapter is to study the impact of tariffs on the manufacturing-supply chain considered. Thus, different tariff rates are imposed on imported raw materials. As the tariffs increase, the purchased materials costs increase as well. As a result, the purchase of raw materials is shifted from global to local suppliers when high tariff rates are imposed. Further, this chapter shows that the total products costs including raw materials, labor, and overhead costs are increased, which eventually affect the selling prices of finished products. Consequently, the demand values of the finished products, determined by demand function, are dropped due to the increase of selling prices. This leads to impact the profits, generated by the manufacturing company. It is observed that the profits of products act differently in which some products generate more profits and others lose profits when the tariffs increase. Moreover, in this chapter, the impact of the trade tariffs on the manufacturing system design is studied considering the number of

O. Alhawari (✉)
College of Business, Ohio University, Athens, OH, USA
e-mail: alhawari@ohio.edu

G. Süer
Department of Industrial and Systems Engineering, Ohio University, Athens, OH, USA

manufacturing cells open to meet the demand. Results show that demand impacts the manufacturing system design in which the number of cells decreases as the demand decreases. It is concluded that although the market share has been impacted in terms of the demand, the manufacturer sustains the business by making satisfactory profits.

1 Supply Chains: Management and Risk

Business organizations pay a great deal of attention to their supply chains to survive in the competitive local and global environments. Stevenson (2019) stated that organizations including their facilities, functions, and activities work together to build a sequence that begins with raw materials supply and extend all the way to final customer. Facilities such as suppliers, plants, warehouses, distribution centers, and retailers are linked together as a chain to help achieve customer requirements. The sequence implies that different nodes act as suppliers and customers where each node is a customer of the preceding one and a supplier to the proceeding one. Generally, a supplier represents the point of origin, and a consumer is considered as the point of consumption. Figure 1 shows the typical manufacturing-based supply chain network. In supply chains, the manufacturer interacts directly with suppliers of tier-1 type. The suppliers can be either local or global supplier. Based on the demand information, manufacturers procure the required raw materials from suppliers. The raw materials are converted into finished products through the manufacturer's operations and eventually delivered to consumers who may exist in local or global market.

Based on Fig. 1, it is noticeable that suppliers are strategically important in the functioning of supply chain. Thus, the issue of supplier selection and evaluation has been very crucial for the success of the entire supply chain. Companies pay attention to raw material costs due to their contribution to the final product cost. Burton (1988)

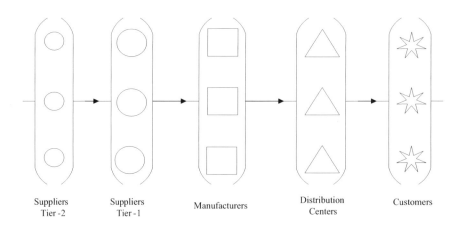

Fig. 1 A manufacturing-based supply chain network

mentioned that the procurement cost of raw materials required by US companies is about 40–60% of the unit cost of a product. Ávila et al. (2012) indicated that supplier selection process consists of three phases: qualification, supplier selection, and evaluation. Süer and Huang (2012) discussed the supplier problems and proposed a long-term sourcing framework that evaluates the supplier dynamics with multiple criteria such as cost, quality, and delivery. Bhutta and Huq (2002) stated that evaluating suppliers effectively requires considering qualitative and quantitative criteria. Patton (1996) addressed some supplier evaluation criteria such as the purchase price, quality, and the delivery time of the materials required. Further, the author mentioned that sales support and technology rate are factors considered for supplier's evaluation. From a manufacturer's perspective as a buyer, material costs and delivery times should be minimized, whereas the quality should be maximized to gain competitive advantage in the marketplace.

The strategic coordination between organizations involved in supply chain is required to integrate supply and demand management effectively and efficiently. According to Stevenson (2019), the need for managing supply chains has been essential for organizations to consider important issues such as:

- The need to improve their operations
- Increasing their level of outsourcing
- Increasing transportation costs
- The awareness of competitive pressure
- The need for increasing globalization
- The need for managing inventories
- The dynamic nature of supply chain
- Advance in IT and e-business environment

Supply chain management (SCM) is defined as managing the network of facilities that performs a sequence of functions starting from procurement of raw materials to production of intermediate and finished products and distribution to customers (Ganeshan & Harrison, 1995). Also, SCM is the integration of activities by facilities who are responsible for procuring raw materials, transforming them into finished products, and delivering them to customers utilizing distribution systems (Lee & Billington, 1995).

Supply chains experience trade tariffs, stochastic demand, seasonal demand, production yield variability, inventory level variability, capacity shortages, defective products, and inspection errors. In supply chains, uncertainty and risk are used interchangeably; however, there is a slight difference in which risk is defined as the probability of occurrence of a defined magnitude of the occurrence (Zsidisin, 2003). The probability is known for the risk, but unknown for the uncertainty. For instance, a supplier risk is the probability that an incident associated with inbound supply from a supplier occurs where its outcomes result in not meeting customer requirement by the purchasing company (Siegel, 2005). In the literature, uncertainties in supply, process, and demand influence the manufacturing function (Wilding, 1998). Also, uncertainties have an impact on decision-makers resulting in ineffectiveness and inefficiency if not handled properly which in return affects the organizational performance (Van Der Vorst & Beulens, 2002).

2 Trade Tariffs in Supply Chains

In global supply chains, the global trade benefits from the use of outsourcing and exploiting opportunities beyond the local market. According to Stevenson (2019), some of the drivers for global supply chain are considered as follows:

- The product design requires inputs (e.g., raw materials) from other countries
- Expand the market share by selling products globally
- Outsource work to countries where labor and material costs are lower

Figure 2 shows the local supply and demand curves in a country, where (P^*) and (Q^*) show the equilibrium status without global trade. However, when this country is open to global trade, (P_{global}) is a given global price in a competitive global market. According to Fig. 2, the local suppliers supply their market with a quantity called (Q_{LS}) which is less than the original quantity supplied (Q^*), whereas the quantity ordered by the local consumer is (Q_{LC}) which is greater than the original quantity ordered (Q^*). For that, the quantity imported to the country is the difference between the local supplied quantity and the local ordered quantity (i.e., quantity imported (Q_I) = Q_{LC} - Q_{LS}). This scenario harms the local suppliers in the global competitive market. Consequently, the local suppliers approach their government to

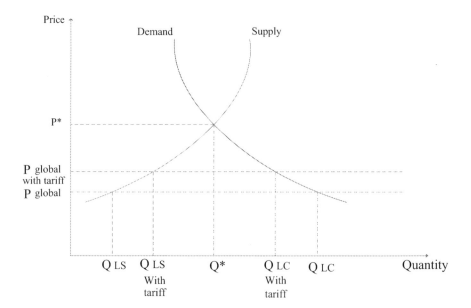

Fig. 2 Supply and demand curves considering tariffs

find solutions for their losses in global trade. The local suppliers cannot afford to produce and sell with the same prices as global suppliers; subsequently, their revenues are lost, and their business might be shut down. To react to this situation, governments usually impose a tariff rate for every unit entering the country to increase the amount supplied by local supplier and decrease the amount imported. A tariff is a tax imposed on imports or exports between countries and the new global price with tariffs is denoted by ($P_{\text{global with tariff}}$). For that, the quantity imported to the country is decreased with tariffs consideration (i.e., quantity imported with tariff (Q_1) = $Q_{\text{LC with tariff}} - Q_{\text{LS with tariff}}$). The trade position of a country, in the case of tariffs, will be different in terms of that the local supply is improved, the quantity ordered globally is decreased, and a government is helped financially by the tariffs revenues.

Nowadays, rival companies advantage from the tariffs imposed (Ungureanu, 2019). Rodrik (2018) indicated that large countries have the power to control their market demand by imposing tariffs on the imported items; therefore, the market demand for imported items is reduced. This influences the world price of the imported items. Further, in this regard, each country imposes their own optimal tariffs. However, the word trade agreements such as the old General Agreement on Tariffs and Trade (GATT) and the World Trade Organization (WTO) include negotiations to minimize tariffs imposed.

3 System Studied

The system studied, in this chapter, considers data used by Alhawari et al. (2021). It includes a family that is composed of 12 similar products with their processing requirements. It is inspired by jewelry manufacturing industries that are mostly labor-intensive systems associated with high operator involvement. Processing time is considered as the time needed to process a unit of a product by a machine and considered as the bottleneck processing time. This product family requires opening five manufacturing cells to meet the expected demand. Each cell includes 13 machines where each machine needs one operator. A cell has a capacity of 40 h/week considering that an operator works for 8 h/day (1-shift consideration) and 5 days per week (i.e., 8 h/day * 5 days/week = 40 h/week). Table 1 shows the expected demand and processing times (in minutes) of the products in the family.

In the manufacturing system, the capacity allocated to meet the expected demand for each product, as shown in Table 1, is determined by parameters of both expected demand and processing time (in minutes) of each product. For example, given the expected demand in units for product 1 (P1) is 613 units and the processing time in hours per unit is 0.025, the capacity allocated for P1 is 15.33 h and computed as follows:

Table 1 Expected demand, processing times, and capacities allocated to the products in family (Alhawari et al., 2021)

Product i	P1	P2	P3	P4	P5	P6	P7	P8	P9	P10	P11	P12
Processing time (bottleneck)	1.5	2.0	1.7	1.6	1.8	1.9	1.4	1.3	0.9	1.2	0.8	1.0
Expected demand	613	337	311	616	166	110	279	706	366	301	495	298
Capacity allocated in hours	15.33	11.23	8.81	16.43	4.98	3.48	6.51	15.30	5.49	6.02	14.85	4.97

Fig. 3 The structure of product 1

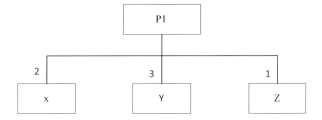

Capacity allocated for product 1 = 613 units * 0.025 hours/unit = 15.33 hours

Each product requires raw materials to be processed and turned into finished product. In this study, it is assumed that there are three types of materials needed and they are X, Y, and Z. For example, product 1 (P1) requires 2 units of material X, 3 units of material Y, and 1 unit of material Z as shown in Fig. 3.

4 Total Product Cost

The product cost consists of direct and indirect costs. Direct costs are the major portion, and they are raw materials and labor costs, whereas the indirect manufacturing costs are incurred when a product is manufactured such as electricity, administrators, and equipment depreciation. In this chapter, the total product cost includes materials, labor, and overhead costs.

4.1 Raw Materials Procurement Cost

It is assumed, in this chapter, that some of the suppliers of the raw materials are in North America (USA; suppliers 4 and 5) and others are in China (suppliers 1, 2, and 3). The procurement costs of raw materials including purchase price and transportation cost are shown in Table 2. Raw materials of X, Y, and Z are supplied by Chinese suppliers due to the lowest procurement cost. The minimum procurement costs for materials X, Y and Z are $1.80, $2.30, and $2.10, respectively.

According to the product structure, the materials costs of each product in the family are provided in Table 3.

4.2 Product Labor Cost

The total labor cost of a manufacturing system is determined based on Eq. (1). In this chapter, the approximate average of hourly wages (Pay Rate) in US manufacturing is

Table 2 Procurement costs of materials

Location	Supplier	Material type: X			Material type: Y			Material type: Z		
		Purchase price/unit	Transportation cost/unit	Procurement cost/unit	Purchase price/unit	Transportation cost/unit	Procurement cost/unit	Purchase price/unit	Transportation cost/unit	Procurement cost/unit
Asia (China)	Supplier 1	$1.80	$0.30	$2.10	$2.00	$0.30	**$2.30**	$1.90	$0.30	$2.20
	Supplier 2	$1.90	$0.30	$2.20	$2.10	$0.30	$2.40	$1.80	$0.30	**$2.10**
	Supplier 3	$1.50	$0.30	**$1.80**	$2.20	$0.30	$2.50	$2.00	$0.30	$2.30
North America (USA)	Supplier 4	$2.70	$0.10	$2.80	$3.40	$0.10	$3.50	$3.20	$0.10	$3.30
	Supplier 5	$3.00	$0.10	$3.10	$3.50	$0.10	$3.60	$2.80	$0.10	$2.90

Bold values are the minimum procurement costs for each material type

Table 3 Total materials costs

Product i	Raw materials required			Raw material cost			Total raw material costs
	X	Y	Z	X	Y	Z	
P1	2	3	1	$3.60	$6.90	$2.10	$12.60
P2	3	2	1	$5.40	$4.60	$2.10	$12.10
P3	1	2	2	$1.80	$4.60	$4.20	$10.60
P4	3	2	1	$5.40	$4.60	$2.10	$12.10
P5	1	2	2	$1.80	$4.60	$4.20	$10.60
P6	2	1	2	$3.60	$2.30	$4.20	$10.10
P7	2	2	1	$3.60	$4.60	$2.10	$10.30
P8	3	1	1	$5.40	$2.30	$2.10	$9.80
P9	1	2	2	$1.80	$4.60	$4.20	$10.60
P10	2	2	1	$3.60	$4.60	$2.10	$10.30
P11	2	1	2	$3.60	$2.30	$4.20	$10.10
P12	1	1	3	$1.80	$2.30	$6.30	$10.40

considered to be $21.37. This average hourly wage value was also used by Alhawari et al. (2021), in their published work.

$$\text{Total labor cost of the system} = \text{Number of operators} * \text{Pay Rate} * \text{Available Capacity}. \quad (1)$$

As the system studied includes five manufacturing cells, the total weekly capacity available is 200 h given that each cell is available for 40 h/week. Additionally, there are 13 operators in each cell. The total labor cost of the system is computed as follows:

$$\text{Total labor cost of the system} = 13 * \frac{\$21.37}{\text{hour}} * 200 \text{ hours} = \$55,562.$$

Having determined the total labor cost of the system, the labor cost per unit of a product is determined by Eq. (2).

$$\text{Labor Cost}_i/\text{unit} = \left(\frac{\text{Capacity allocated } i}{\sum_{i=1}^{12} \text{Capacity allocated } i} \right) * \frac{\text{Total labor cost of the system}}{\text{Expected demand}_i}. \quad (2)$$

For example, according to the expected demand and capacity allocated of products as shown in Table 1, the labor cost per unit of product 1 is determined as follows:

$$\text{Labor Cost}_{\text{product 1}}/\text{unit} = \left(\frac{15.33}{113.39}\right) * \frac{\$55,562}{613} = \$12.25.$$

Additionally, in this chapter, the overhead cost is assumed to be 5% of the total direct costs for labor and materials. The total product cost (TC$_i$) is determined by summing up the raw materials, labor, and overhead costs as shown in Table 4.

5 The Selling Prices of Products

The selling price (SP$_i$) per unit of product i consists of two terms: profit margin per unit (PROFIT MARGINi) and total cost per unit (TC$_i$). It is determined based on Eq. (3).

$$SP_i = TC_i\,(1 + \text{PROFIT MARGIN}_i). \tag{3}$$

In this chapter, the profit margin per unit of each product is assumed to be 20% of the total cost. The selling prices of products are provided in Table 5.

6 Tariffs Impacts

6.1 Tariffs Impact on Raw Materials Imported from Global Suppliers

As local governments impose trade tariffs on the imported items, local industry and local suppliers compete with global suppliers. In this section, the local government imposes a tariff rate on the imported raw materials from the global suppliers in China. Different tariff rates are assumed on the purchase price of raw materials, and they are 25%, 50%, and 75%.

6.1.1 A Tariff Rate of 25% on the Materials Purchase Price

The global suppliers are impacted by tariffs imposed on their shipped raw materials; however, the local suppliers are not. Considering the tariff rate of 25% on the purchase price, a new cost term is added to the procurement costs of the raw material per unit. The procurement cost after the tariff rate is computed according to Eq. (4).

$$\text{Procurement cost} = \text{Purchase price} + \text{Tariff cost} + \text{Transportation cost.} \tag{4}$$

Table 4 Total product cost per unit

Product i	P1	P2	P3	P4	P5	P6	P7	P8	P9	P10	P11	P12
Raw material cost	$12.60	$12.10	$10.60	$12.10	$10.60	$15.52	$11.43	$10.62	$7.35	$9.80	$14.70	$8.17
Labor cost	$12.25	$16.33	$13.88	$13.07	$14.70	$1.28	$1.09	$1.02	$0.90	$1.01	$1.24	$0.93
	$1.24	$1.42	$1.22	$1.26	$1.27							
Total product cost TC_i	$26.09	$29.86	$25.71	$26.43	$26.57	$26.90	$22.82	$21.44	$18.85	$21.11	$26.04	$19.50

Table 5 The selling price per unit of products

Product i	P1	P2	P3	P4	P5	P6	P7	P8	P9	P10	P11	P12
SP_{i1}	$31.31	$35.83	$30.85	$31.71	$31.88	$32.28	$27.38	$25.73	$22.62	$25.33	$31.25	$23.39

Table 6 shows the procurement costs of the materials when a tariff rate of 25% of the purchase price is added. Despite the increased costs of material X due to the additional tariffs of $0.45 per unit, it is still imported from global supplier 3 at a cost of $2.18 per unit. As for material Y, it is still shipped or imported from the global supplier 1 at total procurement costs of $2.80 per unit. Material type Z is shipped from supplier 2 with procurement costs of $2.53 per unit. It is concluded that imposing a tariff rate of 25% on the purchase price of raw materials X, Y, and Z increases the prices but does not lead to any change to global suppliers.

6.1.2 A Tariff Rate of 50% on the Materials Purchase Price

The government strategy is to increase the tariff rate to reach a point where the local supplier can benefit. In this regard, the government increases the tariff rate to 50% on the purchase price. By doing so, the total procurement cost of raw material X is increased, but the global supplier 3 still provides this material at a cost of $2.55 per unit as the lowest costs among all local and global suppliers. With the increase in the tariff rate, the procurement costs of raw material Y go up to the point that local supplier 4 offers $3.50, which is a better price than global supplier 3 does. However, global supplier 1 provides the lowest cost at $3.30; hence, this material is still imported globally at a cost of $3.30. The trade tariff rate of 50% makes raw material Z to be switched from being globally to locally shipped. Local supplier 5 supplies a unit of material Z at a cost of $2.90. This is very critical to global supply chains as the global suppliers suffer by losing portion of their market share. It can be concluded that the impact of the trade tariff rate of 50% on purchase price has been clear when the supply of materials Z switched from global suppliers to the local suppliers. However, other raw materials X and Y are still imported from global suppliers as shown in Table 7.

6.1.3 A Tariff Rate of 75% on the Materials Purchase Price

In this section, the government's trade policy is to support the domestic industry to compete on all the types of raw materials. The government increases the trade tariff rate to 75% on all imported materials. As the import costs go up, the local producers increase their market share. Considering this scenario, raw material X is produced and sold locally by supplier 4 for $2.80. None of the global suppliers can compete with the local price. However, global supplier 3 can provide a unit of material Z for $2.87, which is considered close to the local supplier as shown in Table 8. Material type Y is also supplied locally instead of globally. Local supplier 4 provides material Y to the company at a cost of $3.50 as the minimum cost available. Local supplier 5 is selected due to the minimum cost offered of $2.90 per unit of raw material Z. The costs of materials Y and Z are increased for all the global suppliers. If global suppliers still want to be competitive, they need to lower the purchase price to keep it the lowest even after a tariff is imposed. For instance, this case happens when global

Table 6 Procurement costs of the materials considering a tariff of 25% on purchase price

Location	Supplier	Material type: X				Material type: Y				Material type: Z			
		Purchase price/unit	Tariff cost	Transportation cost/unit	Procurement cost/unit	Purchase price/unit	Tariff cost	Transportation cost/unit	Procurement cost/unit	Purchase price/unit	Tariff cost	Transportation cost/unit	Procurement cost/unit
Asia (China)	Supplier 1	$1.80	$0.45	$0.30	$2.55	$2.00	$0.50	$0.30	$2.80	$1.90	$0.48	$0.30	$2.68
	Supplier 2	$1.90	$0.48	$0.30	$2.68	$2.10	$0.53	$0.30	$2.93	$1.80	$0.45	$0.30	$2.53
	Supplier 3	$1.50	$0.38	$0.30	$2.18	$2.20	$0.55	$0.30	$3.05	$2.00	$0.50	$0.30	$2.80
North America (USA)	Supplier 4	$2.70	$0.00	$0.10	$2.80	$3.40	$0.00	$0.10	$3.50	$3.20	$0.00	$0.10	$3.30
	Supplier 5	$3.00	$0.00	$0.10	$3.10	$3.50	$0.00	$0.10	$3.60	$2.80	$0.00	$0.10	$2.90

Table 7 Procurement costs of the materials considering a tariff of 50% on purchase price

Location	Supplier	Material type: X				Material type: Y				Material type: Z			
		Purchase price/unit	Tariff cost	Transportation cost/unit	Procurement cost/unit	Purchase price/unit	Tariff cost	Transportation cost/unit	Procurement cost/unit	Purchase price/unit	Tariff cost	Transportation cost/unit	Procurement cost/unit
Asia (China)	Supplier 1	$1.80	$0.90	$0.30	$3.00	$2.00	$1.00	$0.30	$3.30	$1.90	$0.95	$0.30	$3.15
	Supplier 2	$1.90	$0.95	$0.30	$3.15	$2.10	$1.05	$0.30	$3.45	$1.80	$0.89	$0.30	$2.97
	Supplier 3	$1.50	$0.75	$0.30	$2.55	$2.20	$1.10	$0.30	$3.60	$2.00	$1.00	$0.30	$3.30
North America (USA)	Supplier 4	$2.70	$0.00	$0.10	$2.80	$3.40	$0.00	$0.10	$3.50	$3.20	$0.00	$0.10	$3.30
	Supplier 5	$3.00	$0.00	$0.10	$3.10	$3.50	$0.00	$0.10	$3.60	$2.80	$0.00	$0.10	$2.90

Table 8 Procurement costs of the materials considering a tariff of 75% on purchase price

Location	Supplier	Material type: X				Material type: Y				Material type: Z			
		Purchase price/unit	Tariff cost	Transportation cost/unit	Procurement cost/unit	Purchase price/unit	Tariff cost	Transportation cost/unit	Procurement cost/unit	Purchase price/unit	Tariff cost	Transportation cost/unit	Procurement cost/unit
Asia (China)	Supplier 1	$1.80	$1.35	$0.30	$3.45	$2.00	$1.50	$0.30	$3.80	$1.90	$1.43	$0.30	$3.63
	Supplier 2	$1.90	$1.43	$0.30	$3.63	$2.10	$1.58	$0.30	$3.98	$1.80	$1.34	$0.30	$3.42
	Supplier 3	$1.50	$1.35	$0.30	$2.87	$2.20	$1.65	$0.30	$4.15	$2.00	$1.50	$0.30	$3.80
North America (USA)	Supplier 4	$2.70	$0.00	$0.10	$2.80	$3.40	$0.00	$0.10	$3.50	$3.20	$0.00	$0.10	$3.30
	Supplier 5	$3.00	$0.00	$0.10	$3.10	$3.50	$0.00	$0.10	$3.60	$2.80	$0.00	$0.10	$2.90

Table 9 Procurement costs when purchase price offered by supplier 2 is reduced (75% tariff rate)

Suppliers	Material type: Z			
	Purchase price/ unit	Tariff cost	Transportation cost/ unit	Total procurement cost/ unit
Supplier 1	$1.90	$1.43	$0.30	$3.63
Supplier 2	$1.45	$1.09	$0.30	$2.84
Supplier 3	$2.00	$1.50	$0.30	$3.80
Supplier 4	$3.20	$0.00	$0.10	$3.30
Supplier 5	$2.80	$0.00	$0.10	$2.90

supplier 2 reduces the purchase price of raw material Z from $1.80 to $1.45. According to that, supplier 2 provides this material at a cost of $2.84 as the lowest option available among all suppliers, as shown in Table 9.

6.2 Impact of Tariffs on Total Cost of Finished Product

In this section, the impact of the imposed tariffs on the total cost of finished product is discussed. According to the product structure of the products shown in Table 3, the cost of the finished product is impacted due to the increased cost of raw materials required X, Y, and Z. The import tax rates of 25%, 50%, and 75% impact the costs of the raw materials imported from global suppliers as shown in Table 10. For example, the trade tariff rate of 25% on the purchase price increases the costs of the raw materials X, Y, and Z to reach $4.36, $8.40, and $2.53, respectively. The total materials cost of product 1 (P1) is $15.29 and that is the summation of materials X, Y, and Z. Additionally, the total materials cost for a unit of P1 is $17.82 when a tariff rate of 50% is imposed; however, the cost reaches $19.00 when a tariff rate of 75% is imposed.

Having determined the impact of tariff rates on material costs of finished products, the total cost of the finished product is impacted as well. The total cost of a product includes raw materials, labor, and overhead costs. Table 11 shows the total costs of finished product considering non-import and import tax considerations. The total costs increase as the trade tariff rate increases. For example, the total costs per unit of P1 are $26.09, $28.92, $31.57, and $32.81 when the trade tariff rates are 0% (without tariff consideration), 25%, 50%, and 75%, respectively.

Based on the change in the total costs due to tariffs imposed, the company cannot maintain the original selling price of each product. The numbers in **bold** and *italic* in Table 11 refer to cases where the company loses money when the original selling price is kept unchanged. For instance, the original selling price of P1 is $31.31 based

Table 10 Material costs of products considering the imposed trade tariff rates of 25%, 50%, and 75%

Product	Raw material cost considering trade tariff rate of 25% on purchase price			Total materials costs/ unit (25% tariff)	Raw material cost considering trade tariff rate of 50% on purchase price			Total materials costs/ unit (50% tariff)	Raw material cost considering trade tariff rate of 75% on purchase price			Total materials costs/ unit (75% tariff)
	X	Y	Z		X	Y	Z		X	Y	Z	
P1	$4.36	$8.40	$2.53	$15.29	$5.02	$9.90	$2.90	$17.82	$5.60	$10.50	$2.90	$19.00
P2	$6.54	$5.60	$2.53	$14.67	$7.53	$6.60	$2.90	$17.03	$8.40	$7.00	$2.90	$18.30
P3	$2.18	$5.60	$5.06	$12.84	$2.51	$6.60	$5.80	$14.91	$2.80	$7.00	$5.80	$15.60
P4	$6.54	$5.60	$2.53	$14.67	$7.53	$6.60	$2.90	$17.03	$8.40	$7.00	$2.90	$18.30
P5	$2.18	$5.60	$5.06	$12.84	$2.51	$6.60	$5.80	$14.91	$2.80	$7.00	$5.80	$15.60
P6	$4.36	$2.80	$5.06	$12.22	$5.02	$3.30	$5.80	$14.12	$5.60	$3.50	$5.80	$14.90
P7	$4.36	$5.60	$2.53	$12.49	$5.02	$6.60	$2.90	$14.52	$5.60	$7.00	$2.90	$15.50
P8	$6.54	$2.80	$2.53	$11.87	$7.53	$3.30	$2.90	$13.73	$8.40	$3.50	$2.90	$14.80
P9	$2.18	$5.60	$5.06	$12.84	$2.51	$6.60	$5.80	$14.91	$2.80	$7.00	$5.80	$15.60
P10	$4.36	$5.60	$2.53	$12.49	$5.02	$6.60	$2.90	$14.52	$5.60	$7.00	$2.90	$15.50
P11	$4.36	$2.80	$5.06	$12.22	$5.02	$3.30	$5.80	$14.12	$5.60	$3.50	$5.80	$14.90
P12	$2.18	$2.80	$7.59	$12.57	$2.51	$3.30	$8.70	$14.51	$2.80	$3.50	$8.70	$15.00

Table 11 The total costs of products considering various tariff rates

Products	Without trade tariff rate consideration	Trade tariff rate of 25% on materials purchase price	Trade tariff rate of 50% on materials purchase price	Trade tariff rate of 75% on materials purchase price	Original selling price
P1	$26.09	$28.92	*$31.57*	*$32.81*	$31.31
P2	$29.86	$32.55	$35.03	*$36.37*	$35.83
P3	$25.71	$28.06	$30.23	*$30.96*	$30.85
P4	$26.43	$29.12	$31.60	*$32.94*	$31.71
P5	$26.57	$28.92	$31.09	$31.82	$31.88
P6	$26.90	$29.12	$31.12	$31.94	$32.28
P7	$22.82	$25.12	$27.25	*$28.28*	$27.38
P8	$21.44	$23.61	$25.56	*$26.69*	$25.73
P9	$18.85	$21.20	*$23.37*	*$24.10*	$22.62
P10	$21.11	$23.40	*$25.54*	*$26.57*	$25.33
P11	$26.04	$28.27	$30.26	$31.08	$31.25
P12	$19.50	$21.77	*$23.81*	*$24.33*	$23.39

on the cost of $26.09 when a tariff is not imposed. However, under tariff rates of 50% and 75%, the company will lose money. As the cost goes higher according to the increase in tariffs, the selling price should be adjusted accordingly to avoid losses. This effect is discussed in the following section.

6.3 Impact of Tariffs on Selling Price of Finished Product

The selling price per unit of each product is impacted by the trade tariffs. As the total costs per unit of product are determined according to the different rates of the trade tariff, the selling prices are determined according to Eq. (3). Table 12 shows how selling prices increase as import taxes or tariff rates increase. The selling price is determined considering the 20% profit margin. For example, when the import tax is not considered, the selling price of P1 is $31.31. However, as the import tax increases by 25% on the purchased materials, the selling price increases to reach $34.70. Additionally, as the import taxes are increased to 50% and 75% on the purchased materials from the global suppliers, the selling prices of P1 are $37.89 and $39.38 per unit, respectively.

6.4 Impact of Tariff on Demand for Finished Product

According to the law of demand, the demand is a function of the selling price of each product. Sana (2011) considered that demand is a function of selling price P_j of each time period j as shown in Eq. (5).

Table 12 The selling prices of products considering various tariff rates

Products	Without trade tariff rate consideration	Trade tariff rate of 25% on materials purchase price	Trade tariff rate of 50% on materials purchase price	Trade tariff rate of 75% on materials purchase price
P1	$31.31	$34.70	$37.89	$39.38
P2	$35.83	$39.06	$42.04	$43.64
P3	$30.85	$33.67	$36.28	$37.15
P4	$31.71	$34.95	$37.92	$39.52
P5	$31.88	$34.70	$37.31	$38.18
P6	$32.28	$34.95	$37.34	$38.33
P7	$27.38	$30.14	$32.70	$33.94
P8	$25.73	$28.33	$30.68	$32.03
P9	$22.62	$25.44	$28.05	$28.92
P10	$25.33	$28.09	$30.64	$31.88
P11	$31.25	$33.92	$36.31	$37.30
P12	$23.39	$26.13	$28.57	$29.19

Table 13 β values for each product in family

Product (i)	D_i	SP_i	β_i
P1	613	$31.31	831
P2	337	$35.83	619
P3	311	$30.85	523
P4	616	$31.71	839
P5	166	$31.88	392
P6	110	$32.28	341
P7	279	$27.38	448
P8	706	$25.73	856
P9	366	$22.62	484
P10	301	$25.33	447
P11	495	$31.25	712
P12	298	$23.39	424

$$D(P_j) = \alpha - \beta P_j - \gamma P_j^2. \qquad (5)$$

In this study, the demand for product i, D_i, is assumed to be a function of the selling price of product i, SP_i, based on Eq. (6).

$$D_i = \beta_i - 0.7\, SP_i - 0.2\, SP_i^2 \qquad (6)$$

Based on expected demand and the selling price of each product without considering trade tariffs, β_i, value for each product is determined according to Eq. (6) and shown in Table 13. For example, given that the selling price and the demand values of P1 are $31.31 and 613, respectively, β equals 831.

As the selling price values of each product are determined by the considerations generated of the trade tariff rates, the demand for each product is impacted in which

Table 14 Demand values of product according to the increase in selling prices

Product i	Without trade tariff rate consideration		Trade tariff rate of 25% on materials purchase price		Trade tariff rate of 50% on materials purchase price		Trade tariff rate of 75% on materials purchase price	
	SP_i	D_i (unit)	SP_i	D_i (unit)	SP_i	D_i (unit)	SP_i	D_i (unit)
P1	$31.31	613	$34.70	566	$37.89	517	$39.38	493
P2	$35.83	337	$39.06	286	$42.04	236	$43.64	207
P3	$30.85	311	$33.67	273	$36.28	234	$37.15	221
P4	$31.71	616	$34.95	571	$37.92	525	$39.52	499
P5	$31.88	166	$34.70	126	$37.31	87	$38.18	73
P6	$32.28	110	$34.95	72	$37.34	36	$38.33	20
P7	$27.38	279	$30.14	245	$32.70	211	$33.94	194
P8	$25.73	706	$28.33	676	$30.68	647	$32.03	629
P9	$22.62	366	$25.44	337	$28.05	307	$28.92	297
P10	$25.33	301	$28.09	270	$30.64	238	$31.88	221
P11	$31.25	495	$33.92	458	$36.31	423	$37.30	408
P12	$23.39	298	$26.13	269	$28.57	241	$29.19	233

it decreases as the selling price increases. Based on Eq. (6), the demand values of products according to each consideration are shown in Table 14.

For example, the demand value of P1 is 613 units when each unit of P1 is sold by $31.31. As the selling price increases to reach a value of $34.70, the demand value declines to 566. Also, the demand values are dropped to 517 and 439 units when the selling prices go up to $37.89 and $39.38, respectively.

6.5 Profit Analysis Considering Tariffs Impact

In this chapter, the company (manufacturer) analyzes the impact of the imposed trade tariffs on the purchased materials. Given that the total costs are increased, the selling prices of each product are increased as well. The market share is affected according to the dropped values of the product demand. The total profit, TP_i, generated by product i is given by Eq. (7).

$$TP_i = TR_i - TC_i = \{D_i * SP_i\} - \{D_i * C_i\}, \tag{7}$$

where
TR_i total revenues generated by product i
TC_i total costs incurred on product i
D_i demand for product i
SP_i selling price of product i
C_i cost per unit incurred on product i

For example, considering imposing trade tariffs of 25% on the purchased raw materials, the selling price and costs per unit of P1 are $34 and $28.92, respectively. The new demand value is 566 units. According to Eq. (7), the total profit generated by P1 is $3273 and calculated as follows:

$$TP_{p3} = TR_{p3} - TC_{p3} = D_{p3} * SP_{p3} - D_{p3}{}^*C_{p3}$$
$$= 566^*\$34.70 - 566^*\$28.92 = \$3,273.$$

Table 15 provides the total profits generated by products considering the different scenarios of the imposed tariffs on the purchased raw materials.

Some observations are captured on the total profits generated by the set of products. Products are impacted in which their profits are either declined or inclined as the trade tariffs are increased. Product 8 (P8) kept the growth of the profits despite the demand inclination according to the different trade tariffs scenarios. For instance, the profits generated by P8 are $3030, $3192, $3307, and $3357 considering no tariffs, 25% tariffs, 50% tariffs, and 75% tariffs scenarios, respectively. However, the profits generated by the products P2, P3, P5, P6, and P10 are declined. Other products such as P1, P4, P9, P11, and P12 behave differently in which the profits generated reveal inclination and deterioration; however, the profits under the imposed tariff rates are higher than the profits without tariffs. For instance, the profits generated by P1 are $3198 when the tariffs are not imposed. Higher profits are generated at a value of $3273 when 25% tariffs imposed. However, the profits decline to reach values of $3267 and $3238, when tariffs rates of 50% and 75% are imposed, respectively. These findings depend on that all products have the same profit margin percentage when selling prices of products are determined, besides the demand function considered.

6.6 Tariffs Impact on Manufacturing System Design (Number of Manufacturing Cells)

The manufacturing system includes five manufacturing cells. As the manufacturer experiences the new demand values in its local market, the manufacturing system design is impacted accordingly: in other words, the number of cells to cover the affected demand. The manufacturer determines the capacity requirements in hours according to the demand values in units. For that, the mean, variance, and standard deviation values of capacity requirements are determined based on Eqs. (8) and (9).

$$\text{Mean Capacity Req.(hours)} = \text{Expected demand(unit)} * \frac{(\text{Bottleneck Processing Time})^2}{60}. \quad (8)$$

Table 15 Total profits generated by products considering tariff rate considerations

Product i	Raw material cost without considering trade tariff rate on purchase price			Raw material cost considering trade tariff rate of 25% on purchase price			Raw material cost considering trade tariff rate of 50% on purchase price			Raw material cost considering trade tariff rate of 75% on purchase price		
	TR_i	TC_i	TP_i	TR_i	TC_i	TP_i	TR_i	TC_i	TP_i	TR_i	TC_i	TP_i
P1	$19,193	$15,995	$3198	$19,636	$16,363	$3273	$19,602	$16,335	$3267	$19,425	$16,188	$3238
P2	$12,075	$10,061	$2014	$11,184	$9320	$1864	$9920	$8267	$1653	$9052	$7543	$1509
P3	$9594	$7995	$1599	$9179	$7649	$1530	$8500	$7084	$1417	$8207	$6839	$1368
P4	$19,533	$16,278	$3256	$19,940	$16,617	$3323	$19,914	$16,595	$3319	$19,731	$16,442	$3288
P5	$5292	$4410	$882	$4389	$3657	$731	$3249	$2707	$541	$2800	$2333	$467
P6	$3551	$2959	$592	$2525	$2104	$421	$1343	$1119	$224	$782	$652	$130
P7	$7639	$6367	$1272	$7393	$6161	$1232	$6911	$5759	$1152	$6584	$5487	$1097
P8	$18,165	$15,135	$3030	$19,154	$15,962	$3192	$19,840	$16,533	$3307	$20,140	$16,783	$3357
P9	$8279	$6898	$1381	$8571	$7143	$1429	$8616	$7180	$1436	$8579	$7149	$1430
P10	$7624	$6353	$1272	$7573	$6311	$1262	$7287	$6072	$1214	$7061	$5884	$1177
P11	$15,469	$12,890	$2579	$15,547	$12,955	$2591	$15,362	$12,802	$2560	$15,212	$12,677	$2535
P12	$6970	$5810	$1161	$7028	$5856	$1171	$6872	$5727	$1145	$6800	$5666	$1133

Table 16 The mean and variance of capacity requirements considering 25% tariff rate

Products in family	Bottleneck processing time (min)	Demand (unit)		Capacity requirements (h)		
		Mean	Standard deviation	Mean	Variance	Standard deviation
P1	1.5	566	142	14.15	12.51	3.54
P2	2	286	72	9.53	5.68	2.38
P3	1.7	273	68	7.74	3.74	1.93
P4	1.6	571	143	15.23	14.49	3.81
P5	1.8	126	32	3.78	0.89	0.95
P6	1.9	72	18	2.28	0.32	0.57
P7	1.4	245	61	5.72	2.04	1.43
P8	1.3	676	169	14.65	13.41	3.66
P9	0.9	337	84	5.06	1.60	1.26
P10	1.2	270	68	5.40	1.82	1.35
P11	1.8	458	115	13.74	11.80	3.44
P12	1.0	269	67	4.48	1.26	1.12

Table 17 Parameters of capacity requirements of product family considering 25% tariff rate

Capacity requirements (hours) of product family		
Mean	Variance	Standard deviation
101.75	69.57	8.34

$$\text{Variance of Capacity Req.(hours)} = (\text{Standard Deviation})^2 * \frac{(\text{Bottleneck Processing Time})^2}{60}. \quad (9)$$

The standard deviation (σ) is assumed to vary between 24.8% and 25.4% of the mean value of each product. Table 16 shows the conversion of the mean and standard deviation of the product demand in units into mean, variance, and standard deviation of the capacity requirements for each product. For example, considering the demand values obtained due to the increase in the selling prices according to the imposed tariffs of 25% on the purchased raw materials, the expected time to produce P1 is 14.15 h with standard deviation of 3.45 h.

Further, the mean capacity requirements for the entire family are determined by the summation of men capacity requirements of all products which is 101.75 h. However, the standard deviation of capacity requirement for the entire family is calculated by taking the square root of the summation of the variance of capacity requirement of products. Table 17 shows the parameters of capacity requirements of the product family.

Having determined the mean and standard deviation values of capacity requirements, the number of cells open to cover the demand is determined based on the demand coverage probabilities. The probability that a cell covers the demand is given by Eq. (10).

A Manufacturing-Supply Chain Risk Under Tariffs Impact in a Local Market

Table 18 Number of cells open to meet demand for product family considering 25% tariff rate

Mean capacity req.	Standard deviation capacity req.	One cell	Two cells	Three cells	Four cells
101.75	8.34	6.66E-14	0.004563	0.98568	1.0

Probability (Capacity Required ≤ Capacity Available)

$$= \text{NORMSDIST}\left(\frac{\text{Capacity Available} - \text{Mean Capacity Requirements}}{\text{Standard Deviation of Capacity Requirements}}\right) \quad (10)$$

According to Eq. (9), the capacity available for one cell is 40 h/week. Given that the mean of the capacity requirements is 101.75 h and the standard deviation is 8.34 h as shown in Table 17, the following computations for one cell and four cells are shown as follows:

- The probability that one cell covers the demand

$$P(CR \leq 40) = \text{NORMSDIST}\left(\frac{(40 - 101.75)}{8.34}\right) = 6.6E - 14$$

- The probability that the four cells cover the demand

$$P(CR \leq 160) = \text{NORMSDIST}\left(\frac{(160 - 101.75)}{8.34}\right) = 1.0$$

Based on the above computations, four cells are open to fully cover the demand for products in family (100% demand coverage probability). Table 18 shows the number of cells open and their demand coverage probabilities.

Table 19 shows the summary of the demand values affected by tariff rate consideration: trade tariff rate of 0% (without tariff rate), 25%, 50%, and 75% on purchased raw materials

The number of cells open to cover the demand values is affected by tariff rates considered. Figure 4 shows the results of the revised number of cells open to meet the demand according to tariff rates considered. Based on the demand that occurs when the tariff rate is not imposed, five cells are open to cover the demand. However, when the demand is dropped, due to the increase of the selling prices of the finished products, based on the tariff rates of 25% and 50% on the purchased raw materials, four cells are open to cover the demand, whereas three cells are open to meet the demand required due to imposed tariff rate of 75%.

In this chapter, it is intended to only show the number of cells open to meet the updated demand based on the demand coverage probabilities. Also, the optimal number of cells based on the profits and cost calculation is not considered. The manufacturer realizes the impact of the increased selling prices due to the tariff rates

Table 19 Demand values affected by tariff rate considerations

Products	Demand with a trade tariff rate is not considered	Demand with a trade tariff rate of 25% on materials purchase price	Demand with a trade tariff rate of 50% on materials purchase price	Demand with a trade tariff rate of 75% on materials purchase price
P1	613	566	517	493
P2	337	286	236	207
P3	311	273	234	221
P4	616	571	525	499
P5	166	126	87	73
P6	110	72	36	20
P7	279	245	211	194
P8	706	676	647	629
P9	366	337	307	297
P10	301	270	238	221
P11	495	458	423	408
P12	298	269	241	233

Fig. 4 Number of cells open vs trade tariff considerations

imposed. The original manufacturing system design is affected due to the changes in the market demand. Although the market share has been impacted in terms of the demand, the manufacturer sustains the business by making satisfactory profits.

7 Conclusions

This chapter shows the impact of imposed tariffs on both the manufacturer and its suppliers in a manufacturing-supply chain network. The manufacturer, who is in the USA, purchases raw materials of types X, Y, and Z from two local and three global suppliers in the USA and China, respectively. According to the capacity of the manufacturing system including five cells, the finished products are priced considering raw materials, labor, and overhead costs. Given that raw materials are purchased from global suppliers, the notion followed in this chapter is that US government imposes tariffs on the imported raw materials. This, in turn, helps local suppliers compete with global suppliers. There are different tariff rates assumed on raw materials' prices, and they are 25%, 50%, and 75%. The 50% tariffs imposed make the purchase of raw material Z to be switched to local suppliers. However, imposing 75% tariffs enables local suppliers to offer the lowest prices and sell all types of raw materials. Accordingly, the total costs of the finished product are increased as well, which eventually increases the selling prices. This study points out to the products making the company lose money in case the original selling price is kept unchanged. Additionally, given that the market demand is a function of the selling price of each product, the increase in the selling price causes the demand value to decrease according to the law of demand. Hence, the market share declines. Due to the drop in the demand values, the generated profits of products are updated accordingly. Results show that the profits of products behave differently in which some products generate more profits while others lose when the tariffs increase. It is worth saying that these findings are based on the assumption of having the same profit margin percentage when the product selling price is set up, besides the demand function considered. Furthermore, the impact of tariffs is shown on the manufacturing system design due to the decline in demand. When the tariff rates of 25% and 50% on the purchased raw materials are considered, the number of open manufacturing cells is dropped from five to four to cover the demand. However, three cells are open to meet the demand when a tariff rate of 75% is imposed. The contributions of this chapter are represented by showing how tariffs can be crucial to global supply chains, particularly, when the sales of global suppliers are deteriorated. Hence, global suppliers lose a slice of their market share; however, local suppliers become the beneficiaries. Additionally, this chapter shows how that low market demand enforces manufacturers to change their system design to absorb the impact of high imposed tariffs.

References

Alhawari, O. I., Süer, G. A., & Bhutta, M. K. (2021). Operations performance considering demand coverage scenarios for individual products and products families in supply chains. *International Journal of Production Economics, 233*, 108012.

Ávila, P., Mota, A., Pires, A., Putnik, G., & Teixeira, J. (2012). Supplier's selection model based on an empirical study. *Procedia Technology, 5*, 625–634.

Bhutta, K. S., & Huq, F. (2002). Supplier selection problem: A comparison of the total cost of ownership and analytic hierarchy process approaches. *Supply Chain Management: An International Journal, 7*(3), 126–135.

Burton, T. (1988). JIT/repetitive sourcing strategies: 'Tying the Knot' with your supplies. *Production and Inventory Management Journal, 29*(4), 38.

Ganeshan, R., & Harrison, T. (1995). *An introduction to supply chain management.* Department of Management Sciences and Information Systems, Pennsylvania State University.

Lee, H. L., & Billington, C. (1995). The evolution of supply-chain-management models and practice at Hewlett-Packard. *Interfaces, 25*(5), 42–63.

Patton, W., III. (1996). Use of human judgment models in industrial buyers' vendor selection decisions. *Industrial Marketing Management, 25*(2), 135–149.

Rodrik, D. (2018). What do trade agreements really do? *Journal of Economic Perspectives, 32*(2), 73–90.

Sana, S. (2011). Price-sensitive demand for perishable items – An EOQ model. *Applied Mathematics and Computation, 217*(13), 6248–6259.

Siegel, P. (2005) *Managing agricultural production risk: Innovations in developing countries.* Commodity Risk Management Group, Agricultural and Rural Development Department. World Bank Report, 32727.

Stevenson, W. J. (2019). *Operations management* (14th ed.). McGraw Hill.

Süer, G., & Huang, J. (2012, July). Long-term sourcing model for multi-objective supplier selection under uncertainty. In *42nd Conference on Computers & Industrial Engineering*, Cape Town, South Africa, pp. 15–18.

Ungureanu, H. (2019). Using financial accounting information for evaluation and control. *Academic Journal of Economic Studies, 5*(2), 120–124.

Van Der Vorst, J., & Beulens, A. (2002). Identifying sources of uncertainty to generate supply chain redesign strategies. *International Journal of Physical Distribution & Logistics Management, 32*(6), 409–430.

Wilding, R. (1998). The supply chain complexity triangle: Uncertainty generation in the supply chain. *International Journal of Physical Distribution & Logistics Management., 28*(8), 599–292.

Zsidisin, G. (2003). A grounded definition of supply risk. *Journal of Purchasing and Supply Management, 9*(5–6), 217–224.

Printed in the United States
by Baker & Taylor Publisher Services